Relative astrophysics is a branch of physics that deals with some of the most difficult challenges in our understanding of nature: the birth of the Universe, the origin of cosmic structures, the nature of exotic objects such as quasers, accretion disks and black holes.

Based on a meeting to honour the pioneering work of Igor Novikov, this book presents 19 specially written chapters covering all aspects of relativistic astrophysics. The distinguished team of authors is: Marek Abramowicz, Bernard Carr, Valeri Frolov, Werner Israel, Bernard Jones, Lev Kofman, Edward ('Rocky') Kolb, Malcolm Longair, Phil Lubin, Donald Lynden-Bell, Igor Novikov, Martin Rees, Sergei Shadarin, George Smoot, Roland Svensson, Gustav Tammann, Kip Thorne, Michael Turner and J. Craig Wheeler.

This volume provides researchers with a broad overview of modern relativistic astrophysics and offers an ideal source of supplementary reading for graduate students in astrophysics, cosmology and physics. It is a succinct yet engaging review of what is today one of the most active and fruitful areas of research in physics.

CAMBRIDGE CONTEMPORARY ASTROPHYSICS

Relativistic Astrophysics

CAMBRIDGE CONTEMPORARY ASTROPHYSICS

Series editors
José Franco, Steven M. Kahn, Andrew R. King and Barry F. Madore

Titles available in this series

Gravitational Dynamics, *edited by O. Lahav. E. Terlevich and R. J. Terlevich* (ISBN 0 521 56327 5)

High-sensitivity radio astronomy, *edited by N. Jackson and R. J. Davis* (ISBN 0 521 57350 5)

Relativistic Astrophysics, *edited by B. J. T. Jones and D. Marković* (ISBN 0 521 62113 5)

Advances in Stellar Evolution, *edited by R. T. Rood and A. Renzini* (ISBN 0 521 59184 8)

Relativistic Astrophysics

Proceedings of the *Relativistic Astrophysics* Conference,
in honour of Professor I. D. Novikov's 60th birthday
held in Copenhagen, Denmark
January 10–13, 1996

Edited by
B. J. T. JONES
Theoretical Astrophysics Center, Copenhagen, Denmark

D. MARKOVIĆ
Theoretical Astrophysics Center, Copenhagen, Denmark

CAMBRIDGE
UNIVERSITY PRESS

PUBLISHED BY THE PRESS SYNDICATE OF THE UNIVERSITY OF CAMBRIDGE
The Pitt Building, Trumpington Street, Cambridge CB2 1RP United Kingdom

CAMBRIDGE UNIVERSITY PRESS
The Edinburgh Building, Cambridge CB2 2RU, United Kingdom
40 West 20th Street, New York, NY 1001–4211, USA
10 Stamford Road, Oakleigh, Melbourne 3166, Australia

First published 1997

Printed in the United Kingdom at the University Press, Cambridge

A catalogue record for this book is available from the British Library

ISBN 0 521 62113 5 hardback

Contents

Epilogue

Participants

Abramowicz, Marek	Chalmers University (Sweden)
Adamson, David	Gordon and Breach Publishers (USA)
Andersson, Nils	Washington University (USA)
Antonuccio, Vincenzo	University of Catania (Italy)
Artemova, Ioulia	TAC, Copenhagen
Bertschinger, Edmund	MIT (USA)
Beskin, Vasily	Lebedev Physical Inst., Moscow (Russia)
Bjørnholt, Peter	Aarhus University (Denmark)
Björnsson, Gunnlaugur	University of Iceland
Bormann, Karsten	Niels Bohr Inst., Copenhagen
Brinchmann, Jarle	Inst.of Astronomy, Cambridge (UK)
Carr, Bernard	Queen Mary and Westfield Coll. (UK)
Chechetkin, Valery	Keldysh Institute (Russia)
Christensen, Per-Rex	TAC, Copenhagen
Christensen-Dalsgaard, Jørgen	TAC, Aarhus (Denmark)
de Felice, Fernando	Dept. of Physics, G. Galilei (Italy)
Demianski, Marek	University of Warsaw (Poland)
de Oliveira-Costa, Angelica	Max-Planck-Inst. für Astrophysik (Germany)
Diener, Peter	TAC, Copenhagen
Doroshkevich, Andrei	TAC, Copenhagen
Einasto, Jaan	Tartu Observatory (Estonia)
Flin, Piotr	Kielce Pedagogical University (Poland)
Frolov, Valery	University of Alberta (Canada)
Ginzburg, Vitaly L.	Lebedev Physical Inst., Moscow (Russia)
Gudmundsson, Einar	University of Iceland
Gurvits, Leonid	Joint Inst. for VLBI in Europe (Netherlands)
Gutierrez, Carlos M.	Inst.de Astrofisica de Canarias (Spain)
Hannestad, Steen	Aarhus University (Denmark)
Helbig, Phillip	Hamburger Sternwarte (Germany)
Hindmarsh, Mark	University of Sussex (UK)
Hivon, Eric	TAC, Copenhagen
Hjelmeland, Svend E.	University of Oslo (Norway)
Homeier, Derek	Hamburger Sternwarte (Germany)
Icke, Vincent	Sterrewacht, Leiden (Netherlands)
Illarionov, Andrei	Lebedev Physical Inst., Moscow (Russia)
Israel, Werner	University of Alberta (Canada)
Istomin, Yakov	Lebedev Physical Inst., Moscow (Russia)
Jensen, Bjørn	University of Oslo (Norway)
Jones, Bernard	TAC, Copenhagen
Jørgensen, Henning	Astronomical Observatory, Copenhagen
Kardashev, Nikolay	Astro-Space Center of Lebedev (Russia)
Khanna, Ramon	Landessternwarte, Heidelberg (Germany)
Khlopov, Maxim	Cent. for Cosmo-Particle Phys. (Russia)
Khokhlov, Alexei	University of Texas, Austin (USA)
Knutsen, Henning	Stavanger College (Norway)
Kofman, Lev	University of Hawaii (USA)
Kolb, Edward	Fermilab, Batavia (USA)
Komberg, Boris	Astro-Space Center, Moscow (Russia)

Kompaneets, Dmitri	Astro-Space Center, Moscow (Russia)
Kotok, Eleonora	TAC, Copenhagen
Lacey, Cedric	TAC, Copenhagen
Larsson, Stefan	Stockholm Observatory (Sweden)
Levin, Janna	Cent. for Particle Astrophysics, Berkeley (USA)
Liddle, Andrew	University of Sussex (UK)
Lilje, Per B.	University of Oslo (Norway)
Lindberg Christensen, Lars	TAC, Copenhagen
Lipunov, Vladimir	Shternberg Astronomical Inst.. Moscow (Russia)
Longair, Malcolm	Cavendish Laboratory, Cambridge (UK)
Lubin, Philip	UC Santa Barbara (USA)
Lukash, Vladimir	Astro-Space Center, Moscow (Russia)
Lynden-Bell, Donald	Inst. of Astronomy, Cambridge (UK)
Madsen, Jes	TAC, Aarhus (Denmark)
Markovic, Draza	TAC, Copenhagen
Massacand, Christophe	University of Tromsö (Norway)
Matarrese, Sabino	Dept. of Physics, G. Galilei (Italy)
Misner, Charles W.	University of Maryland (USA)
Nadyozhin, Dmitri	Inst. of Exp. and Theor. Phys., Moscow (Russia)
Naselsky, Pavel	Rostov State University (Russia)
Nobili, Luciano	Dept. of Physics, Padova (Italy)
Nordlund, Aake	TAC, Copenhagen
Novikov, Dmitri	Astro-Space Center, Moscow (Russia)
Novikov, Igor	TAC, Copenhagen
Novikova, Elena	Geophex, Raleigh (USA)
Olesen Larsen, Peder	The Danish Nat.Research Found.
Oscoz, Alejandro	Inst. de Astrofisica de Canarias (Spain)
Otzen Petersen, Jørgen	TAC, Copenhagen
Padoan, Paolo	TAC, Copenhagen
Pagel, Bernard	NORDITA, Copenhagen
Pethick, Christopher	NORDITA, Copenhagen
Pietilä, Harri	University of Turku (Finland)
Pogosian, Dmitri	University of Toronto (Canada)
Polnarev, Alexander	Queen Mary and Westfield College (UK)
Poutanen, Juri	Stockholm Observatory (Sweden)
Raevsky, Tatiana	(private guest)
Rebolo, Rafael	Inst. de Astrofisica de Canarias (Spain)
Rees, Martin	Cambridge University (UK)
Refsdal, Sjur	Hamburg Observatory (Germany)
Rose, Joanna	Forskning og Framsteg (Sweden)
Roxburgh, Ian W	Queen Mary and Westfield College (UK)
Rugh, Svend E.	Los Alamos Nat. Laboratory (USA)
Ryde, Felix M. F.	Stockholm Observatory (Sweden)
Sa, Paulo M.	University of the Algarve (Portugal)
Sato, Katsuhiko	University of Tokyo (Japan)
Schmid, Christoph	Inst. for Theoretical Physics, Zurich (Switzerland)
Schneider, Raffaella	University of Rome (Italy)
Schnopper, Herbert W.	Danish Space Research Inst. (Copenhagen)
Schutz, Bernard	U. of Wales, Cardiff (UK)
Schwarz, Dominik	Inst. for Theoretical Physics, Zurich (Switzerland)

Sciama, Dennis	SISSA, Trieste (Italy)
Shandarin, Sergei	University of Kansas (USA)
Skagerstam, Bo-Sture	Chalmers University (Sweden)
Smoot, George	UC, Berkeley (USA)
Sommer-Larsen, Jesper	TAC, Copenhagen
Stabell, Rolf	University of Oslo (Norway)
Starobinski, Alexei A.	Landau Inst. (Russia)
Stelmach, Jerzy	University of Szczecin (Poland)
Stern, Boris	Inst. for Nuclear Research (Russia)
Svensson, Roland	Stockholm Observatory (Sweden)
Tammann, Gustav	University of Basel (Switzerland)
Tegmark, Max	Max-Planck-Inst. für Physics (Germany)
Thacker, Robert John	University of Alberta (Canada)
Thorne, Kip	Caltech (USA)
Turner, Michael S.	University of Chicago (USA)
Turolla, Roberto	University of Padova (Italy)
Varshalovich, Dmitry	Ioffe Physical-Technical Inst. (Russia)
Vedel, Henrik	TAC, Copenhagen
West, Richard M.	ESO, Garching (Germany)
Wheeler, J. Craig	University of Texas, Austin (USA)
Wheeler, John A.	Princeton (USA)
Xella, Stefania	University of Bologna (Italy)
Zane, Silvia	SISSA, Trieste (Italy)

Prof. Igor D. Novikov

Preface

In the preface of his book 'Whither does the river of time flow?' Igor Dimitrevich Novikov writes: "Life can offer so much; it is many-sided, marvelous and sometimes terrible. Yet I have never known anything more wonderful than the striving to unravel the mysteries of the universe. And this is not an abstract craving, nor an idle 'philosophical' pondering of the essence of being (I realised early on that it is all but nonsence and often just a form of indolence and absorption with every contortion of one's thought), but a hard and happy work."

For over four decades Igor Dimitrevich has relentlessly pursued Universe's mysteries. Never too willing to give away any of her secrets, the old Sphynx has responded to all his insights by hastily shrouding herself with new veils. Nevertheless, that we have advanced so far in our understanding of the Universe and the marvels it is studded with is due in an unusual measure to the inexhaustable enthusiasm of Igor Dimitrevich.

Between the 10th and 13th January, 1996, many of Igor Novikov's friends and colleagues, former and current collaborators and students assembled in Copenhagen to review and discuss the current stage of the quest. The happy occasion was the *Relativistic Astrophysics* conference held in honour of Igor's 60th birthday.

This book is a collection of the invited lectures given at the Copenhagen conference. Each of the speakers had made vital contributions to the subject reviewed in his talk and did his best to present end explain it so that an uninitiated audience could gain a basic understanding of the physics involved as well as an appreciation of the unresolved issues. Put together, the lectures give a broad overview of modern relativistic astrophysics and offer an almost ideal source of supplemental reading material for the student of astrophysics and relativity.

We have divided the lectures into two main parts: *the Universe at very large redshifts* and *The Physics and Astrophysics of Relativistic Objects.* With this we have an introductory overview by Bernard Carr of all the material covered at the conference and in this book. This introduction serves as a succinct and engaging introduction to the vast variety of science covered in these Proceedings.

Bernard Jones, Draza Marković
Theoretical Astrophysics Center, Copenhagen
December, 1996

Acknowledgements

The conference was generously sponsored by the Danmarks Grundforskningsfond (Danish Foundation for Basic Research) through its support for the Theoretical Astrophysics Center. It was also co-sponsored by Nordita. We are grateful to both these institutions for making this meeting possible.

The scientific organising committee comprised of M. Demianski, N. Kardashev, B. Komberg, B. Pagel, M.J. Rees, and K. Thorne. The local organising committee comprised P.R. Christensen, E. Kotok and A. Uhl. We thank these people for their help in organising this conference. Anette Uhl also acted as conference organisation secretary and in that role coordinated the thousands of tasks that had to be performed to get such a show on the road. Without her this would indeed have been a daunting task and to her we express a special vote of thanks.

The preparation of this book was helped by the efforts of Nora Kotok and Karlis Berzins, who together managed to collect and format Igor's bibliography.

Conference Overview

By BERNARD CARR

Astronomy Unit, Queen Mary & Westfield College, London, E1 4NS, UK

1. Introduction

It is very nostalgic for me to give the overview talk at a conference in honour of Igor Novikov because Igor was the chairman at the first conference presentation I ever gave. This was at an IAU Symposium on Cosmology at Cracow in 1973, organized by Malcolm Longair. I remember being filled with trepidation because I was disagreeing with a paper Igor had written with Zeldovich on primordial black holes. Fortunately, he did not seem to hold this against me — indeed I remember his being wonderful company at a party we both attended the same evening. Today I am filled with trepidation because the chairman is my current boss, Ian Roxburgh, but of course I would not dare to disagree with one of his papers!

The dynamism of Novikov and Zeldovich at the conference in Cracow made a big impression on me. Both of them were forever jumping up and down — on the conference floor by day and on the dance floor by night — and it is gratifying to find that Igor is just as dynamic today (23 years later). When I was first asked to give this summary, I was slightly depressed because one of my friends opined that receiving such invitations is a sign of having reached middle age. On the other hand, by no stretch of the imagination could somebody as energetic as Igor be described as middle-aged and I'm even younger! Indeed seeing so many other distinguished participants who have recently had 60th birthday conferences (Werner Israel, Donald Lynden-Bell, Charles Misner, Bernard Pagel) makes it clear that such a state is no cause for despondency. It is also a special pleasure to see John Wheeler here since in 1972 he too celebrated his 60th birthday with a conference in Copenhagen, the proceedings of which became the famous "Magic Without Magic" volume. On that occasion the conference banquet was held at Carlsberg castle, where John exhibited his famous sense of mischief by letting off fire-crackers!

2. Summarizing Igor

In giving this talk, I am charged with two tasks: firstly, to summarize the highlights of Igor's own career and scientific contributions; secondly, to summarize the talks themselves. These tasks are closely related since, at one time or another, Igor has worked on almost all of the topics covered at the meeting — a reflection in part of Igor's versatility and in part of the organizers' skill in selecting speakers. Igor recalls some of his research in these topics in his own talk, and many other speakers also refer to his contributions, so my main purpose here is to place his work in a broader context

Let me start by recording the bare facts of Igor's academic career. He was an undergraduate at Moscow State University, reading astronomy, from 1954-1959. In 1959 he began his PhD in general relativity at Moscow State University. In 1963 he moved to the Institute of Applied Mathematics in Moscow and there began his famous collaboration with Zeldovich. In 1974 he became Head of the Department of Relativistic Astrophysics at the Space Research Institute in Moscow, remaining there until 1990. Following his split with Zeldovich, he became Professor at the Pedagogical University of Moscow in 1979, Professor of Moscow State University in 1986, and Head of the Department of Theoretical Astrophysics at the Lebedev Physical Institute in 1990. He relinquished the first two positions in 1991 but he still holds the third. In 1991 he moved to Denmark, becoming Professor of Astrophysics at Copenhagen University and Adjoint Professor at NORDITA. In 1994 he became Director of the Theoretical Astrophysics Center.

To glean more personal insights into Igor's life - to breath fire into these dry facts - one could do no better than to read Kip Thorne's engrossing book "Black Holes and Time Warps." Kip records a number of interesting episodes in Igor's career and I would like to recount some of these here. The first episode concerns his wife Eleonora Kotok (Nora), whom he married in 1959, the same year he began his PhD. Obviously Nora has played a crucial role in Igor's personal life but it is perhaps not appreciated how crucial her role has been in his professional life. For without Nora's intervention, Igor would probably never have worked with Zeldovich and he may not even have pursued a career in relativistic astrophysics. Thorne explains:

Igor was completing his kandidat degree with a specialty in general relativity and his wife Nora, also a physicist, was being warned by friends that relativity was a backwater with no relevance to the real Universe. Her husband, for the sake of his career, should leave it. Nora, worried by the warnings, seized an opportunity at a physics conference in Estonia to get advice from the famous physicist Yakov Borisovich Zeldovich. She sought Zeldovich out and asked whether he thought general relativity was of any importance. Zeldovich replied that relativity was going to become extremely important for astrophysics research. Nora then described an idea on which her husband was working, the idea that the implosion of a star to form a black hole might be similar to the big bang origin of the Universe but with time turned around and run backwards. As Nora spoke, Zeldovich became more and more excited. He himself had developed the same idea and was exploring it. A few days later Zeldovich barged into the office that Igor Novikov shared with many other students and began grilling him about his research. Here was an ideal match of talents, Zeldovich realized. Novikov, as a master of general relativity, would be an ideal member of the team. When Novikov, happy at Moscow University, hesitated to sign up, Zeldovich exerted pressure. He went to Mstislav Keldysh, the director of the Institute of Applied Mathematics where Zeldovich's team was being assembled; Keldysh telephoned Ivan Petrovsky, the Rector of Moscow University, and Petrovsky sent for Novikov. Petrovsky was unequivocal: "Maybe you now don't want to leave the University to work with Zeldovich but you will want to." Novikov signed up, and despite some difficult times, never regretted it.

The second episode concerns Igor's first visit to the West in 1964, when he attended a relativity conference in London. (Igor recalls this meeting in his own talk.) It was also where he first made contact with Kip Thorne, who records:

As a matter of course, the Soviet Union was now sending a small delegation of scientists to every major international conference; such delegations were important not only for maintaining the strength of Soviet science, but also for demonstrating the Soviets' strength to Western delegates. Thus it was that Zeldovich, having arranged an invitation from London for Novikov to give one of the major lectures of the Relativity Conference, found it easy to convince the bureaucrats to include his young colleague in the Soviet delegation. Novikov had many impressive things to report; he would create a very positive impression of the strength of Soviet physics. In London, Novikov presented a one-hour lecture to an audience of 300 of the world's leading relativity physicists. His lecture was a tour de force: a series of major contributions to our understanding of relativistic gravity, neutron stars, stellar implosion, black holes, the nature of quasars, gravitational radiation, and the origin of the Universe. As I sat there in London listening to Novikov, I was stunned by the breadth and power of the Zeldovich team's research. I had never before seen anything like it.

The special relationship between Novikov and Zeldovich is legendary. In the Soviet system each great mentor tended to become the centre of a research team and Zeldovich's team soon became the top theoretical astrophysics group in the world. To a large extent this resulted from the combination of Zeldovich's deep physical insight and Novikov's expertise as a relativist. If Zeldovich was the spark, Igor was the engine and between their first joint paper in 1964 (on black holes as the energy sources for quasars) and their last one in 1976 (on gravitational-wave-induced fluctuations in the microwave background), their collaboration led to some of the most important developments in the history of astrophysics. However, Zeldovich cannot have been an easy task-master. We got a tantalizing glimpse of this from Igor's own talk — with the wonderful story of how Zeldovich reprimanded his five team members because they could not equal the output of just three Western theorists: Hawking, Rees and Silk! As is well-known, Zeldovich's group split up in 1978 but, traumatic as that must have been, it at least permitted Igor to become the talented leader of an independent research team.

I must complete this account of Igor's career with a brief resume of his publications. His first paper was published in 1952 when he was only 17 (two years before going to University) on the variable star VZ Cygni. Since then he has published more than two hundred further papers. I cannot summarize them all here but a list of the topics they cover serves to indicate the broad range of his contributions: the cosmic microwave background (1964), the "no hair theorem" (1964), supermassive black holes and quasars (1964), white holes (1965), X-rays from compact objects in binary systems (1966), primordial black holes (1966), galaxy formation from primordial fluctuations (1967), relativistic accretion onto black holes (1972), electromagnetic detectors of gravitational waves (1973), quantum effects of black and white holes (1975), microwave background anisotropies (1977), the origin of large-scale structure (1978), supermassive black holes in active galactic nuclei (1983), spacetimes with closed timelike curves (1989), tidal disruption of stars by black holes (1991), black hole entropy (1993), dark matter and microlensing (1995), advection in accretion (1995). I have given the date of his first publication in each topic since this indicates his foresight in anticipating so many important areas of research.

Igor has also written fifteen books, six of them in English. Some of these, such as the two volumes on "Relativistic Astrophysics" (coauthored with Zeldovich) and "The Physics of Black Holes" (coauthored with Frolov), are now classic textbooks. But he has also been keen to communicate to a wider lay audience. Indeed he might be regarded as Russia's answer to Stephen Hawking, his popular books including "Evolution of the Universe," "Edwin Hubble: Life and Work," and "Black Holes and the Universe." To give a glimpse of Igor's sense of humour, I have included several cartoons from the last book in this summary.

3. Summarizing the Conference

The contributions at this conference have touched on almost every field of relativistic astrophysics and cosmology. For the purpose of this review, I will group them under eight broad headings: the microwave background, the early universe, large-scale structure, dark matter, gravitational lensing, gravitational waves, galaxies, and black holes. Many speakers are mentioned under several headings. Indeed what is so exciting - and what in a sense brings us all together — is the intimate link between each of these topics. It is no longer possible to consider any of them in isolation because advances in one topic invariably have implications for the others.

Figure 1 illustrates this rich network of interconnections. Every talk at this conference has emphasized at least one of the links indicated in this Figure. We have learned, for example, how the nature of the dark matter has crucial implications for the microwave background fluctuations, the formation of large-scale structure, the structure of galactic halos, and processes in the early Universe. We have also seen how the signatures of black

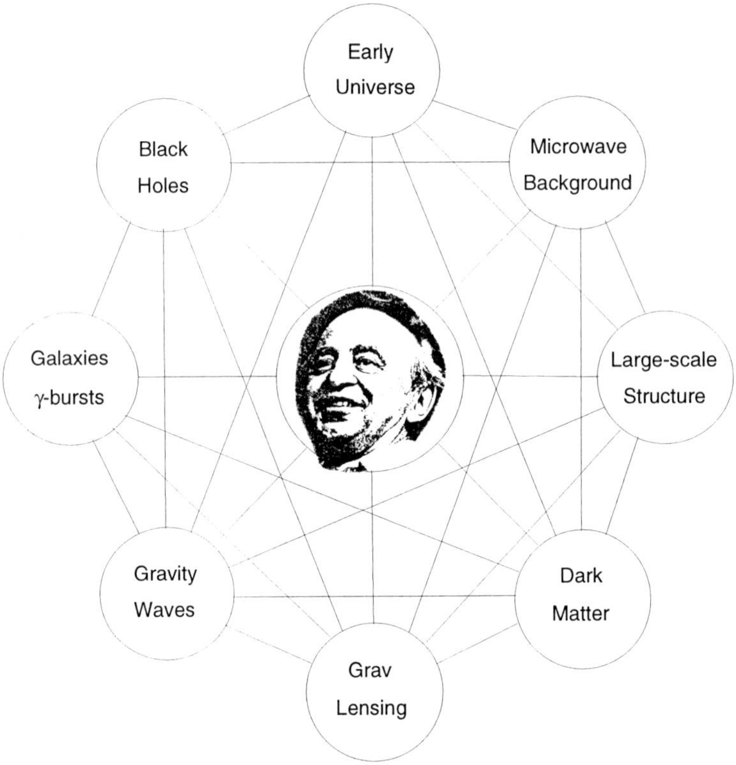

FIGURE 1.

holes include activity in galactic nuclei, gravitational lensing effects and the generation of gravitational waves. I will not attempt to list all the other connections explicitly but the point should be clear: relativistic astrophysics and cosmology involve the cross-fertilisation of many different branches of mathematics and physics and this has born fruit abundantly.

The other side of the coin, of course, is that people involved in our subject must be very versatile and possess expertise in many different branches of physics. One needs to be part astrophysicist, part cosmologist, part nuclear physicist, part high energy physicist, and part relativist. One also needs to be aware of the vast stream of data pouring in from a wide range of particle physics experiments and astronomical programmes. Above all, one needs to be able to span the gulf between astrophysics and relativity. Very few people have been able to bridge this gap and, in Thorne's assessment, no one except perhaps Chandrasekhar has done so as successfully as Igor. This emphasizes another link between the topics in Figure 1: as a brief perusal of his publication list makes clear, Igor has worked on all of them. He provides a central connection, so it is very appropriate to put him in the middle!

4. Early Universe

Most of the early Universe talks focussed on the inflationary scenario. As Mike Turner vividly conveyed, this still provides the best resolution of the horizon and flatness problems but the variety of inflationary scenarios is now bewilderingly large. He drew a distinction between simple toy models (eg. chaotic and power-law inflation), natural models (eg. superstring inflation), complicated models (eg. GUT and electro-weak inflation) and non-standard models (eg. scalar-tensor and multiple inflation). Some of these models may even permit the cosmological density to deviate from the critical value, so it is no longer clear which predictions of inflation are really generic. Alex Starobinsky showed that many inflationary scenarios produce a background of gravitational waves and this may turn out to be the most useful signature of inflation.

A key demand on any model for the early Universe is that it must generate the density fluctuations which produce the observed large-scale structure. These are usually characterised by the exponent n in the wave-number dependence of the power spectrum: $P(k) \propto k^n$. Various observational constraints on scales from quasars to the particle horizon, summarized by Starobinsky, require that n be close to 1 (corresponding to "scale-invariant" fluctuations) and this is usually taken to be a feature of the standard scenario. The simplest inflationary scenarios produce fluctuations in which $P(k)$ has a logarithmic dependence on k (which is nearly scale-invariant) but more general inflationary scenarios can give values for n somewhat different from 1. In this context, I would like to emphasize the importance of studying constraints on inflation imposed by the formation of primordial black holes (a subject which Igor helped to pioneer). If one has "blue" fluctuations (with $n > 1$), then one might expect primordial black holes to form prolifically at the end of inflation. The requirement that one not produce too many of them, together with the COBE quadrupole measurement, constrains the spectrum of fluctuations over 57 decades of mass and already implies $n < 1.3$.

Inflation usually ends when the inflaton field begins to oscillate in the bottom of the inflaton potential, these oscillations leading to reheating and reversion to the standard Friedmann model. We heard from Lev Kofman about a new class of scenarios which involve reheating via the creation of particles associated with a second scalar field. If one has a broad parametric resonance, the reheating is very rapid, so one has a period of non-equilibrium; this means that baryogenesis is easier and may even occur below 1 GeV. If reheating is slow, so that the inflatons never completely decay, then they would today constitute a dark matter candidate but their density would usually be much larger than critical. In particular, slow reheating below 10^{14} GeV is excluded by this argument, so a theory introduced to solve one relic problem (the overabundance of monopoles) may introduce another one!

5. Microwave Background

One of the key challenges in cosmology is to measure the anisotropies in the temperature of the cosmic microwave background (CMB). The dipole anisotropy gives information about our peculiar motion relative to the cosmic rest frame, while the form of the anisotropies on smaller scales — characterised by a number C_l for each multipole l — gives information about the spectrum of the primordial density fluctuations. It is now three years since COBE first detected anisotropies on scales above 10°. At this meeting we were fortunate to hear George Smoot reveal for the first time (and a few days before the official announcement) the results of the 4-year COBE data. After subtracting the contribution from the Galaxy and using a new set of orthonormal functions to analyse the small-scale anisotropies, he reported a spectral index $n = 0.9 - 1.5$ and a quadrupole anisotropy of $4 - 28\,\mu$K (both at 95% confidence level). Whereas the 2-year data were characterised by statistical noise, the signal-to-noise ratio in the 4-year map exceeds two, so COBE is now observing real features in the microwave sky. This suggests a new means of funding future CMB research: since members of the public already pay considerable sums to have stars named after them, perhaps we can also sell off features in the CMB!

FIGURE 2.

It is not only space experiments which are yielding positive detections. Both ground and balloon-based experiments are also doing so. In particular, Phil Lubin told us how five years of data from South Pole balloon experiments have now found clear signals at 0.5° and 1°. (It seems there may even have been a detection in 1991.) He also reviewed the host of future anisotropy experiments - some of them already approved — in space (COBRAS/SAMBA, MAP, FIRE, PSI, RELICT), on balloons (QMAP, MSAM, MAXIMA, BOOMERANG, HACME, ACE) and on the ground (CAT, CBI, VSA). These experiments will be many times more sensitive than COBE and able to probe much higher values of l. It often strikes me how lucky we are to be alive in that small window of time between the first detection of the CMB, when one had to deal with sort of mundane sources of noise illustrated in Figure 2, and ten years hence when there may be no more useful data to be extracted from it!

Bernard Jones and Mike Turner previewed the wealth of cosmological information which we may gain from the CMB data. It seems that almost all the important cosmological parameters may leave their imprint in the form of C_l. In particular, the position and amplitude of the 1st Doppler peak give the total density parameter Ω and the baryonic density parameter Ω_b respectively. As Lubin pointed out, the combination of the South Pole and Cambridge (CAT) data may already allow one to identify the 1st peak and this suggests a high value for Ω. Starobinsky emphasized that high-l anisotropies also contain contributions from gravitational waves and topological defects, so experiments like VSA (whose funding has just been approved and which will probe $l = 100 - 2000$ modes) will provide a vital source of information about the early Universe. Indeed, as illustrated in Figure 3, contemplating the scalar and tensor contributions to C_l made me appreciate the true significance of Smoot's claim to have seen "the Face of God" in the CMB data!

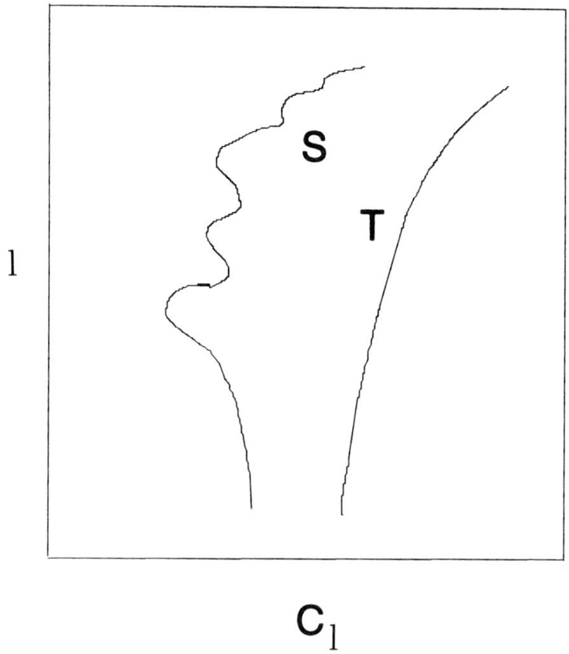

FIGURE 3.

6. Large-Scale Structure

A first step to quantifying large-scale structure is an accurate determination of distance scales since this leads to the value of the Hubble constant H_o. Gustav Tamman described how recent observations of Cepheids in four galaxies in the Virgo cluster seem to yield rather high values ($H_o = 80\,\mathrm{km/s\,Mpc}$), which may make the Universe younger than its oldest stars. However, the velocity of these four galaxies is considerably smaller than the mean velocity of all the galaxies in Virgo, and Tamman argued that this is because they are all on the near side of the cluster and therefore moving inwards. Data from a larger sample of Virgo galaxies - using a variety of methods - give a distance of 21.6 Mpc and a velocity relative to the microwave background of $1179\,\mathrm{km\,s^{-1}}$, corresponding to $H_o = 57 \pm 5\,\mathrm{km/s\,Mpc}$. This is also compatible with a determination of H_o using Type Ia supernovae. Doubtless Gerrard de Vaucouleurs, who died last year, would have disagreed but (to my mind) the argument was rather convincing. I am reminded of a conference on the Hubble constant a few years back: when a member of the audience lamented that "we need some young genius to solve this problem," de Vaucouleurs (already in his 70s) immediately jumped up and cried "I am that young genius!" Sadly de Vaucouleurs is no longer with us but it is gratifying that we have another young genius in the form of Gustav Tamman. Figure 4 illustrates a rather more direct method for determining the distances to other galaxies!

The next few years will see a huge increase in 3-dimensional large-scale structure data. In particular, Jones drew attention to the ESO key project ENACS (which will yield the redshifts of all cluster galaxies within 10-100 Mpc), while Turner emphasized the forthcoming SLOAN digital survey (which will yield a million galaxy redshifts). Jones stressed that proper analysis of all this new data will require the introduction of more sophisticated statistical techniques. There is a "vast, boring and complex" literature on statistics which still needs to be mastered! This will be necessary, for example, if one

FIGURE 4.

wishes to use measures such as skewness, kurtosis and genus to test for deviations from Gaussianity, as advocated by Starobinsky.

Two key ingredients to explaining large-scale structure are the form of the primordial density fluctuations and the nature of the dark matter. As emphasized by Turner, the standard Cold Dark Matter scenario (with $n = 1$) seems to produce too little power on large scales. However, there are various resolutions of this problem, so there is no need to give up the Big Bang picture itself or even the inflationary scenario: one can invoke a cosmological constant, a "tilted" spectrum of fluctuations with $n \neq 1$, a mixture of hot and cold dark matter, or an open cosmological model. To discriminate between these different possibilities, we need to determine how primordial density fluctuations evolve into large-scale structure. This can be done either through N-body simulations or analytically using the Zeldovich approximation. The latter breaks down in the non-linear regime but Sergei Shandarin explained how the truncated Zeldovich approximation (which neglects small wavelength fluctuations and uses the gravitational adhesion approximation) is adequate for many purposes.

A less conventional approach to understanding the development of large-scale structure was advocated by Ed Bertschinger in his discussion of "gravitomagnetic" effects, a topic also touched on by Donald Lynden-Bell. One can divide a gravitational field into an electric part (which has a scalar component corresponding to Newtonian action at a distance) and a magnetic part (which has a vector component corresponding to gravitomagnetic effects). This formalism has applications in such diverse areas as quantum gravity, numerical relativity, the growth of cosmological perturbations and — most relevant for present purposes — the collapse of a gas cloud in an expanding background. Although the last application should be an improvement on the Zeldovich approximation, Ed's earlier work (based on the assumption that the local magnetic part is zero) led to the surprising conclusion that collapse should usually lead to filament formation. This attracted some scepticism, so it is reassuring to learn that this prediction was erroneous; the magnetic part is non-zero even in the Newtonian limit and pancake collapse is now predicted to be generic. The Zeldovich picture is thus correct, although it underestimates the collapse time. Ed should not be too embarrassed by this turn of events: errors are always elucidating, as appreciated by Buddha who used to urge his disciples to do things the wrong way first because that would lead to deeper understanding!

7. Dark Matter

As Turner emphasized, there is now firm evidence for dark matter in galactic halos and clusters of galaxies. In units of the critical density, the associated densities are $\Omega_{\rm H} = 0.03\,h^{-1}(R_{\rm H}/100\,{\rm kpc})$ and $\Omega_{\rm C} = 0.1 - 0.2$, respectively, where $R_{\rm H}$ is the (very uncertain) radius of a typical halo and $h = H_o/100\,{\rm km\,s^{-1}\,Mpc^{-1}}$. Some of the best evidence for dark matter in halos and clusters comes from lensing data (as discussed below). One might also expect there to be background dark matter with almost exactly the critical density since the simplest inflationary scenarios predict $\Omega = 1$. Lynden-Bell argued that Ω might even exceed 1 since only closed cosmological models are compatible with Mach's principle. However, the direct evidence for either $\Omega = 1$ or $\Omega > 1$ is far from convincing. In principle, studies of large-scale streaming motions using theoretical tools such as POTENT and the microwave background dipole anisotropy should yield the value of Ω. However, the density parameter normally appears in conjunction with the bias parameter b and one needs more sophisticated analyses to get Ω and b separately. We have seen that, ultimately, determining the 1st Doppler peak of the CMB may lead to a better estimate of Ω.

If the Universe really does have the critical density, then most of its mass must be non-baryonic (i.e. in the form of relics from the Big Bang) since the cosmological nucleosynthesis picture implies that the total baryonic density can only be $\Omega_{\rm b} \sim 0.01\,h^{-2}$. A number of arguments suggest that this is probably "cold" (in the sense that it can clump inside gravitational potential wells), in which case the halo and cluster dark matter could also be non-baryonic. However, the halo dark matter could still be partly baryonic since there is evidence that the density of visible baryons is well below the cosmological nucleosynthesis value. In this case, halos could contain a large number of low mass stars or stellar remnants and, as discussed below, microlensing may already provide evidence for this.

Two of the most popular cold dark matter candidates are the photino and the axion. Rocky Kolb introduced ingenious new twists on both these candidates. The first twist is that the photino may be "light" (with a mass in the range 200 MeV - 1 GeV). Although such photinos would be hard to detect directly, the model predicts that there should also exist two other types of particle: an R^0 particle formed from the gluino and the gluon with a mass of around 1.5 GeV (important in determining the relic photino abundance) which should be detectable in kaon experiments at Fermilab within the next year; and a light Nambu-Goldstone boson, formed from two gluinos, which may be observed in ALEPH jet events. The second twist is that axions might reside in miniclusters, formed from fluctuations in the misalignment angle in the axion field. Simulations suggest that these form as soliton-type configurations comprising strings and domain walls. They would have a mass of around $10^{-9}\,M_\odot$ and a size of around 10^{13} cm. Such clusters would only strike the Earth every 25 million years, so their direct detections would be very rare, but they might still be found by their femtolensing effects.

8. Gravitational Lensing

An exciting new area of astronomy is gravitational lensing and we were fortunate to hear about this from one of the founders of the subject, Sjur Refsdal. He distinguished between macrolensing (the multiple imaging of a distant source such as a quasar by an intervening object) and microlensing (the amplification in the luminosity of the source). For sources at a cosmological distance, macrolensing is detectable for lenses between $10^5 M_\odot$ and $10^{13} M_\odot$, while microlensing is detectable between $10^4 M_\odot$ and $10^{-3} M_\odot$. For lenses in our own halo, one can detect microlensing all the way down to $10^{-6} M_\odot$, so lensing can probe almost the entire mass range between $10^{-6} M_\odot$ and $10^{13} M_\odot$. This makes it an extraordinarily powerful astronomical tool, providing information on such diverse issues as the background geometry of the Universe, the nature of the dark matter, the value of the Hubble parameter, and the structure of quasars.

Macrolensing already places strong constraints on the density of compact objects in the mass range $10^6 M_\odot$ and $10^{13} M_\odot$. These limits depend on the fact that the probability of a given source being lensed is roughly the density parameter of the compact objects, while the angular separation between the images is determined by their mass. For example, Refsdal mentioned that a sample of 469 quasars contains 5 which are lensed with separations in the range 0.5-5 arcsec; this implies a lens density $\Omega_L = 0.01$ (comparable to the density of galaxies) for masses in the range $5 \times 10^{10} - 10^{12} M_\odot$. He emphasized that macrolensing already provides incontrovertible evidence for dark matter in clusters: the dark matter potential distorts background galaxies into arcs and arclets in a way which allows the potential to be mapped. There is also evidence for dark objects in our own Galactic halo from the detection of microlensing of stars in the LMC. The 1st year data from the MACHO group suggested that the likely mass of the dark objects is around $0.05\,M_\odot$ (which is in the brown dwarf range) but the 2nd year data pushes the mass up to $0.5\,M_\odot$ (which is in the white dwarf range). In either case, the frequency of the microlensing events suggests that the halo fraction is below 1, suggesting that one also needs some non-baryonic component.

Another application of macrolensing mentioned by Refsdal is the possible determination of the Hubble constant. If one has an image-doubled quasar, intrinsic luminosity variations should appear at different times in the two images. The expected time delay depends on the distance and can hence be used to determine H_o. This effect may have already been observed for the quasar 0957+561, with different groups obtaining delays t in the range 1.1 to 1.5 y. In this case $H_o = (58 - 64)(1 - \kappa)\,(t/1.5\,\mathrm{y})$ where κ is the contribution from the cluster of which the lens galaxy is a member. Observers are currently expecting a variation observed in one image about a year ago to appear in the second image, so how long they have to wait is a direct measure of H_o. The inferred value of H_o^{-1} increases by 40 y for every second, so advocates of large H_o will doubtless become increasingly uncomfortable as time proceeds!

Lensing, of course, is a consequence of General Relativity's prediction that light should bend in a gravitational field, an effect first confirmed during the total solar eclipse of 1919. I was fortunate enough to witness a total solar eclipse in Vietnam last October and I was struck by the fact that this crucial test of General Relativity is possible only because the angular sizes of the Sun and Moon are almost identical. As an occasional advocate of the Anthropic Principle, I cannot but wonder whether some beneficent Creator had gravitational-lensing astronomers in mind when he made the world!

9. Gravitational Waves

Another new branch of astronomy with diverse applications will shortly open up with the advent of gravitational wave detectors. Nobody has been a more effective advocate of this than Kip Thorne, so we were lucky to have him update us on recent developments. He started off by describing the various interferometric detectors currently under construction (LIGO in the USA, VIRGO in Italy, GEO in Germany, TAMA in Japan). All of these are sensitive to waves in the frequency range $10 - 10^4$ Hz. For example, the 40 m interferometer at Caltech (a prototype of the 4 km LIGO interferometer) is already sensitive to waves with amplitude down to 2×10^{-19} in the 200 - 2000 Hz band. The future development of space-based interferometers such as LISA could also access the lower frequency range $0.1 - 10^{-4}$ Hz.

Thorne emphasized the importance of anticipating the signal from various astrophysical sources: coalescing binaries, supernovae, pulsars, cosmic strings and black holes. He focussed especially on coalescing neutron stars because, in this case, one can anticipate the time-dependence of the signal very precisely. With a sufficiently accurate template, neutron star coalescences should be detectable out to 200 Mpc and the final merger could give unique information about the neutron star equation of state, as well as astronomical data. Black hole binaries could be detected at even greater distances — all the way back to a redshift of 2 — but one needs supercomputers to calculate their expected waveforms accurately. Detectors like LIGO could detect stellar black holes in the mass range $15 - 10^3\, M_\odot$ but LISA could detect supermassive black holes in the range $3 \times 10^4 - 10^8 M_\odot$. Thorne also reviewed a variety of more exotic gravitational waves sources (boson stars, soliton stars and choptuons), as well as the various gravitational wave backgrounds (discussed by Starobinsky and Turner) which may have been left over from the early Universe.

10. Galaxies and Gamma-Ray Bursts

Malcolm Longair reviewed the evidence for the evolution of the galactic luminosity function. He mentioned that he first became famous for discovering the seventh quasar but I should point out that he also discovered the fifth and the sixth! The Hubble Space Telescope (HST) and Keck telescope have produced dramatic developments in this field. We can now see galaxies all the way back to a redshift $z = 1.6$ and evidence for a young host galaxy has even been found in the spectrum of a quasar at $z = 3.2$. Normal galaxies would be hard to see at such redshifts, so galaxies must have been more luminous in the past. This is confirmed by K, I and B observations of a sample of 367 galaxies, which show that galaxies with $1 < z < 1.6$ are ten times as luminous as those with $0.5 < z < 0.7$. Longair also compared the HST and radio data for the brightest radio galaxies in the redshift range $0.6 < z < 1.6$, his remarkable slides giving clear evidence for jet-induced star formation.

Martin Rees brought us up to date with γ-ray bursts, one of the oldest and most intriguing problems in astrophysics. The Gamma Ray Observatory (GRO) has now observed over a thousand of these and the fact that the number counts have a slope different from 1.5 indicates that they do not have a uniform distribution. There are two possible models: either the sources are located in an extended Galactic halo with a radius of at least 30 kpc or they are cosmological with most of them being at $z \sim 1$ to allow non-Euclidean effects. In the halo model, the sources need a velocity exceeding 700 km s^{-1} (if originating in the disc) and must be active for 10^9y; the energy requirement (10^{46} ergs) would suggest a subset of pulsars — either neutron stars with superstrong magnetic fields ($B > 10^{15}$ G) or neutron stars impacted by asteroids. In the cosmological model, the events are rare (10^{-5} per year per galaxy) and the energy requirement (10^{51} ergs) would be consistent with either coalescing binaries or "failed" supernovae which evolve to black holes with accretion tori (the γ-rays being associated with relativistic outflow at the poles). To decide between these models, Rees emphasized that we need positions to better than a few degrees and this can only come from deep space detectors like Ulysses. In the meantime important clues could come from studies of the Andromeda halo or from searches for lens-induced γ-ray echoes.

11. Black Holes in Astrophysics

Several contributors focussed on astrophysical aspects of black holes. I had an early taste of this subject when I met Igor for the first time at the Les Houches Summer School on "Black Holes" in 1972. I was too junior to attend the whole School but, thanks to Stephen Hawking, who was about to become my PhD supervisor, I was invited to attend for one week and this happened to coincide with Novikov and Thorne's lectures on black hole astrophysics. These lectures are now regarded as the standard work on the topic and the first treatment to include general relativistic effects in the analysis of black hole accretion.

Craig Wheeler told us about black holes in binary systems. There should be around 10^6 stellar black hole remnants in our Galaxy. Only a few of them are likely to be in binary systems with high mass companions but $10^2 - 10^3$ could have low mass companions. The key signature of the latter would be the X-rays generated by accretion of material from the companions, so he stressed the importance of studying Low Mass X-Ray Transients. There are now seven LMXRT black hole candidates and they all have a mass in the range $3 - 6\,M_{\odot}$. Their light-curves have very similar characteristics, the prototytpe being Nova Muscae 91: a steady soft X- ray component associated with the accretion disc, a transient soft component (which decays on a timescale of 30-40 days) associated with thermal instabilities in the disc, and a hard power-law component which is associated with outflow and includes electron-positron annihilation radiation. Important information can be gained from studying the relative phases of the bursts in different wavebands and the quasi-periodic oscillations associated with some black hole candidates. The fact that VLBA has detected superluminal motion in Nova Sco 94 affords a clear link with similar processes associated with supermassive black holes in radio sources.

Roland Svensson reviewed the "unification" scheme for Active Galactic Nuclei (AGN) and the evidence that these are powered by accretion onto supermassive black holes. X-ray and γ-ray satellites (such as ROSAT, GINGA, GRO, AXAF) have generated a lot of new data in this field. They show that most AGN (except for blazars) are weak γ-ray sources. Seyferts have an average spectral index α=0.9 over the range 2-10 keV with a cut-off at around 100 keV and this can account for the form of X-ray background in this energy range. Blazars can explain the most of the γ-ray background but type Ia supernovae must be invoked to produce the 1 MeV background. In order to compare theory with observation, it is important to understand radiative transfer in a mildly relativistic gas allowing for inhomogeneities, anisotropies and pair-cascade processes. Svensson showed that one can now do this: Non-Linear Monte Carlo and Iterative Scattering Methods give similar results and provide a good fit to the data.

The best evidence that AGN are powered by supermassive black holes comes from studying the innermost regions of the accretion disc. It is well known that, for standard thin accretion discs, radiation-pressure domination in the inner regions can lead to viscous and thermal instabilities, causing the inner disc to expand and form an optically

FIGURE 5.

thin stable region. Marek Abramowicz discussed a new regime of optically thin hot solutions for which advection is important. These solutions result from a type of thermal instability which arises when the disc is thick and pressure gradients are non-negligible. This means that the inner radius of the disc is no longer defined by the last stable orbit and the resulting non- uniqueness of the solution gives rise to the possibility of transients. Interesting new physics may be relevant at the transition between the hot inner region and cold outer region. We also heard from Rees (reporting on a result from another conference) about recent ASCA observations of one particular AGN where the line- width of the radiation is so broad that one needs to invoke a Kerr (rather than a Schwarzschild) black hole.

12. Black Holes in Relativity

Most of the more mathematical talks focussed on processes which occur inside black holes. In particular, Werner Israel discussed the stability of the inner event horizon in the Reissner-Nordstrom charged black hole solutions. Since this is an infinite blueshift surface and a Cauchy horizon, it has long been suspected that it is unstable. In studying its perturbations, one needs to include back reaction effects. In the spherically symmetric case, the inflowing radiation tail can lead to the mass at the Cauchy horizon increasing indefinitely (although there is no trace of this externally) and this "mass inflation" generates a light-like singularity. The timelike singularity of the Reissner-Nordstrom solution is therefore turned into a non-timelike singularity, as vividly illustrated in Werner's "jet engine" picture for the structure of spacetime inside the black hole. Some people have argued that this singularity destroys the Cauchy horizon but this is disputed by Werner and colleagues, who find that the singularity never crosses the Cauchy horizon. The same considerations may apply in the Kerr case.

Valeri Frolov spoke about the quantum aspects of black holes and addressed the question of whether a black hole's entropy resides in its surface, its interior or its singularity. He claimed that the dynamical degrees of freedom correspond to states of the quantum fields propagating in the interior and thermally excited by the black hole radiance. This means that the black hole's thermodynamical entropy is not the same as statistical mechanical entropy. Frolov also reported on work by Igor and himself which shows how one might use a wormhole to study the interior of a black hole. He ended with a picture of Igor peering inside a rotating black hole and perceiving something rather more attractive than a naked singularity. However, knowing how Igor's curiosity takes him everywhere, I suspect he would see something even more exotic, as intimated in Figure 5!

13. Final Remark

In concluding this summary, I would like to refer to Igor's recent studies of wormholes and time machines. This work has attracted scepticism is some quarters. However, it must be recalled that General Relativity is full of exotic predictions and many of these were rejected by the luminaries of the day when they were first discovered. For example, one recalls Einstein's rejection of black holes, Eddington's disbelief in the Chandrasekhar mass, Wheeler's initial disagreement with Oppenheimer's picture of gravitational collapse, and Hawking's initial scepticism of Bekenstein's notion of black hole entropy. However, all these ideas eventually gained general acceptance and work by Igor and others on time machines may just be another example of this.

It should be stressed that what makes outrageous ideas acceptable — even before definite evidence has accumulated to support them — is the backing of a few influential protagonists. When Thorne's first paper on time machines was published, he relates how one of his colleagues feared for his sanity. On the other hand, he recalls:

Igor Novikov, by contrast, was ecstatic. Telephoning from Santa Cruz, where he was visiting, Novikov said, "I'm so happy, Kip! You have broken the barrier. If you can publish research on time machines, then so can I!" And he proceeded to do so, forthwith.

It need hardly be said that Igor has also "broken many barriers". Indeed few people have been more influential in making the outrageous predictions of General Relativity become acceptable. It may well transpire that realistic physics precludes time-machines but, if so, we will only discover this by thinking about them. Meanwhile, perhaps the best evidence for their existence is the fact that Igor still looks so young and has published so many papers! So I hope that Igor will continue to follow the trail along which Einstein's equations lead, however fearsome the beasts he encounters en route. For as indicated in Figure 6, even if he finds his path blocked by a wall of ignorance and prejudice, I have every confidence that he will be able to use his incomparable knowledge of physics and astrophysics to peer over it!

FIGURE 6.

The Size and the Age of the Universe

By G. A. Tammann

Astronomisches Institut der Universität Basel, Venusstr. 7, CH-4102 Binningen, Switzerland

Recent determinations of the large-scale value of the Hubble constant H_0 are reviewed in the light of SNe Ia, the Virgo cluster tied into the cosmic expansion field, and field galaxies. Also the evidence from physical distance determinations is reviewed. The conclusion is $H_0 = 55 \pm 8\ \mathrm{km\,s^{-1}\,Mpc^{-1}}$, corresponding to an expansion age of 15.2 ± 2.2 Gy (if $q_0 = 0.1$) and 12 ± 1.7 Gy (if $q_0 = {}^1/_2$). Globular clusters, white dwarfs, and radioactive actinides require a time frame of 13.5 ± 2 Gy. This and the expansion age agree perfectly if the Universe is open; even for a critical Universe with $q_0 = {}^1/_2$ the time agreement is well acceptable within the errors.

1. Introduction

The grandiose title is chosen to match the scope of the work of Igor Novikov whose 60th birthday has brought us here together. The actual purpose of this paper is to discuss recent progress on the Hubble constant H_0 and to discuss the corresponding expansion age of the Universe and to compare it with independent age determinations of globular clusters, radioactive heavy elements, and white dwarfs.

New Cepheid distances from the Hubble Space Telescope (HST) have put the extragalactic distance scale on much stronger footing. Cepheids are the least controversial distance indicators. The zeropoint of their period-luminosity (P-L) relation has moved by only $0.^m07$ over an interval of 30 years (Kraft 1961; Sandage & Tammann 1968; Feast & Walker 1987; Madore & Freedman 1991). The calibration, using only Galactic Cepheids, gives a distance modulus of LMC of $(m - M) = 18.50 \pm 0.10$ (Sandage & Tammann 1968; Feast & Walker 1987), which is independently confirmed to better than $0.^m10$ through the shell of the supernova 1987A in LMC (Panagia et al. 1996), RR Lyr stars (Walker 1993) with the calibration of Sandage (1993a), the red-giant tip (Lee et al. 1993), the Baade-Becker-Wesselink method in BVIJHK for RR Lyr stars (Laney & Stobie 1992), and the I magnitude diameters of Cepheids (di Benedetto 1995).

The slope of the P-L relation taken from LMC is uncritical as long as the available Cepheids cover a sufficient period interval. Metallicity effects on the P-L relation are small (Freedman & Madore 1990; Chiosi et al. 1993; Sandage 1996b); any such effects enter with low weight because no strongly metal-deficient galaxies are considered in the following. Selection bias (Sandage 1988) can be avoided if the Cepheids span a sufficient period interval.

These or similar arguments have led to a general consent that Cepheids, as far as they can reliably be observed, are the most fundamental distance indicators known.

Cepheids are used in two ways to derive H_0: first to determine the distances of nearby galaxies which have produced well observed supernovae of type Ia (SNe Ia) whose absolute maximum magnitude is then used to calibrate the Hubble diagram of SNe Ia out to $30\,000\ \mathrm{km\,s^{-1}}$ (Section 2), and secondly to derive the distance of the Virgo cluster which is well tied into the cosmic expansion field out to $10\,000\ \mathrm{km\,s^{-1}}$; because of the important depth effect of the Virgo cluster its Cepheid distance from only four galaxies is still unsatisfactory and other distance determinations of the cluster are used as well, several of which are again calibrated through Cepheids (Section 3). Tentative results on the distance of the Fornax cluster are discussed in Section 4. Bias-corrected values of H_0 from field galaxies are in Section 5. In Section 6 purely physical distance determinations are compiled which contribute to H_0. The best value of H_0 and the corresponding expansion age are discussed in Section 7. An overview of present ages of the globular clusters, radioactive actinides, and white dwarfs is given in Section 8. The discussion is in Section 9.

2. H_0 from SNe Ia

The most direct way to the "cosmic" (i.e. large-scale) value of H_0 is now provided by SNe Ia. The small scatter of their Hubble diagram proves them to be very useful standard candles at maximum light. An HST program was therefore mounted to calibrate the luminosity of a few nearby SNe Ia through Cepheids. The program team comprises A. Sandage, A. Saha, L. Labhardt, F.D. Macchetto, N. Panagia, and the writer.

The Hubble diagram of all SNe Ia beyond 1100 km s^{-1} with reasonably well determined maximum magnitudes is shown in Fig. 1 (Tammann & Sandage 1995). Only red SNe Ia ($B-V > 0.25$) are omitted to guard against highly absorbed and peculiar objects. Their *intrinsic* luminosity scatter must be considerably less than 0.35 mag, because much of the scatter is expected from observational errors and peculiar motions. Indeed the intrinsic scatter must be very small because even the most distant SNe Ia lie very close to the theoretical Hubble line of slope 0.2. The argument goes as follows. The most distant SNe Ia occupy a volume about 18 000 times larger than that of the local calibrators. The large volume must contain exceptionally luminous SNe Ia – if they existed – and they have a much enhanced discovery chance for two reasons: their apparent magnitude is brighter than average *and* they stay longer above the detection limit. But still, there are no objects significantly above the Hubble line, not even at large distances. This means: The sample of SNe Ia shown in Fig. 1 constitutes a homogeneous class of very luminous and unabsorbed objects.

When in the following the calibration of Table 1 is applied to the SNe Ia in Fig. 1, it should be kept in mind that nearby, non-red, i.e. "Branch normal" SNe Ia (cf. Branch et al. 1993) are compared with the most luminous SNe Ia known. Therefore the resulting value of H_0 can only be, if anything, an *upper* limit.

Forcing a slope of 5 (corresponding to *linear* expansion) to the data in Fig. 1 gives

$$B(\max) = 5\,\log v - (3.186 \pm 0.054)\,, \tag{1}$$

and

$$V(\max) = 5\,\log v - (3.289 \pm 0.055)\,. \tag{2}$$

An easy calculation shows that the constant term C_λ in equations (1) and (2) is determined by

$$C_\lambda = 5\,\log H_0 - M_\lambda - 25\,. \tag{3}$$

In principle H_0 follows from equation 3 to within the scatter of Fig. 1 from M_λ (i.e. $M_B(\max)$ or $M_V(\max)$) of a single SNe Ia. Actually we have calibrated so far the peak luminosity of six SNe Ia. A seventh object has become available through Tanvir's et al. (1995) Cepheid distance of the Leo group. (It can be assumed that the member galaxies of this compact group lie practically at the same distance.) The resulting absolute magnitudes of the seven SNe Ia are shown in Table 1. Detailed discussions of the input parameters are given elsewhere (Sandage et al. 1996; Tammann et al. 1996; negligible differences between these sources are due to a different weighting of individual sources). The agreement to within the errors between the individual luminosities supports the claim that SNe Ia are (nearly) perfect standard candles. Independent confirmation of the luminosities comes from Höflich et al. (1996) who have three SNe Ia in common with

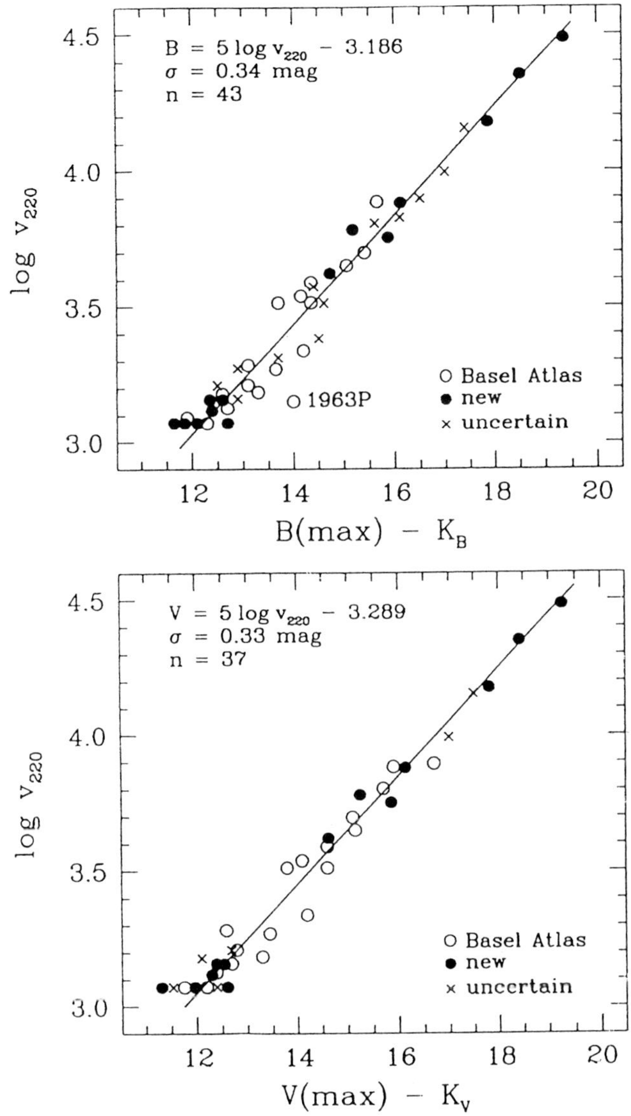

FIGURE 1. Hubble diagrams in B and V of all non-red SNe Ia with known maximum magnitudes. Open circles are from the older archive literature. Closed circles are the modern data used by Phillips (1993) and Hamuy et al. (1995).

Table 1. Their *model* luminosities are the same to within 0.12 ± 0.21 mag. Branch's (1996) model luminosity of SN 1981B agrees fortuitously well with ours, and two SNe Ia of Ruiz-Lapuente (1996) are fainter by only 0.26 ± 0.25 mag judging from their late spectra and the inferred ^{56}Ni mass. Inserting $M_{\rm B}$ and $M_{\rm V}$ from Table 1 into equation (3)

TABLE 1. Absolute Magnitudes of SNe Ia at Maximum.

Supernova	M_B(max)	M_V(max)	Reference[a]
SN 1937C	-19.53 ± 0.15	-19.50 ± 0.17	1
SN 1895B	-19.87 ± 0.22	—	2
SN 1972E	-19.52 ± 0.22	-19.49 ± 0.14	2
SN 1981B	-19.47 ± 0.17	-19.45 ± 0.14	3
SN 1960F	-19.53 ± 0.14	-19.62 ± 0.18	4
SN 1990N	-19.30 ± 0.24	-19.39 ± 0.24	5
SN 1989B	-19.51 ± 0.21	-19.49 ± 0.20	6
unweighted mean	-19.53 ± 0.07	-19.49 ± 0.03	
weighted mean	-19.53 ± 0.07	-19.49 ± 0.07	

[a]References.– (1) Sandage et al. 1992; Saha et al. 1994; (2) Sandage et al. 1994; Saha et al. 1995; (3) Saha et al. 1996a; (4) Saha et al. 1996b; (5) Sandage et al. 1996; (6) Tanvir et al. 1995 .

leads directly to $H_0(B) = 54 \pm 3$ and $H_0(V) = 58 \pm 3$, from which we adopt

$$H_0 = 56 \pm 3 \text{ (internal error).} \tag{4}$$

Equations (1) and (2) are defined out to $30\,000 \text{ km s}^{-1}$. The value of H_0 therefore represents the truly cosmic expansion rate.

At a time when only the very first calibrating SNe Ia were known it was suggested that SN 1972E was overluminous on the basis of its light curve shape (Hamuy et al. 1995; Riess et al. 1995) and that consequently the true value of H_0 was larger. In the light of seven calibrators this argument is now impossible. From first principles of stellar statistics it is known that *seven nearby* objects can on average not be more luminous than a distant, *luminosity-segregated* sample. This point is illustrated in Fig. 2, where the absolute magnitudes of the seven calibrators are compared to the absolute magnitudes of the distant SN sample. The latter are calculated once with $H_0 = 50$ and once with $H_0 = 70$. For $H_0 = 70$ the absurd situation arises that the distant SN Ia are systematically fainter than the nearby calibrators. The firm conclusion from this is that $H_0 < 70$.

A more detailed discussion of all external errors is given elsewhere (Tammann et al. 1996). It yields a confidence range of $44 < H_0 < 64$.

One proviso may be added. While the luminosity segregation of the distant SNe Ia makes the resulting value of H_0 an upper limit, as stated above, this may be offset by another systematic effect: SNe Ia in red or early-type galaxies are somewhat fainter than in late-type spirals (Branch et al. 1996; Sandage & Tammann 1996b). The local calibrators are mainly in late-type spirals whereas the fraction of early-type galaxies increases with distance in the present sample. This tends to make the true value of H_0 larger. It has to be seen which of the two effects is stronger.

Independent of the galaxy type dependence the SN luminosities (again calculated from velocity distances of SNe with $v_{220} > 1000 \text{ km s}^{-1}$ and $(B-V) < 0.25$) depend somewhat on SN color. This dependence is strongest for SNe Ia after 1985, which have presumably the best data:

$$M_B \approx 2.6\,(B-V)_{\text{SN}} + \text{const.} \tag{5}$$

This correction, which has vanishing influence on H_0, is probably intrinsic; it has in fact been predicted by the models of Höflich & Khokhlov (1996). If it were due to absorp-

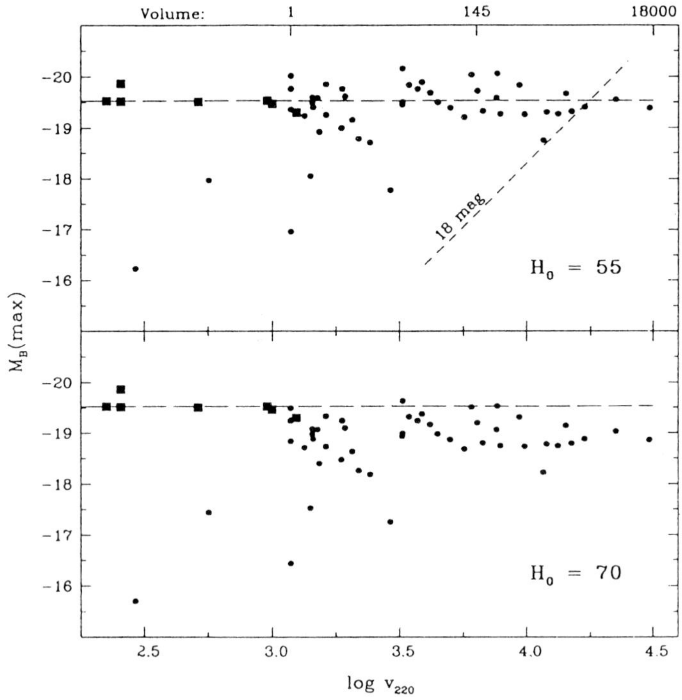

FIGURE 2. The absolute magnitude M_B(max) of all SNe Ia with known m_B(max) versus velocity distance. M_B(max) of the calibrators (squares) is based on their Cepheid distances. For all other SNe Ia M_B(max) is calculated from the recession velocities and $H_0 = 55$ (upper panel) and $H_0 = 70$ (lower panel). Note that for $H_0 = 70$ the impossible case arises that the distant SNe Ia are on average less luminous than the nearby calibrators.

tion the slope in equation (5) would have to be much steeper, i.e. 4.0. Any absorption correction would make the SNe Ia in late-type galaxies still brighter than they already are compared to their counterparts in early-type galaxies. Also the similar luminosity scatter of SNe Ia in late- and early-type galaxies, i.e. $\sigma_M = 0.29$ and 0.21, respectively, speaks against appreciable absorption. It is finally remembered that minimum absorption is expected from strongly luminosity-segregated objects. These arguments refute the suggestion that the distant SNe Ia need substantial absorption corrections (Riess et al. 1996).

3. The Distance of the Virgo Cluster

3.1. *Cepheids*

When the first reliable Cepheid distance of a Virgo galaxy (NGC 4321) became available from *HST* (Freedman et al. 1994) it was precipitately hailed *the* Virgo cluster distance (Mould et al. 1995; Kennicutt et al. 1995), although the value of 17 Mpc was suspiciously low. The next two Virgo galaxies, NGC 4536 (Saha et al. 1996a) and NGC 4496A (Saha et al. 1996b), again had very low distances. Only the fourth galaxy, NGC 4639, revealed the *important depth* of the cluster (note for comparison: the spiral members span ~ 15 deg

TABLE 2. The distance modulus difference between the Leo group and the Virgo cluster

Method	$\Delta(m-M)_{\rm Virgo-Leo}$	Source
Tully-Fisher	1.35 ± 0.20	Federspiel et al. (1996)
Globular clusters	1.47 ± 0.42	Harris (1990)
$D_n - \sigma$	0.97 ± 0.29	Faber et al. (1989)
Planetary nebulae	1.15 ± 0.30	Bottinelli et al. (1991)
Velocities	1.30 ± 0.30	Kraan-Korteweg (1986)
mean:	1.25 ± 0.13	

in the sky!). Its distance is 25 Mpc (Sandage et al. 1996) and yet it is a bona fide cluster member; with a *small* velocity of 800 km s^{-1} it cannot be assigned to the background.

It is no accident that the first three Virgo spirals with Cepheid distances lie on the near side of the cluster. They were selected from Sandage & Bedke's (1988) atlas of galaxies most suited for *HST*; they were thus biased to begin with. NGC 4639 looks much more difficult and would not have been selected had it not produced the archetypal SN 1990N.

It is now clear that it will take at least a dozen Cepheid distances of a randomly selected sample of Virgo members to obtain a meaningful mean cluster distance.

Meanwhile Tanvir et al. (1995) have suggested to step up their Cepheid-based Leo group distance of $(m-M) = 30.32 \pm 0.16$ out to the Virgo cluster by means of *relative* distance determinations. The best available relative distances between the Leo Group and the Virgo cluster are compiled in Table 2. Adding the mean difference of $\Delta(m-M) = 1.25 \pm 0.13$ to the above distance modulus gives a Virgo cluster modulus of $(m-M)_{\rm Virgo} = 31.57 \pm 0.21$. For brevity we will refer to this value in the following as the "Cepheid distance of the Virgo cluster".

3.2. *The Tully-Fisher Relation*

There are now 14 spiral galaxies with Cepheid distances (ten of which come from *HST*) which are useful for the calibration of the relation between absolute magnitude and $\log w$ (w = inclination-corrected 21cm-line width). Two close companions of M101 bring the number of useful calibrators to 16. Two galaxies (M101 and M100) are less inclined than the frequently adopted limit of $i = 45$ deg; yet their inclinations are so well defined by mapping their velocity field that they are still useful as calibrators.

The 16 calibrating galaxies yield the following calibration of the TF relation

$$M_B = -6.60 \log w - 3.42 \qquad (\sigma = 0.40) \tag{6}$$

(Federspiel et al. 1996), where the slope is taken from the Virgo cluster.

An *objective* and *complete* sample of Virgo spirals is defined by the 48 non-peculiar galaxies of type Sab–Sdm from the Virgo Cluster Catalog (Binggeli et al. 1985) with $i \geq 45$ deg and lying within the isopleths of substructures A and B (see Binggeli et al. 1993) or, without changing the result, within the X-ray contour of the cluster (Böhringer et al. 1994). This sample together with equation (6) gives $(m-M)_{\rm Virgo} = 31.70 \pm 0.15$. The application of the *inverse* TF relation does not bring an advantage. An investigation of the TF relation in UVRI magnitudes reveals some systematic effects between field galaxies (calibrators) and Virgo members, which have lower HI surface brightnesses, but this does not change the adopted modulus (Schröder et al. 1996). For the robustness of the result against variations of the input parameters the reader is referred to Federspiel et al. (1996).

3.3. *Other Distances to the Virgo Cluster*

The peak of the luminosity function (LF) of *globular clusters* (GC) has frequently been used as a standard candle. A modern calibration of the GCs in the Galaxy and in M31 combined with a compilation of published GCLFs in five Virgo ellipticals has led to a Virgo modulus of $(m - M)_{\rm Virgo} = 31.75 \pm 0.11$ (Sandage & Tammann 1995). Meanwhile Whitmore et al. (1995) found a very bright peak magnitude in V and I for NGC 4486, which is well determined with *HST* and which corresponds, with our precepts, to a modulus of 31.41 ± 0.28 (Sandage & Tammann 1996a). However, the GCs in NGC 4486 have a bimodal color distribution which is suggestive of age differences and possible merger effects (Fritze-von Alvensleben 1995; Elson & Santiago 1996). Turning a blind eye to this problem and averaging over all available GCLFs in Virgo we obtain $(m-M)_{\rm Virgo} = 31.67 \pm 0.15$. We are aware that the method may still face considerable problems.

The $D_n - \sigma$ *method*, normally applied to ellipticals, was extended to the bulges of S0 and spiral galaxies by Dressler (1987). Using the bulges of the Galaxy, M31, and M81 as local calibrators, one obtains $(m - M)_{\rm Virgo} = 31.85 \pm 0.19$ (Tammann 1988).

Novae are potentially powerful distance indicators through their luminosity-decline rate relation. Using the Galactic calibration of Cohen (1985), Capaccioli et al. (1989) have found the apparent distance modulus of M31 to be $(m - M)_{\rm AB} = 24.58 \pm 0.20$ (i.e. somewhat less than indicated by Cepheids). From six novae in three Virgo ellipticals Pritchet & van den Bergh (1987) concluded that the cluster is more distant by 7.0 ± 0.4 mag than the apparent modulus of M31, implying $(m - M)_{\rm Virgo} = 31.58 \pm 0.45$ (zero absorption is adopted for the Virgo cluster, see Section 3.5). The result carries still small weight, but is interesting because it is based on novae exclusively. *HST* observations, although time-consuming, of novae in the Virgo cluster could much improve this *independent* result.

Theoretical *models of SNe Ia* by various authors converge towards $M_{\rm B} = -19.45 \pm 0.15$ for "Branch normal" objects (Branch 1996; Höflich & Khokhlov 1996; Ruiz-Lapuente 1996). It is true that fainter, nearby SNe Ia are known, but being red and spectroscopically peculiar they can easily be singled out, and they do not contaminate distant, luminosity-selected samples of SNe Ia (cf. Section 2). Eight SNe Ia which have occurred in the Virgo cluster have $\langle m_{\rm B}({\rm max})\rangle = 12.10 \pm 0.15$ mag. This value combined with the theoretical calibration gives $(m - M)_{\rm Virgo} = 31.55 \pm 0.25$. – Had we used instead the empirical calibration of Table 5 below, the Virgo modulus would have become larger by 0.08 mag. We refrain from using this value because the routes towards H_0 in Sections 2 and 3 are to be kept strictly apart.

3.4. *Unreliable Distance Indicators*

The assumption that the LF of the *shells of planetary nebulae* (PN) in the light of the 5007 Å line had a universal cutoff at $M_{5007} = -4.48$ mag has led to a Virgo modulus significantly lower than obtained from the six methods discussed above (Jacoby et al. 1990). Yet it was pointed out that the cutoff magnitude depends on the sample size (i.e. the absolute magnitude of the parent galaxy; Bottinelli et al. 1991; Tammann 1993). Numerically simulated LFs of the shell luminosities confirm indeed the dependence on sample size *and* population age (Méndez et al. 1993). As a consequence the published PN distances deviate systematically from the Cepheid distances. The deviations increase with distance from M81 (Jacoby et al. 1989), NGC 5253 (Jacoby & Ciardullo 1993), and the Leo group (Ciardullo et al. 1989) to reach at the Virgo cluster (Jacoby et al. 1990) 0.74 mag ! – A new method to derive PN distances, allowing for sample size and other effects, has been proposed by Soffner et al. (1995); the first result for the nearby galaxy NGC 300 is encouraging.

TABLE 3. Distance moduli of the Virgo cluster

Method	$(m-M)_{\rm Virgo}$	Galaxy type
Cepheids (via Leo)	31.57 ± 0.21	S
Tully-Fisher	31.70 ± 0.15	S
Globular clusters	31.67 ± 0.15	E
$D_n - \sigma$	31.85 ± 0.19	E/S0
Novae	31.58 ± 0.45	E
Theor. Supernovae	31.55 ± 0.25	E/S0 + S
unweighted mean:	31.65 ± 0.05	
weighted mean:	31.67 ± 0.08	
mean linear distance:	21.6 ± 0.8 Mpc	

Surface brightness fluctuations (SBF) have also been proposed as distance indicators (Tonry & Schneider 1988). The first "test" has remained rather unconvincing, spreading the *elliptical* Virgo cluster members over an interval of 12 to 24 Mpc (Tonry et al. 1990); this interval was interpreted as real although early-type galaxies are known to be concentrated in the *cores* of galaxy clusters. Moreover the individual distances correlate with the Mg_2 index (Lorenz et al. 1993). Finally we note that the SBF distances of NGC 5253 (Phillips et al. 1992), the Leo group (Tonry 1991), and the Virgo cluster (Tonry 1991) are smaller than Cepheid distances by as much as 0.97, 0.48, and 0.56 mag, respectively. For the said reasons we use neither the PN nor the SBF distances.

3.5. *The Adopted Distance of the Virgo Cluster and the Resulting Value of H_0*

The distance determinations of the Virgo cluster in Sections 3.1 to 3.4 are compiled in Table 3. All distance moduli are taken to be true values, i.e. zero absorption is assumed towards the Virgo cluster. If the B-absorption implied by Burstein & Heiles (1984) is applied individually to all galaxies used for the distance determinations, the modulus becomes lower by only 0.06 mag. Even this almost negligible amount may be an overestimation as discussed by Sandage & Tammann (1996).

The resulting Virgo distance of 21.6 ± 0.8 Mpc is the same within the errors of the individual determinations. There is no dependence on galaxy type which suggests that early- and late-type Virgo galaxies are at the *same* distance.

The mean recession velocity of the Virgo cluster has given raise to some controversy. The value of Huchra (1988) is significantly too high; Hubble constants based on it are too high by 10-20% for this single reason! The 364 galaxies with redshifts and listed by Binggeli et al. (1993) to lie within the isopleths of the cluster substructures A and B give $<v_0>= 918 \pm 35$ km s^{-1}(with respect to the centroid of the Local Group). The 361 galaxies within the X-ray contour (Böhninger et al. 1994) give $<v_0>= 983 \pm 39$ km s^{-1}. Taking all 385 galaxies with redshifts in the Virgo survey area (Binggeli et al. 1985), excluding only clear background objects, yields $< v_0 >= 937 \pm 35$ km s^{-1}. From this follows a best cluster velocity of $< v_0 >= 950 \pm 30$ km s^{-1}. There are no significant differences between different Hubble types and between normal and dwarf galaxies. M 87, the first-ranked galaxy in substructure A, has a higher velocity (Binggeli et al.1987), but it should not be given undue weight. It is averaged into the distance determinations of Table 3 (where applicable) as well as into the mean cluster velocity.

To obtain the *cosmic* recession velocity of the Virgo cluster the observed value must still be corrected for the deceleration of the Local Group. We adopt $v_{\rm infall} = 220 \pm 50$ km s^{-1} (cf. Tammann 1996) and find for Virgo $v_{\rm cosmic} = 1170 \pm 61$ km s^{-1}

Yet we prefer a very similar value the rationale of which, however, is quite different. Many authors step up the Virgo distance out to the Coma cluster using the *relative* distance modulus between Virgo and Coma, and find H_0 at the distance of Coma. One can repeat that with any cluster whose distance relative to Virgo is known. In fact there are at least 14 clusters with rather good relative distances and velocities $4000 < v^{\rm CMB} <$ 11 000 km s^{-1} ($v^{\rm CMB}$ is the velocity in the frame of the microwave background). The best cosmic value of H_0 is then the all-sky mean over 14 different H_0 determinations. But more elegant is the reverse method: the relative distances are used to scale down the velocities of the 14 clusters and to predict a mean Virgo cluster velocity, i.e. the $v^{\rm CMB}_{\rm Virgo}$ velocity which the cluster would have in the absence of all local deviations from an ideal expansion field. The result of this procedure is $v^{\rm CMB}_{\rm Virgo} = 1178 \pm 32$ km s^{-1} (Sandage & Tammann 1990, Jerjen & Tammann 1993; Jerjen 1995 for a more rigid error determination). Using only SNe Ia in the Virgo cluster and beyond Hamuy et al. (1996) have recently obtained $v^{CMB}_{Virgo} = 1120 \pm 32$ km s^{-1}. It should be stressed again that these determinations are based on cluster data out to 11 000 km s^{-1} and SNe Ia data out to 30 000 km s^{-1}.

If one adopts a Virgo cluster distance of 21.6 Mpc (Table 3) and a Virgo velocity in the CMB frame of 1150 ± 40 km s^{-1} one obtains a *large-scale* value of

$$H_0 = 53 \pm 2 \qquad \text{(internal error).} \tag{7}$$

4. The Distance of the Fornax cluster

Three well observed SNe Ia have occurred in the Fornax cluster, i.e. SN 1980N and 1981D in NGC 1316 (Sa pec) and SN 1992A in NGC 1380 (S0/a). If their maximum B magnitudes are corrected for the Hubble type of their parent galaxies such that they agree with the mean Hubble type of the calibrators (cf. Section 2) one obtains $<m_B(\max)>= 12.21$. This combined with the calibration in Table 1 gives $(m-M)_{\rm Fornax} = 31.74(\pm 0.15)$ or 22.3 ± 1.7 Mpc. The mean recession velocity of the cluster is $v_{220} = 1338 \pm 50$ km s^{-1}, which leads to $H_0 = 60 \pm 6$.

It is noteworthy that a Cepheid distance of only $(m - M) = 31.30$ has recently been suggested for the Fornax Sbc spiral NGC 1365 (Silberman et al. 1996). If this value is correct at all, it means that NGC 1365 is substantially nearer than the two SN-bearing galaxies of earlier type. This is in fact not unlikely because the late-type spirals are widely distributed about the Fornax cluster and avoid its center. It would be unwise to equate the distance of NGC 1365 with that of the cluster proper.

The literature is abundant with relative distance determinations between the Fornax and the Virgo cluster. Taking 30 determinations between 1973 and 1996 at face value one comes to the conclusion that the late-type galaxies are 0.35 ± 0.09 mag *nearer* than the early-type members. This may be taken as an additional warning against giving too much weight to the distance of NGC 1365.

5. H_0 from Field Galaxies

Distance determinations to field galaxies out to 3000 - 5000 km s^{-1} depend on distance indicators like rotation velocities from optical or 21 cm line widths, the $D_n - \sigma$ relation (diameter versus velocity dispersion), and others. These relations have an intrinsic scatter of $\gtrsim 0.40$ mag. A scatter thus large invites serious selection effects (Malmquist bias) in

TABLE 4. H_0 determinations from field galaxies corrected for Malmquist bias

Method	H_0	Source
Tully-Fisher, distance-limited (local)	48 ± 5	Sandage 1994
Tully-Fisher, flux-limited (distance)	< 60	Sandage 1994
M 101 look-alike diameters	43 ± 11	Sandage 1993b
M 31 look-alike diameters	45 ± 12	Sandage 1993c
Luminosity classes of spirals	56 ± 5	Sandage 1996
M 101, M 31 look-alike luminosities	55 ± 5	Sandage 1996
Tully-Fisher	57 ± 5	Bottinelli et al. 1996
Tully-Fisher (using magn. + diameters)	55 ± 5	Theureau et al. 1996
weighted mean	53 ± 3	

all existing *flux-limited* galaxy catalogs. The effect is that the mean luminosity of the catalogued objects increases with distance and that consequently H_0 spuriously increases with distance as well.

Corrections for Malmquist bias are very difficult because they depend strongly on the size of the *true* scatter (which is difficult to observe), on the fairness of the catalog, on the (unknown) galaxy distribution, and the catalog cutoff. A tutorial on the Malmquist bias is given by Sandage (1995).

Here only a selection is given of H_0 determinations from field galaxies which are carefully corrected for Malmquist bias (Table 4).

6. Purely physical distance determinations

Various methods are emerging which allow extragalactic distances to be determined by purely geometrical or physical methods. They are still not competitive mainly due to obvious or possibly hidden systematic error sources. (An exception is the high-precision distance from the ring of SN 1987A; cf. Section 1). But there is no doubt that some high-weight determinations of H_0 will become available from these methods in the years to come.

It may suffice here to list several methods which hold high promise (Table 5). The reader is referred to the original sources for further explanations.

In principle the gravitational lenses overshoot the problem by determining the Hubble constant at high redshifts where one observes a combination of H_0 and q_0. The effect on H_0 is, however, only $\lesssim 10\%$.

The combined evidence in Table 5 is consistent with $H_0 \approx 50 - 65$.

7. The global value of H_0 and the expansion age

Three determinations of the large-scale value of H_0 have been discussed, i.e. the calibration of the Hubble Diagram of SNe Ia through seven SNe Ia with Cepheid distances giving $H_0 = 56 \pm 3$, the distance of the Virgo cluster tied into the rest frame of the CMB yielding $H_0 = 53 \pm 2$, and purely physical methods suggesting $H_0 \approx 50 - 65$. The evidence for $H_0 = 53 \pm 3$ from field galaxies is fully compatible with these determinations, but the range is restricted to distances of $3000 - 5000\ \mathrm{km\,s^{-1}}$. The tentative distance of the Fornax cluster leading to $H_0 = 60 \pm 6$ has at best only local meaning.

TABLE 5. H_0 from Physical Methods

Method	H_0	Source[a]
Radio Remnant of SN 1979C in Virgo (NGC 4321)	54 ± 20	1
Expanding-photosphere and ^{56}Ni models of SNe Ia	$55 - 70$	2
Expanding-photosphere models of SNe II	73 ± 6	3
	< 50	4
Sunyaev-Zeldovich effect		
cluster A 2218	45 ± 20	5
6 other clusters	60 ± 15	6
cluster A 2163	68 ± 30	7
2 clusters	42 ± 10	8
Gravitational lenses		
QSO 0957 + 561	$\leq 67 \pm 10$	9
B 0218 + 357	~ 60	10
MWB fluctuation spectrum	$30 < H_0 < 50(70)$	11

[a]References.– (1) Bartel 1991; (2) Branch et al. 1996; Höflich & Khokhlov 1996; Höflich et al. 1995; Ruiz-Lapuente 1996; (3) Schmidt et al. 1994; (4) Baron et al. 1995; (5) McHardy et al. 1990; Birkinshaw & Hughes 1994; Jones 1994; Lasenby & Hancock 1995; (6) Rephaeli 1995; Herbig et al. 1995; (7) Holzapfel et al. 1996; (8) Lasenby 1996; (9) Turner 1996; (10) Corbett et al. 1995; Nair 1995; (11) Lasenby 1996.

The conclusion from this is that the best value at present is $H_0 = 55 \pm 8$, where the error is to include also external errors.

The global meaning of this value beyond 5000 km s^{-1} is assured by the Hubble diagram of first-ranked cluster galaxies (Kristian et al. 1978; Hoessel et al. 1980; Lauer and Postman 1992). The expansion field is linear to within $\sim \pm 5\,\%$ as far as it can be tested; beyond $z = 0.15$ the test becomes ambiguous because of uncontrolled effects of luminosity evolution and q_0.

Values larger than $H_0 = 65$ still in the literature can be attributed to a few quite obvious error sources:

(*a*) an unwarrantedly high recession velocity of the Virgo cluster,

(*b*) the unrealistic expectation to fathom the depth of the Virgo cluster with only one well resolved galaxy (NGC 4321),

(*c*) the myth of a sharp, dispersionless cutoff of the luminosity function or planetary nebulae shells, independent of sample size and other factors,

(*d*) the reliance on the suspicious surface brightness fluctuation method, which is in open conflict with Cepheid distances, and/or

(*e*) the Malmquist bias which always artificially increases the value of H_0 from flux-limited samples of field galaxies and incomplete cluster catalogs.

With $H_0 = 55 \pm 8$ the Hubble time becomes $1/H_0 = 18 \pm 2.6$ Gigayears (Gy). This is with $\Lambda = 0$ the maximum expansion age. If we adopt $\Omega_0 = 0.2$ as a realistic minimum value of the matter density, the maximum Friedmann time becomes 15.2 ± 2.2 Gy (Sandage 1961a). The fashionable minimum Friedmann age comes from $\Omega_0 = 1$ ($q_0 = 1/2$), i.e. 12 ± 1.7. Of course, the age could still be smaller than this if q_0 were larger than $1/2$, but nobody seems to favor this solution.

TABLE 6. Recent age determinations of globular clusters

Age [Gy]		Author(s)
14.1 ± 0.3		Sandage 1993a
$11 - 21$	(total range)	Chaboyer 1995
$10 - 14$		Shi 1995
$13(+2, -3)$		Mazzitelli et al. 1995
14.5 ± 1.6		Demarque 1996
< 13		Weiss et al. 1996
$11 - 13$		Caloi et al. 1996

8. The age of the oldest objects in the Galaxy

The age of our Galaxy is well approximated by the age of the globular clusters, which have formed *before* the formation of the Galactic disk, and by the coolest white dwarfs which record the first deaths of stars with less than $\sim 8\,\mathfrak{M}_\odot$ in the Galactic disk. Moreover the onset of the production of radioactive actinides in *short-lived* SNe of type II reflects closely the time of the earliest Galactic star formation.

The fitting of color - magnitude diagrams of globular cluster to theoretical isochrones (e.g. Bergbush & VandenBerg 1992) is sensitive to the adopted distance, reddening, the Fe/H *and* O/Fe ratios, and the mixing length. The distance is usually obtained by fitting the horizontal giant branch (RR Lyr stars) or the main sequence; a new method of fitting the white dwarf cooling sequence has become possible through HST (Renzini et al. 1996). Decreasing distances due to brighter RR Lyr stars (Sandage 1993a) and enhanced O/Fe ratios have tended to decrease the ages over the last years. Recent results are compiled in Table 6.

It appears from the data in Table 6 that a globular cluster age of 13 – 14 Gy is generally acceptable; a value of 12 Gy does not seem to be ruled out.

The number of white dwarfs in the solar neighborhood drops sharply faintwards of $M_V = +16.0$ mag (Winget et al. 1987). From this the authors calculated that the faintest, i.e. oldest white dwarfs have a cooling time of 9 Gy which must be increased by the pre-white-dwarf lifetime of about 0.3 Gy. A rediscussion of the white-dwarf luminosity function by Wood (1992) has given an age range of 7.5 – 11 Gy. Crystallisation processes in white dwarfs, releasing additional energy, tend to increase this age to 9.5 – 12 Gy or even more (Hernanz et al. 1994).

If the radioactive actinides in the Solar System were produced in a single type II SN event, the age of the Galactic disk would be ~ 10.6 Gy. However, it can be shown that several SNe must have contributed to our chemism; in that case the age can only be larger. Allowing for this effect and introducing beta-delayed fission into the r-process network of reactions has increased the age. (Cowan, Thielemann, & Truran 1990). The latest discussion of the input parameters gives an age of 13.8 ± 3 Gy for the onset of heavy-element nucleosynthesis in our Galaxy (Thielemann 1995; Truran 1996).

A Galaxy age of 13 ± 2 Gy would seem like a fair assessment of the above triple evidence. To obtain an age of the Universe this number should still be increased by the gestation time of the first stars setting the clocks for globular clusters, white dwarfs, and the Galactic nucleosynthesis. If the first stars formed at $z \approx 10$ and if $0.1 < q_0 < 0.5$ the gestation time was about 0.5 Gy (Sandage 1961b).

The resulting age of 13.5 ± 2 Gy is derived from only Galactic data. Why is this of cosmic relevance? One can make a string of arguments that there was an epoch of preferred galaxy formation and that most galaxies have comparable ages. Even the gas-rich, young-looking LMC has metal-poor globular clusters which are as old as their Galactic counterparts (Brocato et al. 1996). In any case there is no evidence for a galaxy having formed before the Galaxy.

9. Conclusions

The realistic (Friedmann) expansion time of the Universe has been found to be 12 – 15.2 Gy for $\Omega_0 = 1$ and 0.2, respectively. This range may be increased by the uncertainty caused by the external error of the Hubble constant of ± 8. The possible interval becomes then 10.3 – 17.4 Gy.

The necessary time frame of 13 ± 2 Gy discussed in Section 8 fits astonishingly well into this interval without any necessity of introducing a cosmological constant Λ.

If, however, it is assumed that that $\Omega_0 = 1 (q_0 = {}^1/_2)$ the Friedmann time interval is narrowed down to 10.3 – 13.7 Gy. The upper limit here corresponds to $H_0 = 47$, which is not impossible but at present quite unpopular. Yet a contradiction-free Friedmann time of 13 Gy is obtained with a quite possible value of $H_0 = 51$. If one insists that $H_0 = 55$ the Friedmann time becomes 12 Gy which still cannot be excludes by the data in Section 8. Only if $H_0 = 60$ and $q_0 = {}^1/_2$ the Friedmann time decreases to 11 Gy, and the time agreement becomes so marginal that one may think of introducing a cosmological constant Λ. The situation is illustrated in Figure 3.

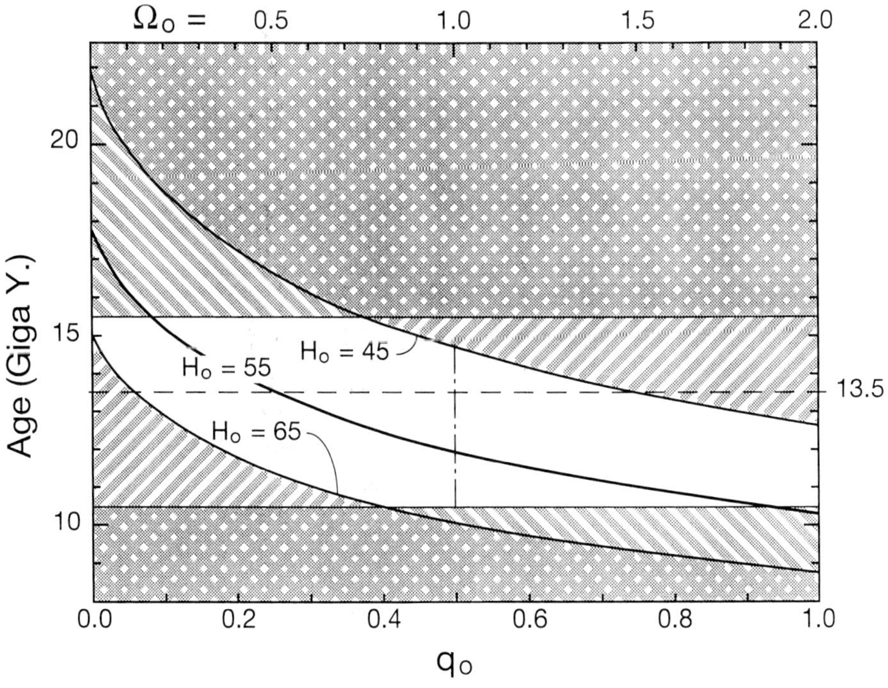

FIGURE 3. Illustration of the interplay of H_0, q_0 and the Friedmann expansion age. Limits are shown for $45 < H_0 < 65$ and for a rather wide interval of the Galactic age of 10.5 – 15.5 Gy. The critical Universe ($q_0 = {}^1/_2$) is marked with a dashed-dotted line.

Acknowledgement: The author thanks Dres. A. Sandage, A. Saha, L. Labhardt, F.D. Macchetto and N. Panagia for a most joyful collaboration in the HST project of the calibration of H_0. He is indebted to Mr. B. Reindl for preparing the manuscript. He acknowledges the support of the Swiss National Science Foundation.

REFERENCES

Baron, E., et al. 1995, ApJ 441, 170

Bartel, N. 1991, in: Supernovae, ed. S.E. Woosley, (New York: Springer), 760

Baum, W.A., et al. 1995, AJ 110, 2537

Bergbush, P.A., & VandenBerg, D.A. 1992, ApJS 81, 163

Binggeli, B., Popescu, C.C., & Tammann, G.A. 1993, A&AS 98, 275

Binggeli, B., Sandage, A., & Tammann, G.A. 1985, AJ 90, 1681

Birkinshaw, M., & Hughes, J.P. 1994, ApJ 420, 33

Böhringer, H., et al. 1994, Nature 368, 828

Bottinelli, L., Gougenheim, L., Paturel, G., & Teerkorpi, P. 1991, A&A 252, 560

Bottinelli, L., et al. 1996, preprint

Branch, D. 1996, in: Thermonuclear Supernovae, eds. R.Canal, P. Ruis-Lapuente, & J. Isern, (Dordrecht: Kluwer), in press

Branch, D., Fisher, A., & Nugent, P. 1993, AJ 106, 2383

Branch, D., Nugent, P., & Fisher, A. 1996, in Thermonuclear Supernovae, eds. R. Canal, P. Ruiz-Lapuente, & J. Isern (Dordrecht: Kluwer Academic Publishers), in press

Branch, D., Romanishin, W., & Baron, E. 1996 ApJ, in press

Brocato, E., et al. 1996, preprint

Burstein, D. & Heiles, C. 1984, ApJS 54, 33

Capaccioli, M., Della Valle, M., D'Onofrio, M., & Rosino, L.A. 1989, AJ 97, 1622

Caloi, V., D'Antona, F., & Mazzitelli, I. 1996, Poster presented at the STScI May Symposium, Baltimore

Chaboyer, B. 1995, ApJ, 444, L9

Chiosi, C., Wood, P., & Capitanio, N. 1993, ApJS 86, 541

Ciardullo, R., Jacoby, G.H., & Ford, H.C. 1989, ApJ 344, 715

Cohen, J.G. 1985, ApJ 292, 90

Corbett, E.A, Browne, I.W.A., Wilkinson, P.N., & Patniak, A.R. 1995, preprint

Cowan, J.J., Thielemann, F.-K., & Truran, J.W. 1990, ApJ 323, 543

Dahle, H., Maddox, S.J., & Lilje, P.B. 1994, ApJl 435, L79

Demarque, P. 1996, Talk presented at the STScI May Symposium, Baltimore

Dressler, A. 1987, ApJ 317, 1

Elson, R.A.W., & Santiago, B.X. 1996, MNRAS, in press

Faber, S.M., et al. 1989, ApJS 69, 763

Feast, M., & Walker, A.R. 1987, ARA&A 25, 345

Federspiel, M., Tammann, G.A., & Sandage, A. 1996, in press.

Freedman, W.L. & Madore, B.F. 1990, ApJ 365, 186

Freedman, W.L. & Madore, B.F. 1994, ApJ 427, 628

Freedman, W.L., et al. 1994, Nature 371, 757

Fritze-von Alvensleben, U. 1995, private communication

Gould, A. 1994, ApJ 425, 51

Hamuy, M., et al. 1995 AJ 109, 1

Harris, W.E. 1990, PASP 102, 966

Herbig, T., Lawrence, C.R., & Readhead, A.C. S. 1995, ApJ 449, L5

Hernanz, M., García-Berro, E., Isern, J., Mochkovitch, R., Segretain, L., & Chabrier, G. 1994, ApJ 434, 652

Höflich, P., & Khokhlov, A. 1996, ApJ 457, 500

Höflich, P., & Khokhlov, A., Wheeler, J.C., Nomoto, K., & Thielemann, F.K. 1995, preprint

Höflich, P., & Khokhlov, A., Wheeler, J.C., Nomoto, K., & Thielemann, F.K. 1996, in: Thermonuclear Supernovae, eds. R. Canal, P. Ruiz-Lapuente, & J. Isern (Dordrecht: Kluwer Academic Publishers), in press

Holzapfel, W.B., et al. 1996, cited by Lasenby (1996)

Huchra, J. 1988, in: The Extragalactic Distance Scale, eds. S. van den Bergh & C.J. Pritchet, (San Francisco: Astronomical Society of the Pacific), 257

Jacoby, G.H., Ciardullo, R. 1993, in: Planetary Nebulae, eds. R. Weinberger & A. Acker, (=IAU Symposium 155), 503

Jacoby, G.H., Ciardullo, R., Booth, J., & Ford, H. C., 1989, ApJ 344, 704

Jacoby, G.H., Ciardullo, R., & Ford, H.C. 1990, ApJ 356, 332

Jerjen, H. 1995, private communication

Jerjen, H., & Tammann, G.A. 1993, A&A 276, 1

Jones, M. 1994, Ap. Lett. Commun., in press

Kelson, D. 1995, private communication

Kennicutt, R.C., Freedman, W.L., & Mould, J.R. 1995, AJ 110, 1476

Kraan-Korteweg, R. C. 1986, A&AS 66, 255

Kraft, R.P. 1961, ApJ 134, 616

Lasenby, A.N. 1996, Talk presented at the STScI May Symposium, Baltimore

Lasenby, A.N., & Hancock, S. 1995, in: Current Topics in Astrofundamental Physics: The Early Universe, eds. N. Sanchez & A. Zichichi, (Dordrecht: Kluwer Academic Publishers), 327

Lauer, T.R., & Postman, M. 1992, ApJ 400, L47

Lorenz, H., Böhm, P., Capaccioli, M., Richter, G. M., & Longo, G. 1993, A&A 277, L15

Madore, B.F., & Freedman, W. L. 1991, PASP 103, 933

Mazzitelli, I., D'Antona, F., & Caloi, V. 1995, A&A 302, 382

McHardy, L.M., Stewart, G.C., Edge, A.C., Cooke, B.A., Yamashita, K., & Hatsukade, I. 1990, MNRAS 242, 215

Méndez, R.H., Kudritzki, R.P., Ciardullo, R., & Jacoby, G.H. 1993, A&A 275, 534

Mould, J.R., et al. 1995, ApJ 449, 413

Nair, S. 1995, preprint

Panagia, N., Gilmozzi, R., Macchetto, F.D., Adorf, H.-M., & Kirshner, R.P. 1991, ApJl 380, L23

Panagia, N. et al. 1996, Poster presented at the Extragalactic Distance Scale STScI May Symposium, Baltimore

Phillips, M.M. 1993, ApJl 413, L105

Phillips, M.M., Jacoby, G. H., Walker, A.R., Tonry, J.L., & Ciardullo, R. 1992, BAAS 24, 749

Pritchet, C. J., & van den Bergh, S. 1987, ApJ 318, 507

Renzini, A., et al. 1996, preprint

Rephaeli, Y. 1995, ARA&A 33, 541

Riess, A. G., Press, W.H., & Kirshner, R.P. 1995, ApJl 438, L17

Riess, A. G., Press, W.H., & Kirshner, R.P. 1996, ApJ, in press

Ruiz-Lapuente, P. 1996, preprint

Saha, A., Labhardt, L., Schwengeler, H., Macchetto, F.D., Panagia, N., Sandage, A., & Tammann, G.A. 1994, ApJ 425, 14 (IC 4182)

Saha, A., Sandage, A., Labhardt, L., Schwengeler, H., Tammann, G.A., Panagia, N., & Macchetto, F.D. 1995, ApJ 438, 8 (NGC 5253)

Saha, A., Sandage, A., Labhardt, L., Tammann, G.A., Macchetto, F.D., & Panagia, N. 1996a, ApJ, in press (NGC 4536)

Saha, A., Sandage, A., Labhardt, L., Tammann, G.A., Macchetto, F.D., & Panagia, N. 1996b, ApJ, in press (NGC 4496A)

Sandage, A. 1961a, ApJ 133, 355

Sandage, A. 1988, PASP 100, 935

Sandage, A. 1993a, AJ 106, 703 and 719

Sandage, A. 1993b, ApJ 402, 3

Sandage, A. 1993c, ApJ 404, 419

Sandage, A. 1994, ApJ 430, 1 and 13

Sandage, A. 1995, in: The Deep Universe, eds. B. Binggeli & R. Buser (Berlin: Springer), 1

Sandage, A. 1996a, AJ 111, 1 and 18

Sandage, A. 1996b, BAAS 28, 52

Sandage, A., & Bedke, J. 1994, Atlas of Galaxies (Washington, D.C.: NASA)

Sandage, A., Saha, A., Tammann, G.A., Labhardt, L., Panagia, N., & Macchetto, F.D. 1996, ApJl, in press

Sandage, A., Saha, A., Tammann, G.A., Labhardt, L., Schwengeler, H., Panagia, N., & Macchetto, F.D. 1994, ApJl 423, L13 (NGC 5253)

Sandage, A., Saha, A., Tammann, G.A., Panagia, N., & Macchetto, F. D. 1992, ApJl 401, L7 (IC 4182)

Sandage, A., & Tammann, G.A. 1968, ApJ 151, 531

Sandage, A., & Tammann, G.A. 1974, ApJ 194, 223

Sandage, A., & Tammann, G.A. 1990, ApJ 365, 1

Sandage, A., & Tammann, G.A. 1995, ApJ 446, 1

Sandage, A., & Tammann, G.A. 1996a, ApJl, in press

Sandage, A., & Tammann, G.A. 1996b, in preparation

Schaefer, B.E. 1995, ApJl 447, L13

Schmidt, B.P., et al. 1994, ApJ 432, 42

Schröder, A.C. 1995, Ph.D. Thesis, Univ. Basel

Shi, X. 1995 ApJ, 446, 637

Silberman, N.A., et al. 1996, Poster presented at the STScI May Symposium, Baltimore

Soffner, T., et al. 1995, A&A, in press

Tanvir, N.R., Shanks, T., Ferguson, H.C., & Robinson, D.R.T. 1995, Nature 377, 27

Tammann, G.A. 1988, in: The Extragalactic Distance Scale, eds. S. van den Bergh & C.J. Pritchet, (San Francisco: Astronomical Society of the Pacific), 282

Tammann, G.A. 1993, in Planetary Nebulae, eds. R. Weinberger & A. Acker, (=IAU Symposium 155), 515

Tammann, G.A., & Sandage, A. 1968, ApJ 151, 825

Tammann, G.A., & Sandage, A. 1995, ApJ 452, 16

Tammann, G.A., Sandage, A., Saha, A., Labhardt, L., Macchetto, F. D., & Panagia, N. 1996, in Thermonuclear Supernovae, eds. R. Canal, P. Ruiz-Lapuente, & J. Isern (Dordrecht: Kluwer Academic Publishers), in press

Theureau, G. et al. 1996, A&A, in press

Thielemann, F.-K., 1996, W.A. Fowler memorial volume, in press

Tonry, J.L. 1991, ApJl 373, L1

Tonry, J.L., Ajhar, E.A., & Luppino, G.A. 1990, AJ 100,1416

Tonry, J.L., & Schneider, D.P. 1988, AJ 96, 807

Truran, J.W. 1996, Talk presented at the STScI May Symposium, Baltimore

Turner, E.L. 1996, Talk presented at the Critical Dialogue in Cosmology Conference, Princeton

Weiss, A., Salaris, M., & Degl'Innocenti, S. 1996, Poster presented at the STScI May Symposium, Baltimore

Whitmore, B.C., et al. 1995, ApJ 454, 773

Winget, D.E., et al. 1987, ApJ 315, L77

Wood, M.A. 1992, ApJ 386, 539

Active Galaxies at Large Redshifts

By Malcolm S. Longair

Cavendish Laboratory, University of Cambridge, Cambridge CB3 0HE, England.

A wide variety of recent observations related to active galaxies at large redshifts is reviewed. Studies of galaxies with the Hubble Space Telescope have shown that active star formation in galaxies has continued over the same redshift interval over which active galaxies display strong cosmological evolution. At redshifts greater than 2, there is evidence that many galaxies underwent bursts of continued star formation and that, at the largest redshifts, $z \sim 3$, the comoving space density of star-forming galaxies is approximately half of that of luminous galaxies ($L \geq L^*$) at the present day. Radio galaxies at redshifts $z \sim 1$ have high optical surface brightness emission regions stimulated by the passage of the radio jets responsible for powering their large-scale radio structures. The significance of these results for understanding the cosmological evolution of active galaxies is discussed.

---

1. Introduction

It is a great pleasure to deliver this review in honour of Igor Novikov's 60th birthday. I got to know Igor when I spent the academic year 1968-9 in Moscow. Seeing so many familiar faces from that period in the audience causes for me a sort of time-warp — one might be back at Zeldovich's seminar at the Shternberg Astronomical Institute. Indeed, I can almost feel the ghosts of Yakov Borisovich Zel'dovich and Iosef Samuilovich Shklovsky hovering over this meeting. These great scientists and teachers influenced everyone who came within their orbit. I am sure they would approve of the outstanding contributions made by Igor and so many of his colleagues.

I propose to discuss the following topics, all of which bear in one way or another upon our understanding the origin and evolution of active galaxies and galaxies in general:

- The cosmological evolution of active galaxies;
- Unification scenarios for radio galaxies and radio quasars;
- HST observations of galaxies at large redshifts;
- Radio galaxies at redshifts $z \sim 1$.

This may seem a somewhat diverse set of topics but they are all, in fact, related — this lecture is distinctly multiply-connected.

2. The Cosmological Evolution of Active Galaxies

One of the first important results of modern astrophysical cosmology was the discovery that the radio luminosity function of radio galaxies and radio quasars has evolved rapidly with cosmological epoch. It took a long time before the form of this evolution could be defined with reasonable precision, largely because it was necessary to identify large complete samples of radio galaxies and radio quasars which extend to large redshifts. The most complete analysis of a wide range of data on complete samples of these radio sources has been carried out by Dunlop & Peacock (1990) with the results shown in Figure 1 for the redshift range $0 \leq z \leq 2$. The upper diagrams show the changes of the comoving radio luminosity function of steep and flat spectrum radio sources which can account for all the available data. The evolution has been described by pure luminosity evolution in which the radio luminosities of all the sources change with redshift z as

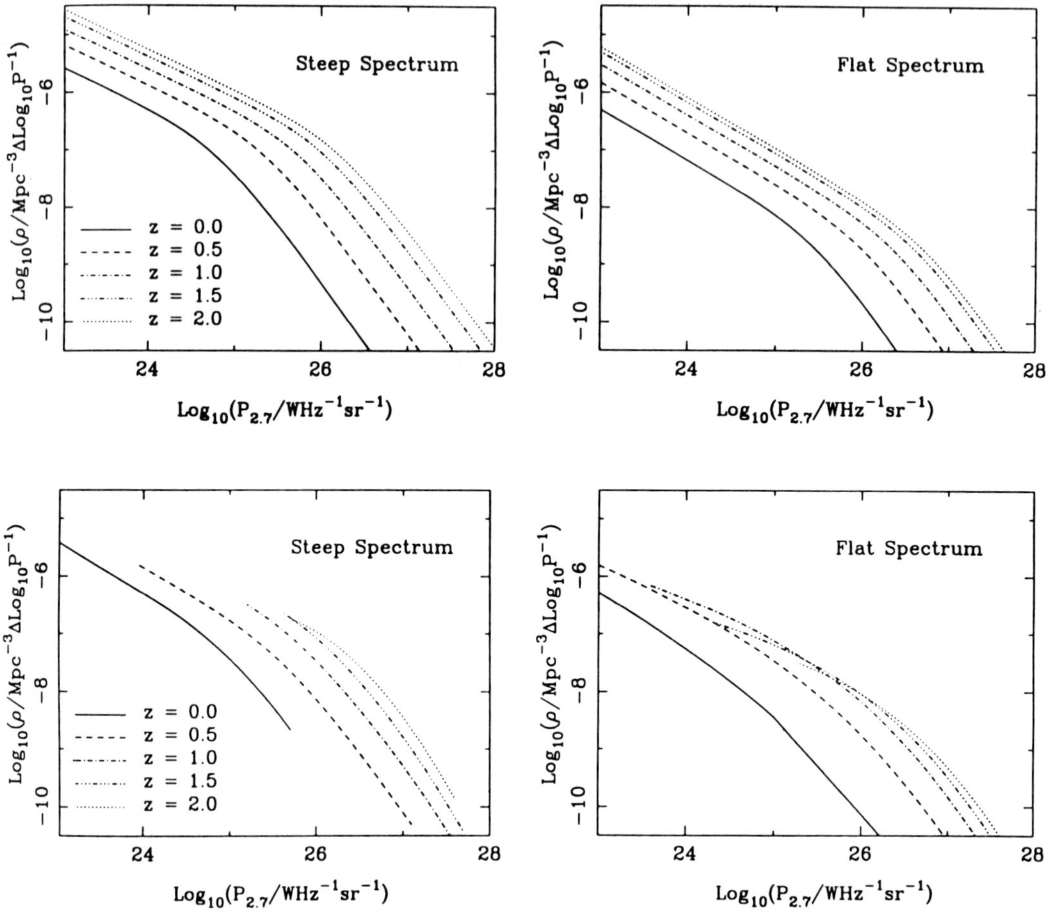

FIGURE 1. Estimates of the evolution of the comoving radio luminosity function of radio galaxies and radio quasars over the redshift interval $0 \leq z \leq 2$. The forms of the comoving luminosity function are shown at redshifts $z = 0,\ 0.5,\ 1.0,\ 1.5,\ 2.0$. In the upper panels, the evolution is parameterised by luminosity evolution in which the radio luminosities of the sources change with redshift as $L(z) = L(0)(1+z)^3$. The evolution of the function is shown separately for sources with steep ($\alpha > 0.5$) and flat ($\alpha < 0.5$) radio spectra, the spectral index α being defined by $S \propto \nu^{-\alpha}$. The lower panels show the same information but restricted to those regions of the luminosity functions for which the comoving space densities are known within a factor of three (Dunlop & Peacock 1990).

$L(z) = L(0)(1+z)^3$ over the redshift interval $0 \leq z \leq 2$. The lower panels show an alternative presentation of the changes of the comoving radio luminosity function for those luminosities for which the comoving space densities are known within about a factor of 3. Within the uncertainties, the radio galaxies and radio quasars exhibit the same form of cosmological evolution.

Similar results have been found for the evolution of the comoving optical luminosity function for optically-selected quasars. Figure 2 shows the forms of function which can describe their evolution over the redshift interval $0 < z < 2$. There are some differences as compared with the evolution of the radio luminosity functions, for example, there appears to little evolution of the most luminous optically selected quasars, but the overall picture

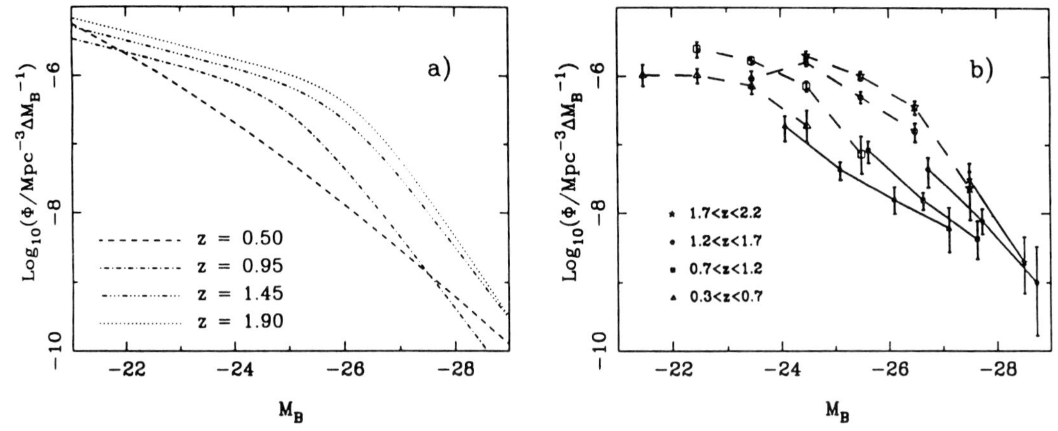

FIGURE 2. (a) The evolution of the comoving optical luminosity function of optically selected quasars at redshifts $z < 2$ as predicted by the modified model of Hewett *et al.* (1993). (b) The same evolution as determined for the Edinburgh UVX survey (solid symbols and lines) and the survey of Boyle, Shanks & Peterson (1988) (open circles and dashed lines) plotted in four redshift intervals (Dunlop 1994, adapted from Miller *et al.* 1993).

is the same — there is a dramatic increase in the comoving space density of quasars out to redshift 2.

At redshifts greater than 2, there is less agreement about the form of the evolution of the comoving luminosity functions. Figure 3 shows the range of possible behaviours of the comoving luminosity functions, ranging from models in which the comoving space density is constant beyond redshift 2 to models in which it decreases abruptly beyond redshift 2, as illustrated by the bell-shaped curve in Figure 3(a) (after Dunlop 1994). The shaded areas in these diagrams illustrate the range of possible variations of the comoving space density of radio sources of radio luminosity 10^{27} W Hz^{-1} sr^{-1} at 2.7 GHz. The reason for the difficulty in determining the behaviour of the luminosity functions at redshifts greater than 2 is that the sources are very luminous and rare. In addition, the comoving volume per unit redshift interval changes from the local Euclidean relation $dV \propto z^2\, dz$ to $dV \propto z^{-1.5}\, dz$ at redshifts $\Omega z > 1$. On this nostalgic occasion, it is perhaps permissible to include in Figure 3(a), the results of the analysis which Andrei Doroshkevich, Yakov Borisevich Zeldovich and I carried out in 1969, when the data were far sparser that they are now (dotted line in Figure 3a). It can be seen that we did not do so badly (Doroshkevich, Longair & Zeldovich 1970).

The key point is that the rapid evolution of the comoving density of active galaxies cannot continue to large redshifts. Dunlop (1994) has redrawn Figure 3(a) in terms of cosmic epoch rather than redshift (Figure 3b), which illustrates how dramatic the changes in the function are with cosmic epoch. It is evident that the quasars and active galaxies flourished when the Universe was only about 20 - 25% of its present age. The cosmological evolutionary trends of the radio galaxies and the radio quasars are remarkably similar — this is not so surprising if one is a believer in unification scenarios.

3. Unification Schemes for Active Galactic Nuclei

The concept of unification resulted from the realisation that projection effects must be important in the observation of certain types of active galactic nuclei. A number

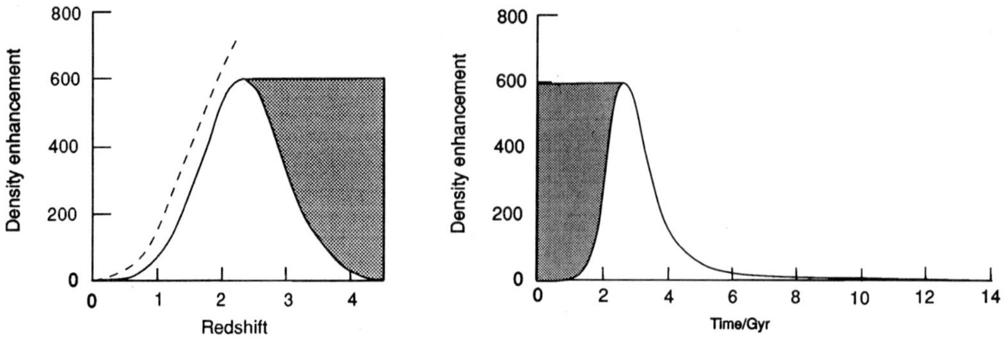

FIGURE 3. (a) Illustrating the evolution of the comoving space density of active galaxies with cosmic epoch. The curve shows the changes of the comoving space density of radio sources with radio luminosity 10^{27} W Hz^{-1} sr^{-1} at 2.7 GHz. The shaded area illustrates the range of possible behaviours of the space density at redshifts $z > 2$. The dotted line shows the results of Doroshkevich, Longair & Zeldovich 1970). (b) The same variation of comoving space density but now plotted as a function of cosmic epoch (Dunlop 1994).

of workers have proposed that much of the diversity of active galactic nuclei can be accounted for in terms of a model in which a single class of active galaxy is viewed at different angles to the line of sight. The intriguing question is the extent to which *all* types of active galactic nuclei can be accommodated within a single scheme.

The idea that projection effects are important in distinguishing between Seyfert I and Seyfert II galaxies was first convincingly demonstrated by Antonucci & Miller (1985) in their classic study of the nearby Seyfert II galaxy NGC 1068. When observed spectropolarimetrically the polarised line emission is as broad as the broad permitted lines observed in Seyfert I galaxies. Antonucci & Miller interpreted these observations in terms of what has become the standard model for the unification of Seyfert I and Seyfert II galaxies. There must be some form of 'obscuring torus' about the nucleus so that, when the galaxy is observed at a small angle to the axis of the torus, the nuclear regions, containing the active nucleus and the broad-line emitting regions, are observed and the galaxy is classified as a Seyfert I galaxy. If the axis of the torus is observed at a large angle to the line of sight, the nuclear regions are obscured and only the narrow-line regions, located further from the nucleus, are observed. The central regions can, however, be observed in the scattered light of the nucleus and this is what is observed spectropolarimetrically. Excellent (and amusing) up-to-date surveys of the unification hypothesis for Seyfert I and Seyfert II galaxies are presented by Antonucci (1993) and Miller (1994). In addition to the spectropolarimetric evidence, other evidence which supports this picture includes:

- Evidence for the beaming of the ionising continuum emission from the nucleus comes from comparing the flux of ionising photons observed from the nucleus with number required to excite the emission line regions. In a number of cases, the observed flux of ionising radiation is less than that required to account for the ionisation and excitation of the gas clouds, suggesting that they are exposed to a more intense radiation field than that which we observe.
- These observations suggest that the radiation from the nucleus is emitted in an 'ionisation cone' and examples of these have been observed by the Hubble Space Telescope. In the case of NGC 1068, the cone structure is entirely consistent with the type of geometry inferred by Antonucci & Miller (1985) and with photon counting analyses (Kinney

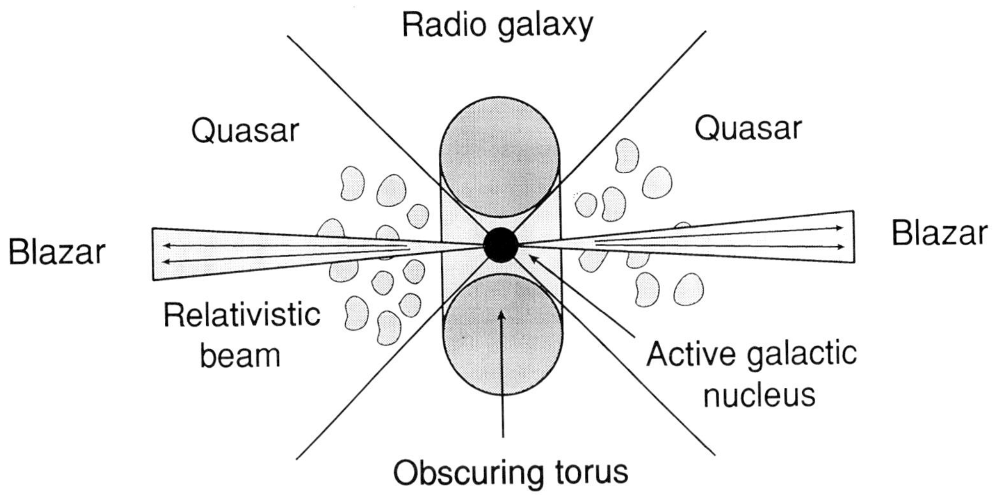

FIGURE 4. Illustrating the unified model for the radio galaxies and radio quasars observed in bright samples of radio sources. Radio quasars are observed when the axis of the radio source lies within about 45° of the line of sight. When observations are made almost along the axis of the radio jet, superluminal radio sources and blazars are observed.

et al. 1991). In the case of the radio galaxy NGC 4261, there seems to be a cone-like structure emanating from the nuclear regions of the galaxy and there is also a remarkable ring about the nucleus (Stockman 1994). The elongated large-scale radio structure lies along the axis of the ring.

• There is a difference in the X-ray absorption properties of Seyfert I and Seyfert II galaxies in the sense that the nuclear X-ray emitting regions of Seyfert I galaxies are observed through much greater column depths of cold gas as compared with the Seyfert II galaxies (Mushotzky *et al.* 1994).

Another major unification industry concerns the strong radio sources as advocated by Barthel (1989, 1994). Barthel's analysis concerned the unification of radio galaxies and radio quasars, by which are meant those extremely luminous radio sources which appear in catalogues of bright radio sources, such as those in the 3CR catalogue (Laing, Riley & Longair 1983) and other bright radio source catalogues (for example, Wall & Peacock 1985). These are very much rarer objects than the Seyfert galaxies and typically have redshifts in the range $0.1 < z < 2$. The advantage in using them is that the double radio structures have well-defined axes and so projection effects can be studied in a straightforward way. The arguments in favour of unification advanced by Barthel (1994) included the following:

• Superluminal motions in compact radio quasars in bright source samples suggest that these sources are observed at a small angle to the line of sight and so projection effects are certainly important for them.

• In some radio quasars, one-sided radio jets are observed emanating from their nuclei, suggesting that relativistic beaming of the radio emission is important.

• Depolarisation asymmetries of the large-scale radio structures are strongly correlated with presence of one-sided radio jets, as expected if the jets were beamed towards the observer (Laing 1988, Garrington *et al.* 1988).

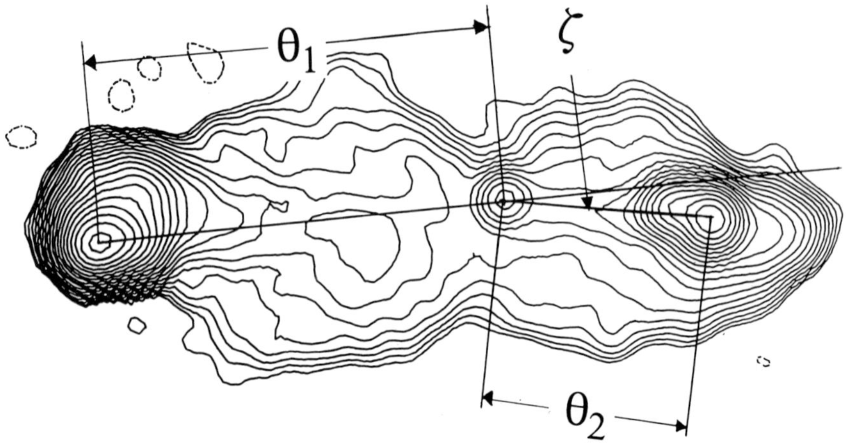

FIGURE 5. Illustrating the typical structure of a Fanaroff-Riley class 2 radio source. The hot-spots lie towards the leading edge of the extended radio structures. The various angles used in the analysis of their structures are shown on the diagram.

- The projected linear sizes of the radio structures of the radio galaxies are greater than those of the radio quasars, as expected if the radio quasars are observed at a smaller angle to the line of sight than the radio galaxies.
- The optical emission observed about a number of radio galaxies is linearly polarised, as expected if the radiation of an obscured nucleus were being scattered into the line of sight, just as in the case of the Seyfert II radio galaxies.

The upshot of these considerations is that a unified picture for radio galaxies and radio quasars, illustrated in Figure 4, has become popular. This picture bears a strong resemblance to the unification scenario for Seyfert galaxies but now radio quasars are observed when the nucleus is observed within a cone of half-angle roughly 45° with respect to the axis of the radio source and a radio galaxy observed when the nuclear regions are hidden by the obscuring torus. Highly collimated radio jets are assumed to be emitted from the nucleus along the axis of the torus and these are responsible for powering the outer radio hot-spots and extended radio lobes. If the radio jet is observed at an angle close to the line of sight, the emission of the relativistic jet is strongly enhanced by relativistic beaming and superluminal motions are observed. The class of objects observed close to the direction of the beam is referred to generically as 'blazars'. This is the orthodoxy advocated by Barthel and we have recently obtained some new results which would be consistent with this picture.

4. Asymmetries in Double Radio Sources

Radio maps of extragalactic radio sources of superb quality are now available and, in particular, those of the brightest radio sources in the sky, for example, those in the 3CR sample, have improved enormously. We have used these data to study asymmetries in the structures of the Fanaroff-Riley Class 2 3CR radio sources (Fanaroff and Riley 1974), that is, sources in which there are hot-spots towards the outer edges of the radio source components (Best, Bailer, Longair and Riley 1995). An example of the structure of these sources is shown in Figure 5.

A convincing case can be made that the 'hot-spots' observed towards the outer edges of these radio structures are continuously supplied with energy by highly collimated jets originating in the active galactic nucleus. The consequent ram pressure of the jet

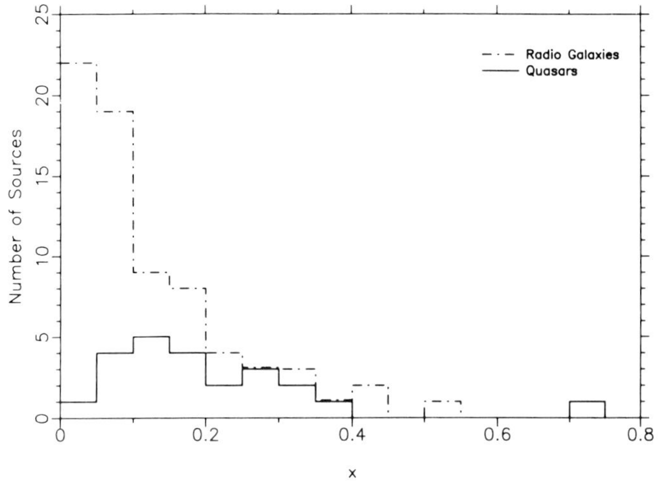

FIGURE 6. The distributions of fractional separation difference x for quasars and radio galaxies in a complete sample of 3CR radio sources. (Best et al. 1995).

causes the hot-spots to 'burn' their way out through the intergalactic medium at a high speed. Thus, asymmetries in the disposition of the hot-spots and the fact that they are generally not perfectly lined up provides information about the dynamics of the jets and their interaction with the surrounding media. To study these asymmetries, we defined three asymmetry properties of the hot-spots using the quantities defined in Figure 5 for all sources in which there are hot-spots in both radio components. These are the *separation quotient* $Q = \theta_1/\theta_2$, the *fractional separation difference* $x = (\theta_1 - \theta_2)/(\theta_1 + \theta_2) = (Q-1)/(Q+1)$, and the *asymmetry angle* ζ. There were 23 quasars and 72 radio galaxies for which the asymmetry parameters could be defined for a complete sample of 3CR radio sources.

First, we assumed that the sources were linear, that is, $\zeta = 0$. The distributions of the fractional separation differences were found to be significantly different for the quasars and the radio galaxies (Figure 6). The radio galaxies have a much narrower distribution of x as compared with the quasars which have a broad distribution and a deficit of sources with $x = 0$. The difference is significant at the 99% level. The x-distributions can be explained naturally within the context of a simple relativistic model in which the hot-spots move outwards from the nucleus at a constant relativistic velocity $v = \beta c$. Because of the light travel time across the source, if the axis of the source lies at an angle θ to the line of sight, the separation quotient $Q = (1 + \beta\cos\theta)/(1 - \beta\cos\theta)$ and the fractional separation difference $x = \beta\cos\theta$ (Ryle and Longair 1967, Longair & Riley 1979). Thus, if quasars are observed at smaller angles to the line of sight than the radio galaxies, they are expected to be more asymmetric and there should be no symmetric quasars at all, unlike the radio galaxies which should have many values of x close to zero.

To test this hypothesis, we have followed the procedure recommended by Banhatti (1980). For a sample of sources with a fixed velocity βc, the probability distribution of x is flat from 0 to β. Therefore, if the probability that the velocity of the hot-spots lies in the range β to $\beta + d\beta$ is $g(\beta)\ d\beta$, the probability distribution for x is

$$p(x) = \int_{\beta=x}^{1} \frac{g(\beta)}{\beta}\ d\beta$$

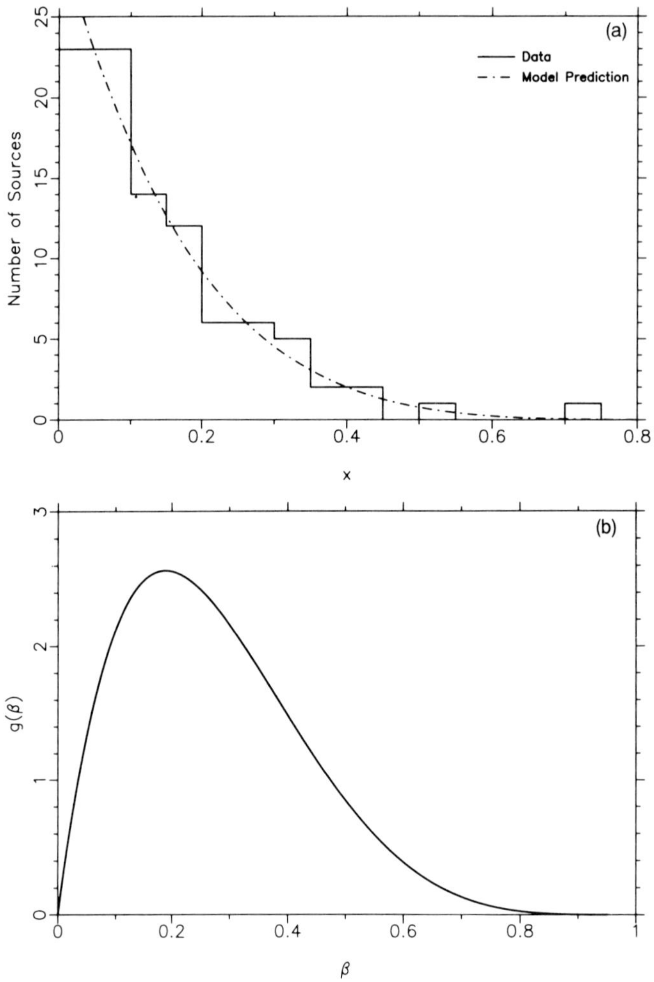

FIGURE 7. (*a*) Illustrating the fit of the function $p(x) = 6.3(1-x)^{5.3}$ to the observed distribution of fractional separation quotients for the complete sample. (*b*) The inferred distribution of hot-spot velocities found by differentiation of the distribution of x shown in (*a*), according to the procedure discussed in the text.

We therefore fitted the distribution $p(x)$ by a suitable function and found $g(\beta)$ by differentiation. Figure 7 shows that there is a broad distribution of inferred velocities with an average value of $\beta \sim 0.2$ but with a distribution which extends to large velocities. Then, assuming that a double radio source is classified as a quasar if $\theta < 45°$ and as a radio galaxy if $\theta > 45°$, we can derive the expected probability distributions of $p(x)$ for the radio galaxies and quasars. This comparison is shown in Figure 8, in which it can be seen that the observed distributions and the simple model are in excellent agreement.

In a second analysis, we investigated the distribution of asymmetry angles ζ for the quasars and the radio galaxies. Figure 9 shows that there is a significant difference between the probability distributions for the quasars and the radio galaxies in the sense that the quasars have a broader and flatter distribution than the radio galaxies. Again, this is exactly as expected if the quasars and radio galaxies have the same intrinsic distribution of asymmetry angles but the quasars are observed preferentially within 45° of the line of sight. To quantify this difference, we constructed a simple model for the

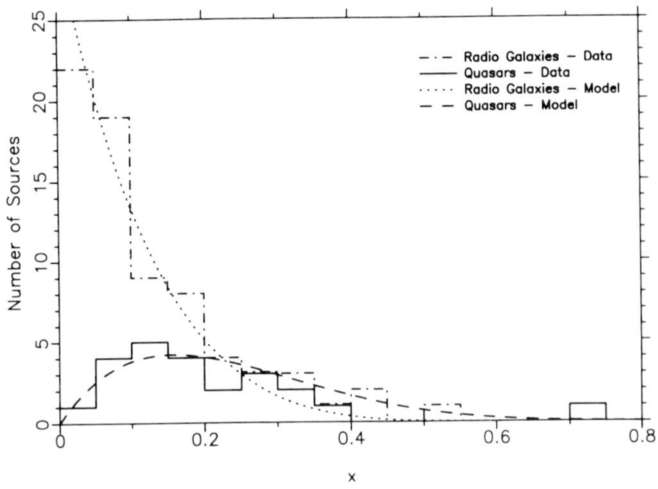

FIGURE 8. Comparison of the observed and predicted distributions of the fractional separation difference for radio galaxies and quasars in the complete 3CR sample of sources.

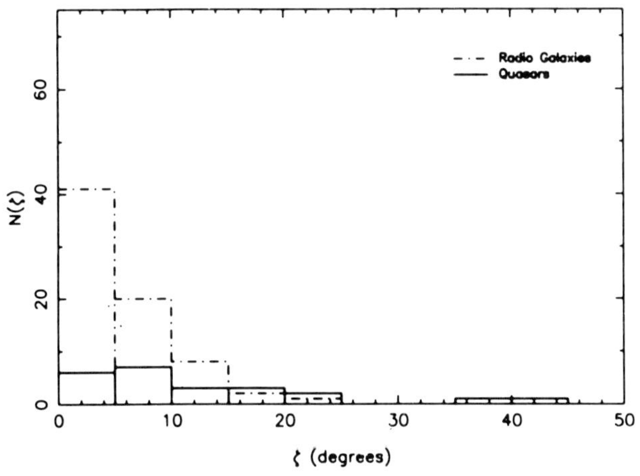

FIGURE 9. The distribution of asymmetry angles ζ for the radio galaxies and radio quasars in the complete 3CR sample.

asymmetry in which one component is ejected at an angle ϕ with respect to the axis of the source. We assumed that the source component is ejected randomly within an angle $\phi_{\max}$. This geometry is illustrated in Figure 10. The predicted distributions for different assumed values of $\phi_{\max}$ are shown in Figure 11, from which it is apparent that typical values of $\phi_{\max}$ of about 10° to 15° can account for the observed distribution of asymmetry angles.

An important aspect of this analysis is that it is *independent* of any assumption about the velocities of the hot spots — the fact that the radio structures of the quasars are found to be more asymmetric than the radio galaxies is simply a projection effect. The intrinsic angular asymmetries are similar for the radio galaxies and the quasars, once they are deprojected. If the hot-spots were ejected randomly within cones in opposite

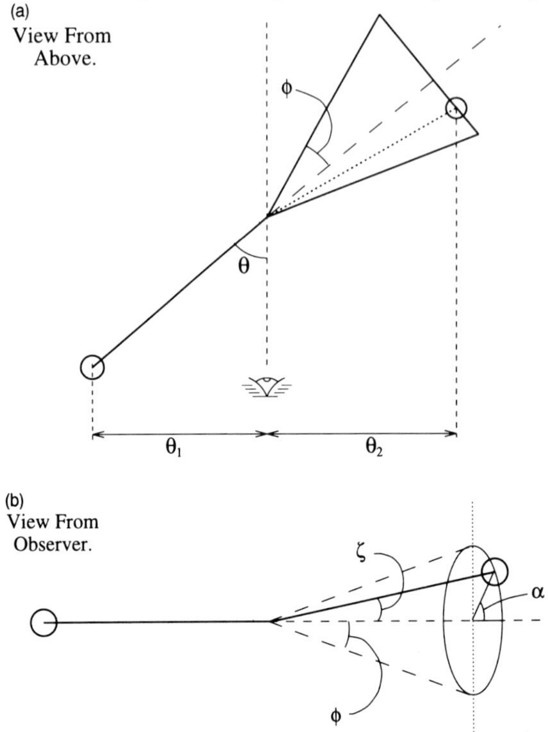

FIGURE 10. Illustrating the simple source geometry used to model the distribution of asymmetry angles observed in the complete 3CR sample of sources.

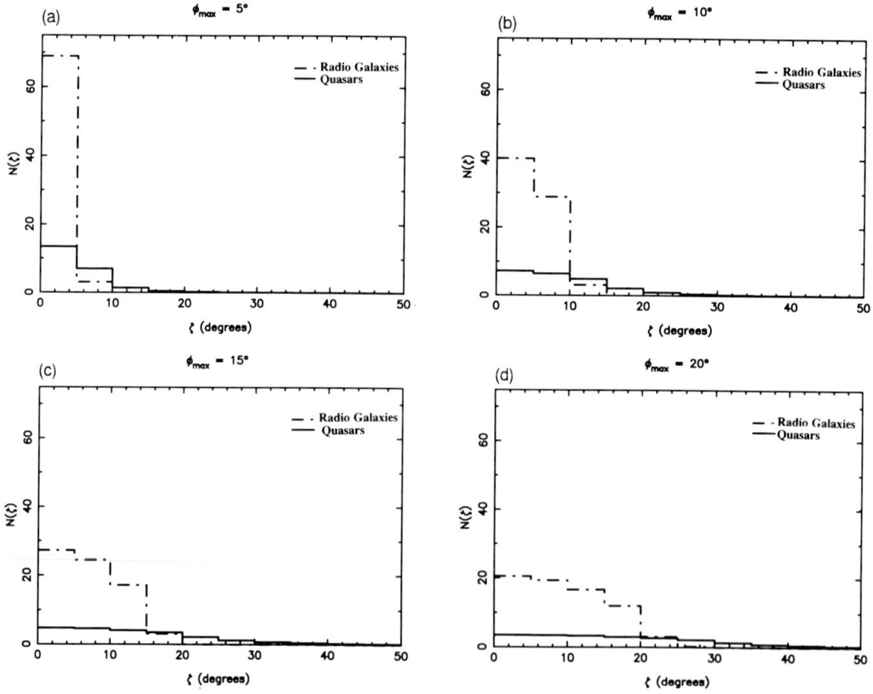

FIGURE 11. The predicted distributions of the asymmetry angle ζ for different assumed values of the intrinsic asymmetry angle $\phi_{\max}$.

directions along the axis of the source, the total solid angle on either side of the source would be the same and so the cones would be somewhat smaller, with $\phi_{\rm max}$ about 7° to 10°.

In our third analysis, we have repeated the first analysis but now assumed that one source component is ejected along the axis of the source and the other randomly within the cone of angle $\phi_{\rm max}$. We found that the asymmetries in x cannot be attributed to the distribution of intrinsic asymmetry angles ϕ but that the x-distributions can be accounted for if the hot-spots move at a significant relativistic velocity, with a velocity distribution similar to that shown in Figure 7(b). The environment of the source may also make a significant contribution to the asymmetry, as noted by McCarthy et al. (1991), but the differences in the distributions of the fractional separation quotients can be most simply attributed to the light travel time effect which implies that the hot-spots move significantly relativistic. Similar high velocities have been found by an entirely independent argument. Liu, Pooley & Riley (1992) determined the rate of advance of the hot spots from the parent galaxy in a sample of double radio sources using synchrotron ageing arguments. Laing (1993) has plotted these velocities as a function of radio luminosity. The sources in our sample all have radio luminosities greater than about 3×10^{27} W Hz^{-1} and the hot-spot velocities $v \sim 0.1 - 0.2$ are similar to those found in our analysis.

The upshot of these considerations is that it is quite plausible that the radio galaxies and radio quasars belong to the same population of objects but observed at different angles to the line of sight, consistent with the identical forms of their evolution with cosmic epoch. If true, these results are important for understanding the physics of active galaxies. For example, if one wishes to study the host galaxies of radio quasars, this has already been done — they are the radio galaxies. They form a remarkably homogeneous class of galaxy as is attested by their tight K-magnitude — redshift relation (Lilly & Longair 1984). Incidentally, we have redetermined that relation for the radio galaxies discussed in Section 8 and find that, with the new data, there is less than 0.5 magnitudes scatter about the mean K-z relation, entirely consistent with the earlier work of over a decade ago (Best *et al.* 1996b).

5. Distant Galaxies Observed by the Hubble Space Telescope

Observations with the Hubble Space Telescope have resulted in important new results for different aspects of these studies. It is important to appreciate exactly what is observed when HST observations of galaxies extend to redshifts $z \geq 0.5$. Giavalisco *et al.* (1996) have simulated the appearances of normal galaxies at large redshifts using observations of nearby galaxies in ultraviolet wavebands at 152 nm and 249 nm. The galaxies M32, M31, M33, M74, M77, M81 and M82 were observed in these wavebands by the Ultraviolet Imaging Telescope of the Astro-1 mission and Giavalisco and his colleagues have used these to determine the morphologies and apparent magnitudes these galaxies would have if observed by the WFPC2 in 4.3 hour integrations. Some of their results are summarised in Table 1.

The figures enclosed in square brackets in Table 1 indicate that, if the galaxies were located at these redshifts, they would not be observable by the HST because they have too low surface brightnesses. NGC 1068 is a special case since it possesses a Seyfert nucleus and so can be observed even at the very large redshift $z = 3$, but the host galaxy would be undetectable. The two other galaxies which are just detectable at redshift $z = 1.4$ are observed with too low signal-to-noise ratio for them to be classified morphologically. These computations make the important point that, if galaxies are to be classified successfully at redshifts $z > 1$ by the HST, they must either have greater

Galaxy		Type	Luminosity	Magnitude at $z = 1.4$ through F606W filter	Magnitude at $z = 3.0$ through F606W filter
NGC221	M32	E2	$\sim 0.1L^*$	[33.2]	[38.1]
NGC224	M31	Sb	$\sim L^*$	[28.9]	[31.0]
NGC598	M33	Sc	$\sim 0.1L^*$	27.5	[29.0]
NGC628	M74	Sc	$\sim L^*$	25.9	[27.8]
NGC1068	M77	SeyII	a few L^*	23.9	26.2
NGC3031	M81	Sb	$\sim L^*$	[27.2]	[30.2]
NGC3034	M82	Irr	$\sim 0.1L^*$	[29.2]	[31.6]

TABLE 1. Local galaxies observed by the Ultraviolet Imaging Telescope of the Astro-1 Shuttle Mission. The table shows the expected magnitudes of the galaxies if they were observed at redshifts $z = 1.4$ and $z = 3.0$ by the HST. Square brackets indicate that the galaxies would be of too low surface brightness to be detected (Giavalisco *et al.* 1996).

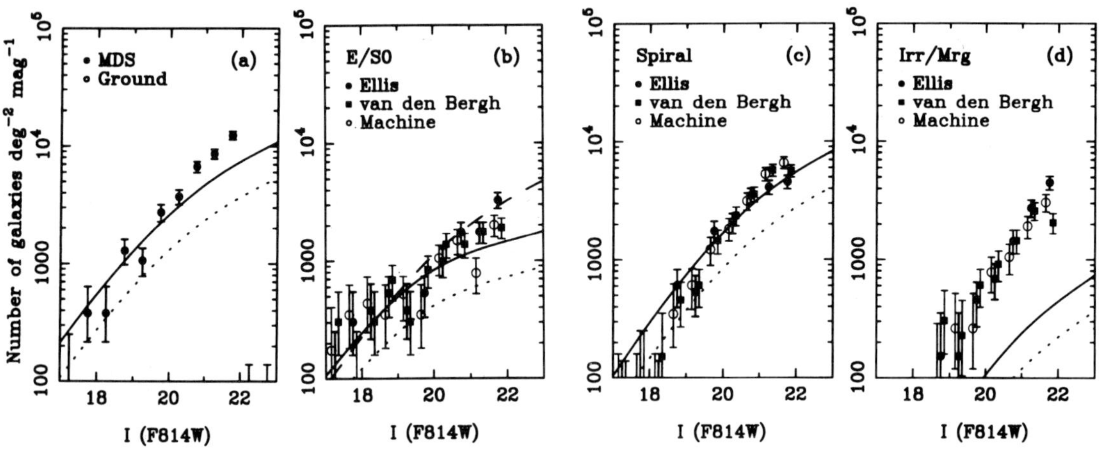

FIGURE 12. The counts of faint galaxies of different morphological types. The counts as determined by van den Bergh, Ellis and according to automatic classification procedures for the same sample of galaxies are indicated by different symbols. The solid lines show the expected galaxy counts for a uniform world model with no evolution. The pronounced excess of irregular/merging galaxies is apparent (Abraham *et al.* 1996).

surface brightnesses or luminosities than the typical galaxies we observe nearby. We will find that there are good observational reasons why this is, in fact, the case. It is also clear that it is possible to make excellent progress in the morphological classification of galaxies out to redshifts of about 0.5, and greater for more luminous galaxies.

Several studies of the morphologies of faint galaxies have been reported (Glazebrook *et al.* 1995(a), Abraham *et al.* 1996, Driver *et al.* 1995, 1996). Figure 12 shows the counts of galaxies of different morphological types by Abraham *et al.* (1996). The first panel shows the total counts of galaxies which exhibits the well-known excess at faint apparent magnitudes. The solid lines show the expected counts for uniform world models with $q_0 = \frac{1}{2}$ and it can be observed that the E/S0 and spiral samples follow closely these expectations. The important result is that the excess blue galaxies are associated with the objects classed as Irregulars/Mergers.

There has been considerable debate about the nature of the excess faint blue galaxy population. It had been expected that there might well be more blue galaxies at faint apparent magnitudes because even passively evolving models of galaxies suggest that the old stellar population should be brighter in the past and there should be more star formation activity. One of the surprises of the first redshift surveys, which extended to apparent magnitudes at which the blue excess was observed, was that the mean redshift of the galaxies at the faintest magnitudes did not increase any more rapidly than would have been expected if the galaxy population had remained unchanged with cosmic epoch (Glazebrook *et al.* 1995b). This result was interpreted as indicating that the excess blue galaxies were associated with a population of dwarf galaxies. There must, however, be more to the story than this.

6. Galaxies at redshifts $z \sim 1$

It is now possible to study fainter samples of galaxies and three examples indicate what is now becoming the state of the art. Schade (1996) has described the most recent results of the Canada-France Redshift Survey (Le Fèvre *et al.* 1995). This magnitude-limited survey contains 943 objects with magnitudes $17.5 \le I_{AB} \le 22.5$. The objects for which redshifts have been measured extend to redshifts $z = 1.2$. HST images have been secured for 32 randomly selected galaxies with redshifts $z > 0.4$ and these display the normal range of morphological types. There are however important differences. The mean rest frame surface brightnesses of the late type galaxies are about 1.2 magnitudes greater than those of nearby galaxies. Some degree of peculiarity/asymmetry is observed in 30% of the objects and 13% show clear signs of mergers or interactions. There are compact blue components in 30% of the galaxies and these occur predominantly in the peculiar systems, but a few of them are also present in normal systems. According to Schade *et al.* (1995), these galaxies are predominantly the cause of the excess of faint blue galaxies and the numbers are consistent with those found in the morphological survey described above.

The second example concerns the recent results of Cowie *et al.* (1995) on the redshift distribution of faint galaxies. The galaxies were selected from very deep surveys in small areas of sky and consisted of all 367 objects in these areas which satisfied the criteria $K < 20$, $I < 22.5$ and $B < 24.5$. Spectra for these galaxies are being obtained with the Keck 10-metre telescope and the programme has been remarkably successful in discovering large redshift galaxies. Among the galaxies, 91 are already known to have redshifts $z > 0.7$ and 40 redshifts $z > 1$. The reason for their success is immediately apparent from the typical spectra of the galaxies for which spectra have been obtained (Figure 13). The large redshift galaxies have strong emission lines, characteristic of regions of star formation. According to their analysis, these are luminous galaxies with absolute B magnitudes, $M_B \sim -21$. The luminosities of the large redshift galaxies in the [OII] line are typically at least an order of magnitude greater than those of a reference sample at redshifts $z < 0.7$. Cowie *et al.* infer that these are star-forming massive galaxies at redshifts $z \sim 1$ and the rates at which the stars are being formed can account for between about 5 and 20% of their present stellar populations. Deep HST images have been obtained for 9 of the galaxies at redshifts $z > 1$ in the I waveband. As they note, the galaxies have 'strikingly unusual morphologies, often consisting of chains or structures of compact blobs, suggesting that they are generally not dominated by uniformly distributed star formation'. These results suggest that among the faint blue galaxies there is a population of distant star-forming massive galaxies in addition to objects at smaller redshifts. Notice that the success of the HST observations depended upon these galaxies having regions of enhanced surface brightness.

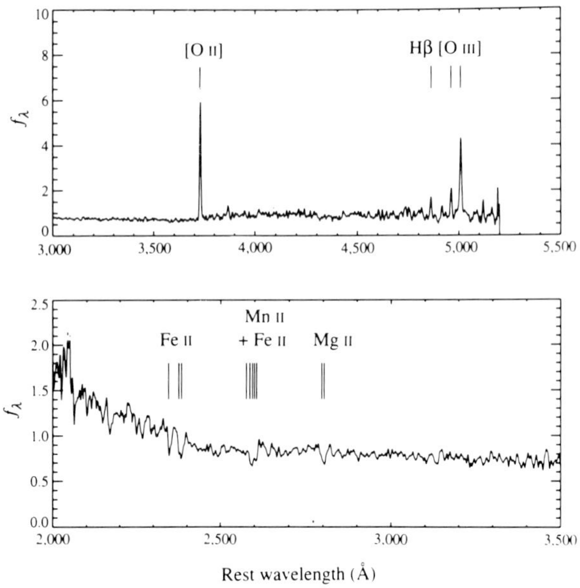

FIGURE 13. The average rest-wavelength spectra of the objects in the deep survey field SSA13. It is apparent that they are strong emission line objects with only weak absorption lines in the ultraviolet region of the spectrum (Cowie *et al.* 1995).

Similar preliminary results were reported by Koo (1996) from a deep survey undertaken with the Keck 10-metre Telescope in the HST Groth strip. The sample extends to $I = 24$ and in this area they already have redshifts for 35 galaxies in the range $0.3 < z < 1.2$, with a mean redshift of about 0.8. Again, the sample contains a large number of unusual galaxies including objects with multiple knots and the types of chain galaxy reported by Cowie *et al.* (1995). In addition, they have found 9 red galaxies which are as red as elliptical galaxies are at the present epoch. The inference is that these galaxies have already had time to form old stellar populations and so must have undergone their last major burst of star formation at redshifts $z > 2$.

7. Galaxies at Large Redshifts

I will discuss two examples of the study of galaxies at very large redshifts. The first results from the study by Windhorst and Keel (1995) of what they refer to as a young 'elliptical' radio galaxy at a redshift $z = 2.390$. Pasarelle (1996) described new observations of the field of this radio galaxy which, by great good fortune, lies at such a redshift that the Lyman-α line is redshifted into the narrow F410W filter of the WFPC2 of the HST. They find evidence for 18 Lyman-α objects at this redshift, all of them with luminosities between about 0.1 and 1 L^*. All of these objects seem to be compact and the sum of their brightness distributions seems to follow the $r^{1/4}$ law. They suggest that this is evidence for the early formation of the bulges of galaxies.

The largest redshift systems which have been identified as young star-forming galaxies have been discovered by searching for the redshifted Lyman limit by multicolour photometry. The technique is similar to that described by Lilly & Cowie (1987) and refined for the detection of 'Lyman-limit galaxies' by Steidel & Hamilton (1992, 1993). The predicted spectrum of a starburst galaxy at a redshift of 3 is illustrated in Figure 14. At

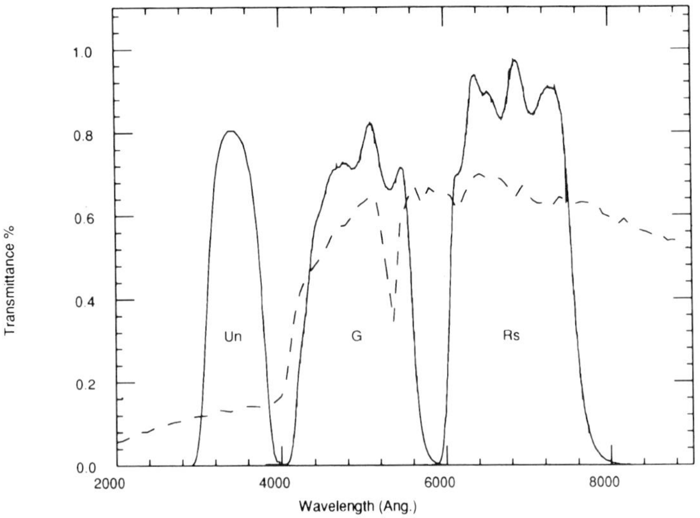

FIGURE 14. (a). Illustrating how three colour photometry in the U_n, G and R_s wavebands can isolate star-forming galaxies at large redshifts. The dashed line shows the spectrum of a star-forming galaxy at a redshift $z = 3$ (Macchetto & Giavalisco 1995).

this redshift, the Lyman limit is shifted to 400 nm and so the characteristic signature of these objects is that roughly equal intensities are observed in the G and R_S wavebands but the intensity in the ultraviolet waveband U_n is low as illustrated in Figure 14.

The story began with the successful attempt to identify the large redshift absorption systems present in the background quasar QSO 0000-262, which has an emission redshift $z = 4.11$ (Steidel & Hamilton 1992, 1993). In this field, Macchetto *et al.* (1993) identified a 'Lyman-α radio quiet galaxy' at a redshift $z = 3.428$. Searches in four other QSO fields are described by Steidel *et al.* (1995). Macchetto & Giavalisco (1995) and Steidel *et al.* (1996) have described further observations of these fields. Macchetto & Giavalisco (1995) described the application of this multicolour technique to the field containing the galaxy at redshift $z = 3.428$ and several objects with the signature of star-forming galaxies were found. Giavalisco (1996) described the exciting result that spectroscopy with the Keck 10-metre telescope has confirmed that these objects are indeed galaxies at redshift $z \sim 3.2$. These galaxies have been imaged by the WFPC2 and, when the images of the galaxies are summed, they are found to follow the standard de Vaucouleurs $r^{1/4}$ dependence of surface brightness upon radius for elliptical galaxies.

Steidel *et al.* (1996) have obtained the spectra of 24 candidate star-forming galaxies selected in both the quasar fields and in random regions of sky and have had great success in measuring redshifts for these with the Keck 10-metre telescope. 17 of the objects have redshifts in the interval $3.01 \leq z \leq 3.43$. The velocity dispersions within the galaxies suggest that they are massive galaxies. They find the important result that the comoving space density of these star-forming galaxies in the redshift interval $3 \leq z \leq 3.5$ is about half that of luminous galaxies with $L \geq L^*$ at the present epoch. The star formation rates correspond to about about $8.5h_{50}^{-2}\ M_\odot\ \mathrm{yr}^{-1}$, similar to the star formation rates per galaxy found by Cowie *et al.* (1995). Steidel *et al.* infer that they have discovered the formation of the spheroidal components of the progenitors of massive galaxies — massive galaxy formation was certainly well underway by a redshift of 3.

The overall trend of these recent results is that it seems beyond doubt that star formation has proceeded actively in normal galaxies from a redshift of at least three onwards. Thus, for the first time, we have direct evidence that there was certainly enhanced star formation activity at the early epochs when the active galaxies were at maximum activity.

8. Radio Galaxies at Redshifts $z \sim 1$

Finally, let me describe some of our recent HST observations of radio galaxies at redshifts $z \sim 1$. The programme consists of WFPC2 imaging observations of a complete sample of 3CR radio galaxies in the redshift interval $0.6 < z < 1.8$. These radio galaxies are of special interest because they exhibit the strong cosmological evolutionary trends observed in the radio source and quasar populations as a whole. It is also known that, in the majority of cases, the optical structures are aligned with the radio axes of these double radio sources (Chambers *et al.* 1987, McCarthy *et al.* 1987). All the radio galaxies were observed with the HST in wavebands corresponding more or less to rest-wavelength U and B wavebands. These observations have been supplemented by infrared observations at 2.2 μm taken with UKIRT, which have angular resolution of about 1 arcsec, and by 8.4 GHz VLA observations with angular resolution 0.18 arcsec. The infrared observations provide images of the old stellar populations in these galaxies and they all resemble standard giant elliptical galaxies.

The high resolution HST images are dramatically different. Virtually all the images show emission regions aligned with the jets which are assumed to be powering the hot-spots in the outer radio lobes. Perhaps the most striking result is the comparison of the maps of all the radio galaxies in our sample in the redshift interval $1 \leq z \leq 1.3$ (Best, Longair & Röttgering 1996a). The radio sources have the same intrinsic luminosities, implying that the jet luminosities are the same for all of them. Figures 15 and 16 show a montage of these 8 radio galaxies in order of increasing separation between the components of the double radio sources. It can be seen that the most remarkable structures are associated with the smaller double radio sources. The optical emission regions associated with the radio galaxies 3C 266, 368, 324 and 280 are all aligned along the axis of the double radio sources. Comparison of the optical and infrared images shows that the optical structures are on more or less the same physical scale as the host galaxy. As the sizes of the double source increase, the optical emission regions become less prominent and, although there is still some alignment with the radio axis, the structures are on a smaller physical scale. According to the theory of double radio sources, the large sources are older than the smaller sources and what is of particular interest is that, for some of them, synchrotron ages are available. These arguments suggest that the radio sources associated with 3C 266 and 280 are a few million years old (Liu, Pooley & Riley 1992).

These observations suggest that the strongly-aligned optical structures are short-lived phenomena, which are stimulated by passage of the radio jet. As we have pointed out already (Longair, Best & Röttgering 1995), it seems that no single theory of these alignments can account for all the observations. The observation of polarised optical emission from some of these sources suggests that scattering of light from an obscured quasar must play some role in accounting for the emission from these optical structures. It seems, however, that the primary cause of the structures must be the interaction of the radio jet with cool interstellar clouds within the parent galaxy. One possibility is that the structures are associated with jet-induced star-formation. It would then be possible to account rather naturally for the change in structure with increasing physical size. The lifetime of the newly formed stars and associated HII regions would amount to only about 10^7 years. After this time, the luminosities of the star-forming regions would decay and the star clusters relax within the potential of the parent giant elliptical galaxy. The

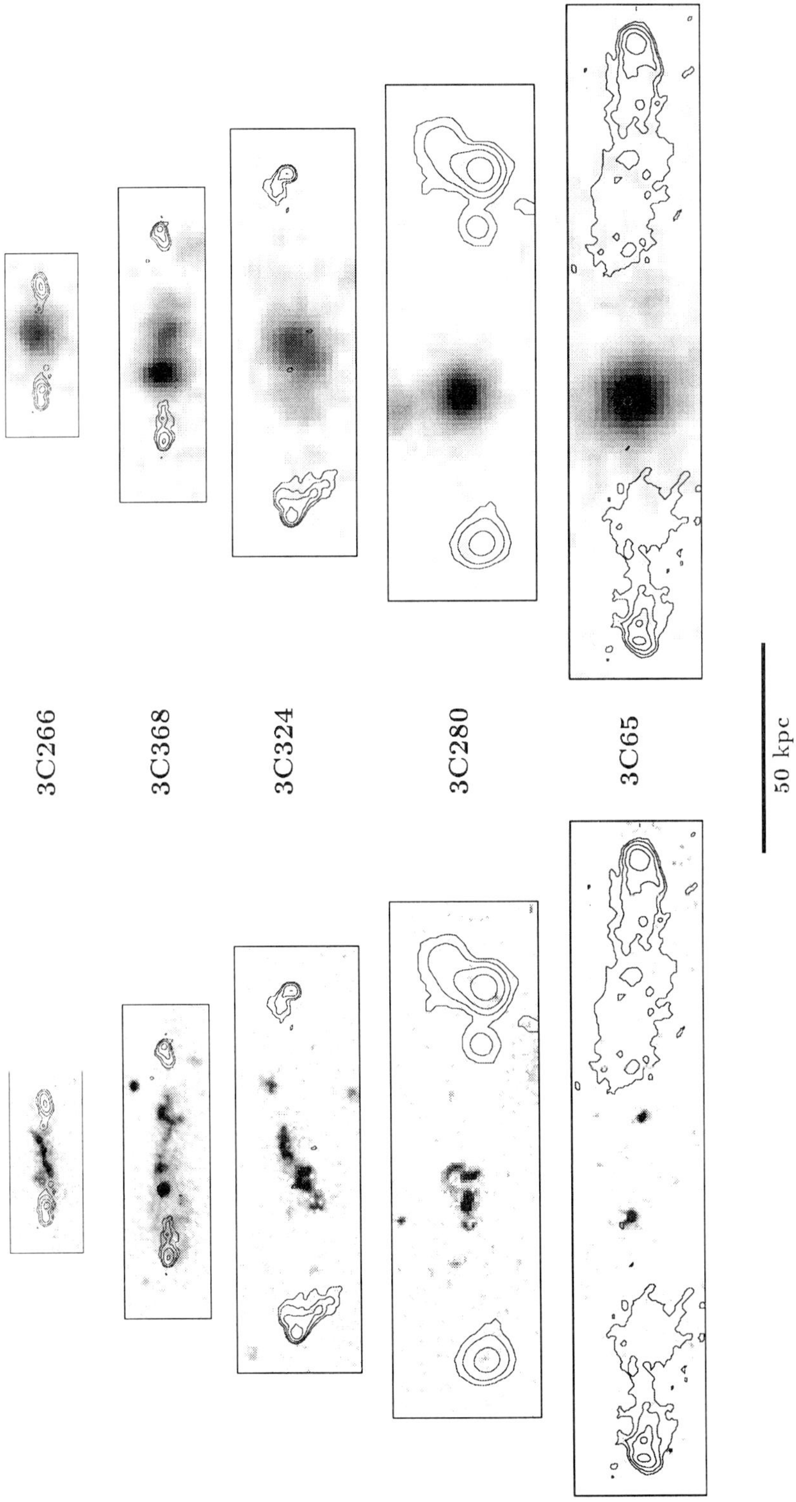

FIGURE 15. HST (left panels) and UKIRT (right panels) images of the radio galaxies 3C 266, 368, 324, 280 and 65 with the VLA radio contours superimposed. The images are drawn to the same physical scale (Best, Longair & Röttgering 1996a).

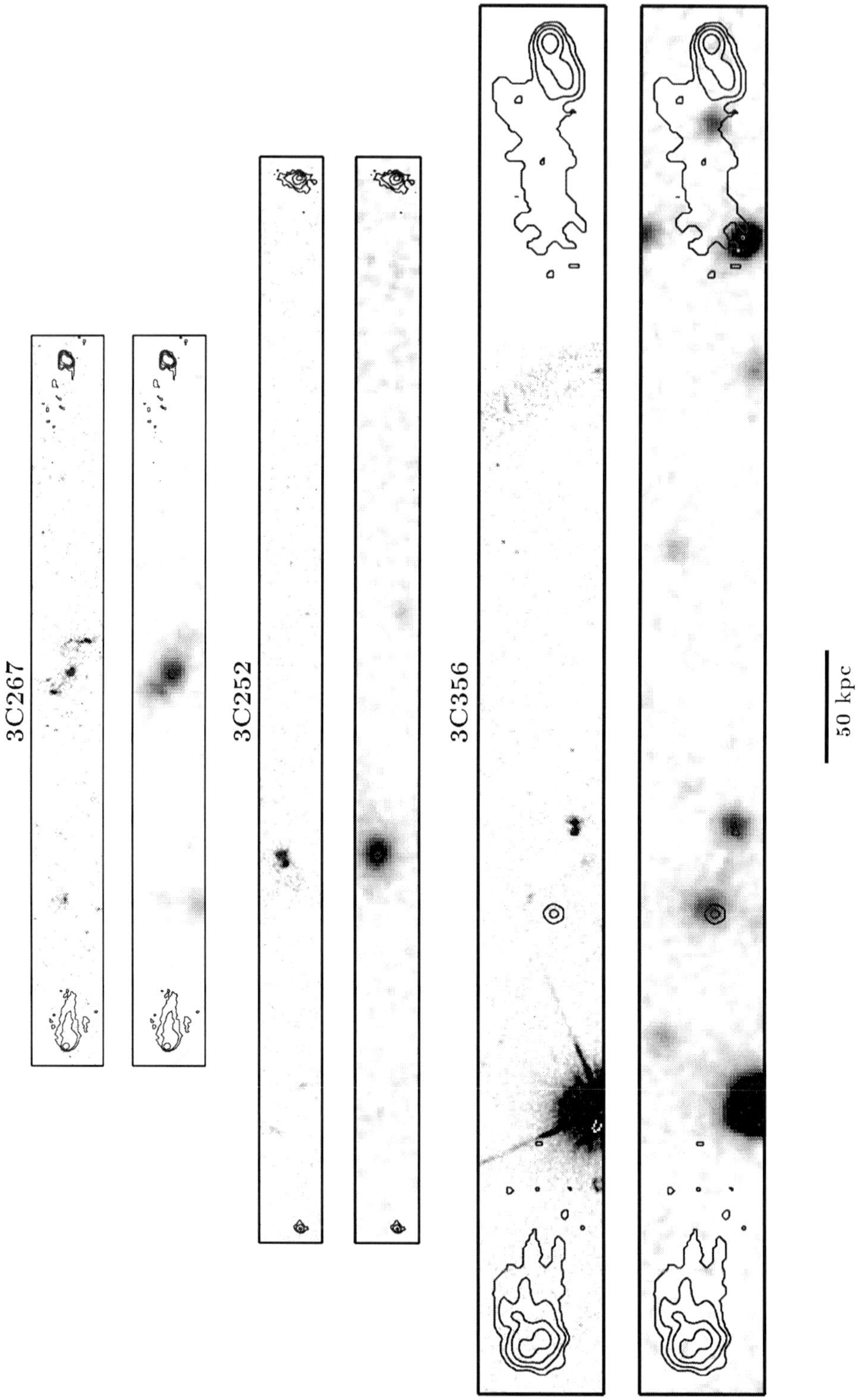

FIGURE 16. HST (upper panels) and UKIRT (lower panels) images of the radio galaxies 3C 267, 252 and 356 with the VLA radio contours superimposed. The images are drawn to the same physical scale (Best, Longair & Röttgering 1996a).

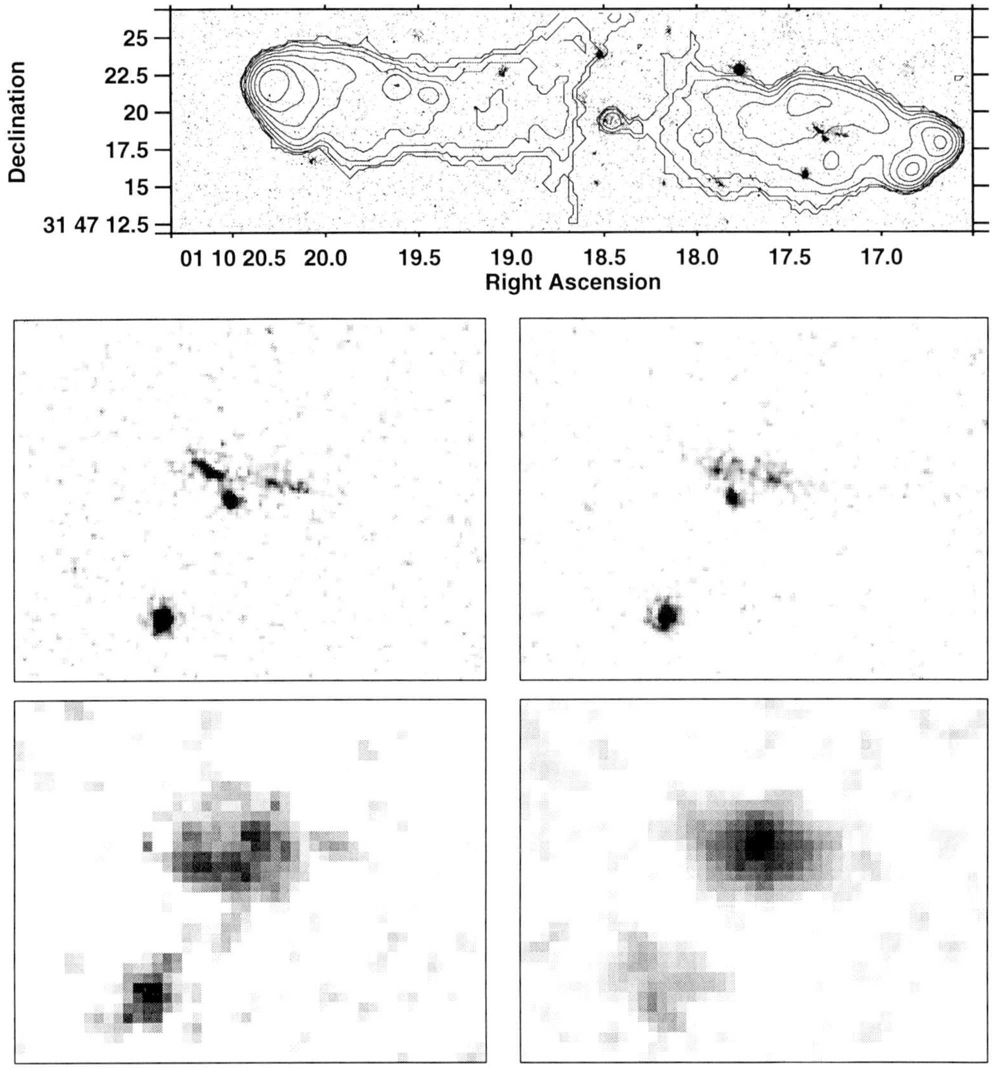

FIGURE 17. (a) The HST image of the radio galaxy 3C 34 superimposed upon which are the VLA radio contours of the double radio source structure. (b) and (c) Images of the structure of the optical 'jet' which lies along the line from the nucleus of the galaxy to the western hot-spot as observed through the f555W and f785LP filters respectively. (d) J and (e) K images of the galaxy associated with the optical jet observed with UKIRT (Best, Longair & Röttgering 1996b)

polarisation of the light would be attributed to the scattering of the light of an obscured quasar by the dust or gas associated with the star-forming regions. A problem with this picture is that there have so far been no reports of young stars in the spectra of the emission regions, but this may well be because the most massive O and B stars have already died by the stage these structures are observed. Alternatively, the alignments may be due to the illumination of pre-existing dust and gas clouds by a central obscured quasar. There remains the problem of accounting for the presence of large amounts of cool gas and dust within the body of the parent galaxy.

One remarkable observation, which may be evidence for jet-induced star-formation, has been found in the field of the radio galaxy 3C 34, which has a redshift of 0.69. In this source, the radio galaxy is rather diffuse with no prominent structures such as those found in the smaller double sources, as can be seen in Figure 17. However, along the axis from the nucleus of the radio galaxy to the brightest western hot-spot, there is a remarkable linear feature which passes close to the galaxy labelled (a). These structures are shown in more detail in Figure 10(b-e). The linear feature is bluer than other objects in the field. Our interpretation of this feature is that it represents the interaction of the radio jet with the interstellar gas in a galaxy which just happens by accident to lie in the path of the jet.

9. Conclusions

The results of this discussion which I believe are important for astrophysical cosmology are:

- The epoch of maximum activity of active galaxies occurred when the Universe was only about 20 – 25% of its present age.
- Active star formation in galaxies and the formation of their stellar populations was already well underway by a redshift of 3.
- It would be remarkable if there were not some relation between the evolution of the population of active galaxies and formation of their stellar populations.
- The radio galaxies at redshifts $z \sim 1$ differ from these at small redshifts in that there must be large amounts of cold gas present within in the body of the galaxies to account for the high brightness structures aligned with the radio jets.
- It is entirely plausible that the optical structures observed in the 3CR radio galaxies are associated with star formation regions which have been created by the incidence of the radio jets.
- A plausible interpretation these phenomena is that they are all associated in different ways with the changing gaseous environments within galaxies as their stellar populations are formed.

REFERENCES

Abraham, R.G., van den Bergh, S., Glazebrook, K., Ellis, R.S., Santiago, B.X., Surma, P. & Griffiths, R.E. 1996. *MNRAS*, in press.

Antonucci, R. 1993. *ARAA*, **31**, 473.

Antonucci, R. & Miller, J. 1985. *ApJ*, **297**, 621.

Banhatti, D.G. 1980. *AA*, **84**, 112.

Barthel, P.D. 1989. *ApJ*, **336**, 606.

Barthel, P.D. 1994. In *First Stromlo Symposium: Physics of Active Galactic Nuclei*, (eds.) Bicknell, G.V., Dopita, M.A. & Quinn, P.J., p. 175. San Francisco: Astron. Soc. Pacific Conf. Series.

Best, P.N., Bailer, D.M., Longair, M.S. & Riley, J.M. 1995. *MNRAS*, **275**, 1171.

Best, P.N., Longair, M.S. & Röttgering, H.J.A. 1996a. *MNRAS*, (in press).

Best, P.N., Longair, M.S. & Röttgering, H.J.A. 1996b. *MNRAS*, (in preparation).

Boyle, B.J., Shanks, T & Peterson, B.A. 1988. *MNRAS*, **238**, 957.

Chambers, K.C., Miley, G.K.& van Breugel, W.J.M. 1987. *Nature*, **329**, 604.

Cowie, L.L., Hu, E.M. & Songaila, A. (1995). *Nature*, **337**, 603.

Doroshkevich, A.G., Longair, M.S. & Zeldovich, Ya.B. 1970. *MNRAS*, **147**, 139.

Driver, S.P., Windhorst, R.A. & Griffiths, S.P. 1995. *ApJ*, **453**, 48.

Driver, S.P., Windhorst, R.A., Ostrander, E.J., Keel, W.C., Griffiths, R.E. & Ratnatunga, K.U. 1995. *ApJ*, **449**, L23.

Dunlop. J.S. 1994. In *Frontiers of Space & Ground-based Astronomy*, (eds) W. Wamsteker, M.S. Longair & Y. Kondo, p. 395. Dordrecht: Kluwer Academic Publishers.

Dunlop, J.S. & Peacock, J.A. 1990. *MNRAS*, **247**, 19.

Fanaroff, B.L. & Riley, J.M. 1974. *MNRAS*, **167**, 31P.

Garrington, S.T., Leahy, J.P., Conway, R.G. & Laing, R.D. 1988. *Nature*, **331**, 149.

Giavalisco, M. 1996. *HST-2*, Proceedings of the Second European Symposium on the Hubble Space Telescope (eds. P. Benvenuto, F. Macchetto & E. Schreier), (in press).

Giavalisco, M, Livio, M., Bohlin, R.C., Macchetto, F.D. & Stecher, T.P. 1996. (preprint)

Glazebrook, K., Ellis, R.S., Santiago, B. & Griffiths, R. 1995a. *MNRAS*, **275**, L19.

Glazebrook, K., Ellis, R., Colless, M., Broadhurst, T., Allington-Smith, J. & Tanvir, N. 1995b. *MNRAS*, **273**, 157.

Hewett, P.C., Foltz, C.B. & Chaffee, F.H. 1993. *ApJ*, L43.

Kinney, A., Antonucci, R., Ward, M., Wilson, A. & Whittle, M. 1991. *ApJ*, **377**, 100.

Koo, D. 1996. *HST-2*, Proceedings of the Second European Symposium on the Hubble Space Telescope (eds. P. Benvenuto, F. Macchetto & E. Schreier), (in press).

Laing, R.D. 1988. *Nature*, **331**, 149.

Laing, R.A. 1993. In *Astrophysical Jets*, (eds) D. Burgarella, M. Livio & C. O'Dea, p 95. Cambridge: Cambridge University Press.

Laing, R.A., Riley, J.M. & Longair, M.S. 1983. *MNRAS*, **204**, 151.

Le Fèvre, O., Crampton, D., Lilly, S.J., Hammer, F. & Tresse, L. 1995. *ApJ*, **455**, 60.

Lilly, S.J. & Cowie, L.L. 1987. In *Infrared Astronomy with Arrays*, (eds. C.G. Wynn-Williams & E.E. Becklin), 473. Honolulu: Institute for Astronomy, University of Hawaii Publications.

Lilly, S.J. & Longair, M.S. 1984. *MNRAS*, **211**, 833.

Liu, R., Pooley, G.G. & Riley, J.M. 1992. *MNRAS*, **257**, 545.

Longair, M.S., Best, P.N. & Röttgering, H.J.A. 1995. *MNRAS*, **275**, L47.

Longair, M.S. & Riley, J.M. 1979. *MNRAS*, **188**, 625.

Macchetto, D & Giavalisco, M. (1995). *ESO Messenger*, September 1995, page 14.

Macchetto, D., Liparo, S., Giavalisco, M., Turnshek, D.A. & Sparks, W.B. (1993). *ApJ*, **404**, 511.

McCarthy, P.J., van Breugel, W.J.M., Spinrad, H. & Djorgovski, S. 1987. *ApJ*, **321**, L29.

McCarthy, P.J., van Breugel, W.J.M. & Kapahi, V.K. 1991. *ApJ*, **371**, 478.

Miller, J.S. 1994. In *First Stromlo Symposium: Physics of Active Galactic Nuclei*, (eds.) Bicknell, G.V., Dopita, M.A. & Quinn, P.J., p. 149. San Francisco: Astron. Soc. Pacific Conf. Series.

Miller, L.S., Goldschmidt, P., La Franca, F. & Cristiani, S. 1993. In *Observational Cosmology*, ASP Conference Series, **51**, 614.

Mushotzky, R.F., Done, C. & Pounds, K.A. 1994. *ARAA*, **31**, 717.

Pascarelle, S.M. 1996. *HST-2*, Proceedings of the Second European Symposium on the Hubble Space Telescope (eds. P. Benvenuto, F. Macchetto & E. Schreier), (in press).

Ryle, M. & Longair, M.S. 1967. *MNRAS*, **136**, 123.

Schade, D. 1996. *HST-2*, Proceedings of the Second European Symposium on the Hubble Space Telescope (eds. P. Benvenuto, F. Macchetto & E. Schreier), (in press).

Schade, D., Lilly, S.J., Crampton, D., Hammer, F., Le Fèvre, O. & Tresse, L. (1995). *ApJ*, **451**, L1.

Steidel, C.C., Pettini, M. & Hamilton, D. (1995). *AJ*, **110**, 2519.

Steidel, C.C., Giavalisco, M., Pettini, M., Dickinson, M. & Adelberger, K.L. (1996). *AJ*, (in press)

Steidel, C.C. & Hamilton, D. (1992). *AJ*, **104**, 941.

Steidel, C.C. & Hamilton, D. (1993). *AJ*, **105**, 2017.

Stockman, H.S. 1994. In *Frontiers of Space & Ground-based Astronomy*, (eds) W. Wamsteker, M.S. Longair & Y. Kondo, p. 87. Dordrecht: Kluwer Academic Publishers.

Wall, J.V. & Peacock, J.A. 1985. *MNRAS*, **216**, 173.

Windhorst, R.A. & Keel, W.C. 1996. *ApJ*, (in press).

Windhorst, R.A., Fomalont, E.B., Kellermann, K.I., Partridge, R.B., Richards, E., Franklin, B.E., Pascarelle, S.M. & Griffiths, R.E. (1995). *ApJ*, **375**, 471.

Observational Cosmology with the CMB

By GEORGE F. SMOOT[1] †,

[1]Lawrence Berkeley National Laboratory & Department of Physics, University of California Berkeley CA 94720 USA

The Cosmic Microwave Background (CMB) encodes information about primoridal perturbations and the geometry of the universe as well as information on critical cosmological parameters such as Ω_0, Ω_B, Ω_Λ, and H_0. It is quite appropriate that we should review this topic on the occassion of the celebration of Igor Novikov's 60th birthday as he was a pioneer in this area. With the new precise measurement of the CMB spectrum and especially with the discovery and early mapping of the CMB anisotropy power spectrum we are finally in a position to carry out the program and tests suggested by Novikov some two decades previously.

By circumstance, during Novikov's 60th birthday celebration, we have released the *COBE* DMR 4-year data and a set of 10 papers presenting the first round of results from the analysis of those 4-year data. As a result I have modified my talk and this paper to emphasize the newly released *COBE* DMR data and results in addition to summarizing briefly the state of CMB observations at present.

---

1. Introduction

The observed cosmic microwave background (CMB) radiation provides strong evidence for the hot big bang model of cosmology. Primordial nucleosynthesis calculations require a cosmic background radiation (CBR) with a temperature $kT \sim 1$ MeV at a redshift of $z \sim 10^9$. Gamow, Alpher & Herman (1948) realized that this CBR was required and predicted its evolution to a faint residual relic radiation with a current temperature of a few degrees.

The CMB was serendipitously discovered by Penzias & Wilson (1965) in 1964 and its existence immediately interpreted by Dicke et al. (1965) as the relic radiation manifestation of the hypothesized primordial CBR. The observed CMB spectrum is well characterized by a 2.73 ± 0.01 K blackbody spectrum. The hot big bang model predicts that the CBR should have a fairly precise thermal spectrum. The thermal coupling between the matter and radiation is extremely good at the time of primordial nucleosynthesis. A non-interacting Planckian spectrum at temperature T_i at redshift z_i transforms with the universal expansion to another Planckian distribution at redshift z_r with temperature $T_r/(1+z_r) = T_i/(1+z_i)$. Hence thermal equilibrium, once established (e.g. at the nucleosynthesis epoch), is preserved by the expansion, in spite of the fact that photons decoupled from matter at early times. Because there are about 10^9 photons per nucleon, the transition from the ionized primordial plasma to neutral atoms at $z \sim 1000$ does not significantly alter the CBR spectrum (Peebles 1993). The large ratio of photons to nucleons means that it takes a very significant release of energy or photons by the matter to have any appreciable effect on the CBR.

The major step in the precision measurement of the CMB spectrum is provided by the *COBE* FIRAS observations. The current observations of the CMB spectrum are shown in Figure 1.

The remarkable precision with which the CMB spectrum is fitted by a Planckian distribution provides limits on possible energy releases in the early Universe, at roughly the

† Present address: Bldg 50-205 LBNL Berkeley CA 94720 USA E-mail: SMOOT@COSMOS.LBL.GOV

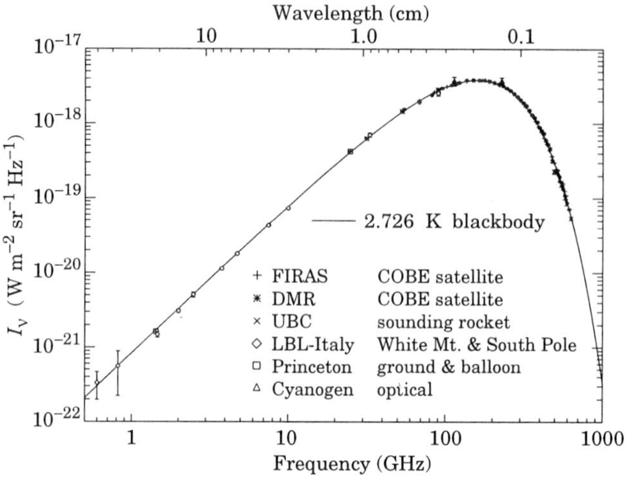

FIGURE 1. Precise measurements of the CMB spectrum. The line represents a 2.73 K blackbody, which describes the spectrum very well, especially around the peak of intensity. The spectrum is less well constrained at frequencies of 3 GHz and below.

fractional level of 10^{-4} of the CBR energy, for redshifts $\lesssim 10^7$ (corresponding to epochs $\gtrsim 1$ year). These observations provide significant constraints on possible cosmologies. Combined with the observed spectrum of the first order dipole anisotropy they also provide us with the knowledge and ability to separate CMB anisotropies from the various foregrounds.

2. Novikov's Influence

When I began my work in cosmology and in particular with the CMB, though still young, Igor Novikov was already well-established, having written a number of papers on the relativistic astrophysics relevant to cosmology and the CMB. At that time I studied his papers, e.g. Doroshkevich, Lukash & Novikov (1973), which discussed the isotropization of anisotropic cosmological models to match the isotropy indicated by the CMB observations and those which indicated what cosmological constraints could be derived from CMB observations. The mathematics was complex (and the notation different from our Western version) so needless to say I was very impressed by the dazzling insight and intellectual abilities of the authors. Novikov became a mythical figure to me as one of the pioneers in the field.

When I first met Novikov at a conference, I was surprised by how young he was – and now still seems. This convinced me all the more that there was much I could learn from him and thus I was quite excited when a couple of years later Novikov came to Berkeley to visit and give a seminar. I still have my notes from that seminar given on 26 January 1978 plus some handwritten drawings and responses to my questions Igor mailed to me shortly afterwards.

Finding out that Igor would be staying through the weekend I jumped at the chance to spend a day with him. Asking him what he might want to do I suggested going out sight seeing since the bay area had much to offer. "Would you like to go see the redwoods, the Marin coast, the wine country, or San Francisco?" I asked. Novikov responded, "They all sound good but I have a first request. I would like to get some new skis for the season." Well it made sense that skiing would be big in the Russian winter and I started to list stores that I knew had skis. Novikov stopped me and said "I am looking for water skis and I know the best store is on San Pablo Avenue." This called for a revision in my opinion of Novikov. Not only was he a relativistic demigod but he water-skied in the Russian winter and he had figured out where the best water-ski shop in the world was located. Soon Igor mentioned that the water skis were mainly for his wife and his daughter, who was a Russian champion water-skier. Whoa - this called for a revision in my opinion of Russian women. Eventually, it came out that there was a large indoor water skiing area in Moscow where water skiing occurred in the winter. Immediately, my world model and the data were congruent and theoretical visions of cold open water in broken lanes of ice and so forth vanished. We went to the shop and then on to Marin and the redwoods, coming back to Berkeley through San Francisco. I got the chance to learn more about Novikov and learn more from him.

I was immediately intrigued and motivated to action by Novikov's seminar and some of our discussions. It concerned work he had done with A.G.Doroshkevich and our session chairman V. N. Lukash (1974) on the anisotropy expected in a homogeneous but anisotropic universe. One must remember that at that time we were still in the age of "classical" cosmology in the sense that though we knew that there were galaxies in the universe and that their progenitor seeds must produce anisotropy in the CMB through the work of Silk (1967) and Sachs & Wolfe (1967), we thought that they would be small-angular-scale (roughly arc minute) and small amplitude ($\sim 10^{-3}$) pock marks on the overall CMB. I hoped that their random appearance would not mask the "interesting" anisotropies from the geometry and dynamics of the universal space-time in which we live. I optimistically anticipated that large scale structure fluctuations would either be sufficiently small in amplitude and frequency or would average out sufficiently on the large angular scale that one could find out about the anisotropy, shear, and vorticity of the Universe. This was not completely wishful thinking in that we did know the causally connected horizon was on the order of a degree and no one promoted a process that would produce large scale perturbations. This situation changed in a couple of years with the advent of the inflation concept and with the realization that the experiments we were doing had begun to push the limits for making galaxies and the other now better understood very large scale structures in the present universe. Since that time and the discovery of anisotropies, the research community has been focused upon models in which the primordial density (and perhaps gravitational wave) perturbations lead to the observed CMB anisotropies.

What Novikov and his colleagues had determined was that in an open universe, $\Omega < 1$, a quadrupolar expansion anisotropy would manifest itself not as a quadrupolar CMB anisotropy but as a 'spot' in the CMB with angular size $\Delta\theta \sim 4\Omega$ radians. This 'spot' shape is unusual and characteristic. It looks very much like the end of a navel orange or bottom of an apple. For obvious reasons I chose to call the 'spot' the navel of the universe rather than the 'spot' when describing attempts to discover it. A sketch of the predicted anisotropy is shown in Figure 2.

As the sketch indicates in a fairly open universe nearly all the sky is fairly well described by a simple dipole anisotropy. At that time we had discovered the dipole anisotropy but the measurements covered only a very limited part of the sky. The fact that the dipole anisotropy did not point in the direction of that expected from the doppler shift caused by the motion of the solar system along with the general rotation of the Galaxy was causing some consternation amongst astronomers. Many thought that the Galaxy must be nearly at rest with respect to the general cosmological expansion frame; otherwise, how could one explain the Hubble law? How could the entire local group of galaxies be moving at such a speed? We now believe that they are but it seemed extreme to many people

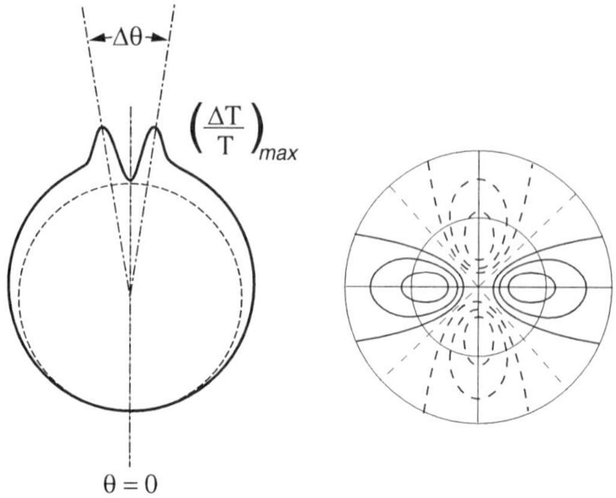

FIGURE 2. Sketch of CMB anisotropy in an open universe with quadrupolar expansion anisotropy. The left hand portion of the figure shows a cross section through the symmetry axis along the plane of minimum expansion rate. The righthand portion of the figure shows a contour plot of the CMB anisotropy levels around the navel 'spot' aligned on the same symmetry axis but with different expansion rates on all axes.

at the time. Perhaps we were confusing anisotropy in the form of a velocity-induced dipole and the pattern of a quadrupolar anisotropic expansion in an open universe. If this were true, then the 'spot' would lie on the great circle containing the velocity vector inducing the doppler dipole and the observed 'dipole' direction. Detecting such a 'spot' or navel for the universe offered the combined chance of both directly measuring Ω and an anisotropy in universal expansion as well as potentially separating the Galactic velocity and rotation.

After Novikov's talk and visit, I went into action to test this idea. Chris Witebsky (then a graduate student) and I designed a small radio telescope capable of detecting the navel ('spot'). We figured that the most likely value for Ω would be either the observed mass density of $\Omega \sim 0.01$ or the big bang primordial nucleosynthesis value of about $\Omega = 0.03 - 0.10$ meaning that we needed to be able to detect the characteristic navel shape with extent $4\Omega \sim 2.2°$ or $4\Omega = 6 - 22°$ across respectively. In a short time we had built a small radio telescope with a beam size of about 0.5° and were using the Earth's rotation to scan it through likely regions on the sky. Needless to say we did not find the navel or a likely 'spot' with the sensitivity ($\delta T/T \lesssim 10^{-3.5}$) possible from the ground at that time. We did, however, check the list of radio sources that I had been compiling from the literature as possible foreground limitations on future CMB anisotropy measurements. We found that the radio flux from larger, extended sources were systematically underestimated on the basis of large radio telecope observations. For example we found that the Orion nebula was emitting about twice the flux reported in the literature. This caused us to revise our estimates for many sources upward but also to have the confidence that our estimates were at the appropriate order of magnitude which was a concern since most radio measurements were made at much lower frequency than the CMB anisotropy measurements had to be made.

There was a second lesson I learned. It was that theorists could come up with scenarios in which missing even a small portion of the sky could provide one with an incorrect conclusion. And though most such models would not be valid, one should avoid the chance of missing such a result or leaving the theorists the wiggle room. This was a critical period in the design of the *COBE* anisotropy experiment and this experience was one of the motivations for my continued insistence on complete sky coverage in as redundant a manner as possible.

I kept a large notebook with the papers, notes, and data from the observations in a book shelf in my office under the headings: Novikov, Lukash, and Doroshkevich; Navel of the Universe. A few years later another graduate student came in my office to discuss what he had been reading in the book "Relativistic Astrophysics" by these people Novikov and Zel'dovich. He was very impressed and regarded them in the same mythical light that I had many years earlier. I told him that I had met the two and that Novikov had come to visit and we had the chance to get acquainted. At that moment the student noticed the book on the shelf and all you could see were the words Novikov and Navel, one above the other like a book title. The student turned and said I guess you really do know him well, you have a whole notebook about his navel. So the mystery and myth of Igor Novikov continues into another generation.

3. *COBE* Four-Year Observations

The Cosmic Background Explorer was the first NASA satellite dedicated to cosmology. It was successful in making very precise measurements of the CMB.

3.1. *Anisotropy Observations with the COBE DMR*

The Differential Microwave Radiometers (DMR) experiment discovered cosmic microwave background (CMB) anisotropies from analysis of its first year of data (Smoot et al. 1992; Bennett et al. 1992; Wright et al. 1992; Kogut et al. 1992). The CMB temperature fluctuations were measured at an angular resolution of 7° at frequencies of 31.5, 53, and 90 GHz. The *COBE* results were confirmed by the positive cross-correlation between the *COBE* data and data from balloon-borne observations at a shorter wavelength (Ganga et al. 1993) and later by comparison of the *COBE* data and data from the ground-based Tenerife experiment (Lineweaver et al. 1995) at longer wavelengths. The positive correlation at both longer and shorter wavelengths provides confidence in the results. The results from analysis of two years of DMR data (Bennett et al. 1994) reconfirmed the results from the first year data. This talk and proceedings summarize the results and cosmological implications obtained from the full *COBE* DMR 4-year data and provide references to further detailed reports about the data and their analysis.

A major finding of the initial *COBE* DMR discovery (Smoot et al. 1992) was that the CMB was anisotropic on all observed angular scales. A key question is what these anisotropies represent. In the early 1970's the observed large scale structure and scaling arguments led to the prediction (Peebles & Yu 1970; Harrison 1970; Zeldovich 1972) that the primordial gravitational potential perturbations must have an equal *rms* amplitude on all scales. This corresponds to a matter perturbation power-law spectrum, $P(k) \propto k^n$, where k is the comoving wavenumber, with $n = 1$. At the time there were no known mechanisms for producing such a scale-invariant power spectrum of fluctuations. It is now known that topological defects are naturally produced in a scale-invariant manner. In 1982 it was found that inflationary models predicted nearly scale-invariant perturbations as a result of quantum mechanical fluctuations at very early times. Even with the proliferation of inflationary models, it is found that essentially all reasonable inflationary models predict $n \approx 1$. The translation from a scale-invariant spectrum of perturbations to the CMB temperature anisotropies depends upon angular scale and the contents of the universe. On large angular scales the results of most models are fairly similar. Including the effects of a standard cold dark matter model, *COBE* DMR should find $n_{eff} \approx 1.1$ for a Peebles-Harrison-Zeldovich $n = 1$ universe. This is consistent with the power spectrum measured by the *COBE* DMR data. The observed power spectrum of fluctuations' amplitude is also consistent with models of large scale structure formation based upon primordial seeds produced by quantum fluctuations or topological defects in the early universe.

Most cosmological models do not predict the exact CMB temperature pattern that would be observed in our sky, but rather predict a statistical distribution of anisotropies. In the context of such models, the CMB temperature observed in our sky is only a single realization from a statistical distribution. Theoretical models most often predict a power spectrum in spherical harmonic amplitudes; as the physics of the models leads to primordial fluctuations that are Gaussian random fields, the power spectrum is sufficient to characterize the results. Observations of the sky are usually expressed as a spherical harmonic temperature expansion $T(\theta,\phi) = \sum_{\ell m} a_{\ell m} Y_{\ell m}(\theta,\phi)$. If the original perturbations are Gaussian random fields, the $a_{\ell m}$ are Gaussianly distributed, and the power at each ℓ is $(2\ell+1)C_\ell/(4\pi)$, where $C_\ell \equiv \langle |a_{\ell m}|^2 \rangle$, is sufficient to characterize the results. For an idealized full-sky observation, the variance of each measured C_ℓ is $[2/(2\ell+1)]C_\ell^2$. This sampling variance (known as cosmic variance) comes about because each C_ℓ is chi-squared distributed with $(2\ell+1)$ degrees of freedom for our observable volume of the Universe (White 1994). Thus, in addition to experimental uncertainties, we account for the *cosmic sample variance* uncertainties due to our observation of a single realization in our analyses of the DMR maps. Cosmic variance exists independently of the quality of the experiment. The power spectrum from the 4-year DMR map is cosmic variance limited for $\ell \lesssim 20$.

This talk summarizes the key results which are detailed in a set of DMR 4-year analysis papers (Banday et al. 1996a, 1996b; Gorski et al. 1996; Hinshaw et al. 1996a, 1996b; Kogut et al. 1996a, 1996b, 1996c; Lineweaver et al. 1996; Wright et al. 1996).

3.2. *OBSERVATIONS*

The DMR consists of 6 differential microwave radiometers: 2 nearly independent channels, labeled A and B, at frequencies 31.5, 53, and 90 GHz (wavelength 9.5, 5.7, and 3.3 mm). Each radiometer measures the difference in power between two 7° fields of view separated by 60°, 30° to either side of the spacecraft spin axis (Smoot et al. 1990). *COBE* was launched from Vandenberg Air Force Base on 18 November 1989 into a 900 km, 99° inclination circular orbit, which precesses to follow the terminator (light dark line on the Earth) as the Earth orbits the Sun. Attitude control keeps the spacecraft pointed away from the Earth and nearly perpendicular to the Sun with a slight backward tilt so that solar radiation never directly illuminates the aperture plane. The combined motions of the spacecraft spin (75 s period), orbit (103 m period), and orbital precession ($\sim 1°$ per day) allow each sky position to be compared to all others through a highly redundant set of temperature difference measurements spaced 60° apart. The on-board processor box-car integrates the differential signal from each channel for 0.5 s, and records the digitized differences for daily playback to a ground station.

Ground data analysis consists of calibration, extensive systematic error analyses, and conversion of time-ordered-data to sky maps (Kogut et al. 1996a). The DMR time-ordered-data include systematic effects such as emission from the Earth and Moon, the instrument's response to thermal changes, and the instrument's response to the Earth's magnetic field. The largest detected effects do not contribute significantly to the DMR maps: they are either on time scales long compared to the spacecraft spin sampling (e.g. thermal gain drifts) or have time dependence inconsistent with emission fixed on the celestial sphere (e.g. magnetic effects). Detected and potential systematic effects were quantitatively analyzed in detail by Kogut et al. (1996a). Data with the worst systematic

Channel	Noise per 0.5 sec Observation (mK)
31A	58.27
31B	58.35
53A	23.13
53B	27.12
90A	39.10
90B	30.76

TABLE 1. DMR Instrument Noise in Antenna Temperature

contamination (lunar emission, terrestrial emission, and thermal gain changes) were not used in the map making process and constitute less than 10% of the data in the 53 and 90 GHz channels. The remaining data were corrected using models of each effect. The data editing and correction parameters were conservatively chosen so that systematic artifacts, after correction, are less than 6 μK (95% confidence upper limit) in the final DMR map in the worst channel. This is significantly less than the levels of the noise and celestial signals.

We subtract a dipole $T_d = 3.356$ mK anisotropy signal (thermodynamic temperature in Galactic coordinates Cartesian components [X,Y,Z] = [-0.2173, -2.2451, +2.4853] mK) from the time-ordered differential data prior to forming the 4-year sky maps to reduce spatial gradients within a single pixel. A small residual dipole remains in the maps from a combination of CMB and Galactic emission. Table 1 shows the instrument noise per half-second observation. The mean signal-to-noise ratios in the 10° smoothed maps are approximately 0.5, 1.5, and 1.0 for 31, 53, and 90 GHz, respectively. For a multi-frequency co-added map the signal-to-noise ratio is ~ 2. This signal-to-noise level is adequate to portray an accurate overall visual impression of the anisotropy. Visual comparison of the full sky maps at each frequency, after averaging the A and B channels, removing the CMB dipole, and smoothing to 10° effective resolution show coincident features. Well off the Galactic plane these are clearly true CMB anisotropy features. Simulated data in combination with the noise appropriate to 1-, 2-, and 4-years of DMR 53 GHz observations show the convergence of the DMR maps with the input simulated data. Increasing years of data result in the emergence of the input large scale features. We can be confident that the large scale features in the 4-year DMR maps are real features rather than confusing noise.

Given the sensitivity of the 4-year DMR maps we have chosen to extend the cut made in our previous analyses to exclude additional Galactic emission. Along with the previous $|b| < 20°$ exclusion zone, we use the *COBE* DIRBE 140 μm map as a guide to cut additional Galactic emission features. The full sky DMR maps contain 6144 pixels. An optimum Galactic cut maximizes the number of remaining pixels while minimizing the Galactic contamination. This cut leaves 3881 pixels (in Galactic pixelization) while eliminating the strongest $|b| > 20°$ Galactic emission. Moderate changes to this cut will cause derived CMB parameters to change somewhat, but this is consistent with the data sampling differences of real CMB anisotropy features and not necessarily Galactic contamination. Likewise, derived CMB parameters also vary by the expected amount when the maps are made in ecliptic rather than Galactic coordinates since about 1/2 of the noise is re-binned.

Kogut et al. (1996b) examine the Galactic contamination of the high Galactic latitude regions of the DMR maps which remain after the Galactic emission cut (described above). No significant cross-correlation is found between the DMR maps and either the 408 MHz synchrotron map or the synchrotron map derived from a magnetic field model (Bennett et al. 1992). This places an upper limit $T_{\rm synch} < 11\ \mu$K (95% confidence) on synchrotron emission at 31 GHz.

A significant correlation is found between the DMR maps and the dust-dominated DIRBE 140 μm map, with frequency dependence consistent with a superposition of dust and free-free emission. This corresponds to a 7° rms free-free emission component of $7.1 \pm 1.7\ \mu$K at 53 GHz and a dust component of $2.7 \pm 1.3\ \mu$K at 53 GHz. Since this emission is uncorrelated with CMB anisotropies it constitutes $< 10\%$ of the CMB power. The amplitude of the correlated free-free component at 53 GHz agrees with a noisier estimate of free-free emission derived from a linear combination of DMR data which includes *all* emission with free-free spectral dependence. The combined dust and free-free emission contribute $10 \pm 4\ \mu$K *rms* at both 53 and 90 GHz, well below the 30 μK cosmic signal. These Galactic signal analyses are consistent with the fact that the fitted cosmological parameters are nearly unaffected by removal of modeled Galactic signals (Górski et al. 1996, Hinshaw et al. 1996), with the notable exception of the quadrupole, which has significant Galactic contamination (Kogut et al. 1996b). A search by Banday et al. (1996a) finds no evidence for significant extragalactic contamination of the DMR maps.

4. Interpretation

Monopole $\ell = 0$: Despite the fact that the DMR is a differential instrument, the known motion of the *COBE* spacecraft about the Earth and the motion of the Earth about the Solar System barycenter provides a means to determine the CMB monopole temperature from the DMR data. The CMB at millimeter wavelengths is well described by a blackbody spectrum (Mather et al. 1994, Fixsen et al. 1996). The Doppler effect from the combined spacecraft and Earth orbital motions creates a dipole signal $T(\theta) = T_0[1 + \beta\cos(\theta) + O(\beta^2)]$, where $\beta = v/c$ and θ is the angle relative to the time-dependent velocity vector. The satellite and Earth orbital motions are well known and change in a regular fashion, allowing their Doppler signal to be separated from fixed celestial signals. We fit the time-ordered data to the Doppler dipole and recover a value for the CMB monopole temperature, $T_0 = 2.725 \pm 0.020$ K (Kogut et al. 1996a).

Dipole $\ell = 1$: The CMB anisotropy is dominated by a dipole term usually attributed to the motion of the Solar System with respect to the CMB rest frame. A precise determination of the dipole must account for Galactic emission and the aliasing of power from higher multipole orders once pixels near the Galactic plane are discarded. One can account for Galactic emission by using a linear combination of the DMR maps or by cross-correlating the DMR maps with template sky maps dominated by Galactic emission (Kogut et al. 1996b). The high-latitude portion of the sky is fitted for a dipole with a CMB frequency spectrum using a pixel-based likelihood analysis (Hinshaw et al. 1996a). Accounting for the smoothing by the DMR beam and map pixelization, the CMB dipole has amplitude 3.353 ± 0.024 mK toward Galactic coordinates $(l, b) = (264^\circ\!.26 \pm 0^\circ\!.33, 48^\circ\!.22 \pm 0^\circ\!.13)$, or equatorial coordinates $(\alpha, \delta) = (11^{\rm h}12^{\rm m}\!.2 \pm 0^{\rm m}\!.8, -7^\circ\!.06 \pm 0^\circ\!.16)$ epoch J2000.

A second analysis approach utilizes a phenomenological estimate of the Galactic foreground by examining the dipole fitted parameters as a function of cuts in Galactic latitude. It was found that the largest source of error in the dipole direction was reduced by using lower Galactic latitude cuts. Using the four year data set from all six channels of the COBE Differential Microwave Radiometers (DMR), the best-fit dipole amplitude $3.359 \pm 0.001 \pm 0.023$ mK in the direction $(\ell, b) = (264^\circ.33 \pm 0^\circ.04 \pm 0^\circ.17,\ +48^\circ.05 \pm 0^\circ.02 \pm 0^\circ.07)$ where the first uncertainties are statistical and the second are estimates of the combined systematics (Lineweaver et al. 1996).

This dipole measurement is consistent with previous DMR and FIRAS results.

Quadrupole $\ell = 2$: On the largest angular scales (e.g., quadrupole), Galactic emission is comparable in amplitude to the anisotropy in the CMB. The quadrupole amplitude is found by a likelihood analysis which simultaneously fits the high-latitude portion of the DMR maps for Galactic emission traced by synchrotron- and dust-dominated surveys and a quadrupole anisotropy with a thermodynamic frequency spectrum (Kogut et al. 1996b, Hinshaw et al. 1996a). After correcting for the positive bias from instrument noise and aliasing, the CMB quadrupole amplitude observed at high latitude is $Q_{rms} = 10.7 \pm 3.6 \pm 7.1\ \mu$K, where the quoted errors reflect the 68% confidence uncertainties from random statistical errors and Galactic modeling errors, respectively. The observed quadrupole amplitude, Q_{rms}, has a lower value than the quadrupole expected from a fit to the entire power spectrum, Q_{rms-PS}, but whether this is a chance result of cosmic variance or reflects the cosmology of the universe cannot be determined from *COBE* data. The 68% confidence interval for the quadrupole amplitude, $6\ \mu\mathrm{K} \leq Q_{rms} \leq 17\ \mu$K, is consistent with the quadrupole normalization of the full power spectrum power-law fit (discussed below): $Q_{rms-PS} = 15.3^{+3.8}_{-2.8}\ \mu$K.

Power spectrum $\ell \geq 2$: The simplest probe of the angular power spectrum of the anisotropy is its Legendre transform, the 2-point correlation function. The 2-point correlation function of the 4-year maps is analyzed by Hinshaw et al. (1996b), where it is shown that the 2-point data are consistent from channel to channel and frequency to frequency. The data are robust with respect to the angular power spectrum. A Monte Carlo-based Gaussian likelihood analysis determines the most-likely quadrupole normalization for a scale-invariant ($n = 1$) power-law spectrum. The results are summarized in Table 2 which also includes the results of 3 additional, independent power spectrum analyses, discussed below. The normalization inferred from the 2-point function is now in better agreement with other determinations than was the case with the 2-year data. The change is due to data selection: with the 2-year data, we only analyzed the 53×90 GHz cross-correlation function; with the 4-year data we have analyzed many more data combinations, including the auto-correlation of a co-added, multi-frequency map. This latter combination is more comparable to the data analyzed by other methods, and the 2-point analysis yields consistent results in that case. The combined 31, 53 and 90 GHz CMB rms is $29 \pm 1\ \mu$K in the 10° smoothed map (Banday et al. 1996b), consistent with the level determined by the 2-point results.

It is also possible to analyze the power spectrum directly in terms of spherical harmonics. However, there is considerable subtlety in this because the removal of the Galactic plane renders the harmonics non-orthonormal, producing strong correlations among the fitted amplitudes. Wright et al. (1996) has solved for an angular power spectrum by modifying and applying the technique described by Peebles (1973) and Hauser & Peebles (1973) for data on the cut sphere. They compute a Gaussian likelihood on these data and calibrate their results with Monte Carlo simulations. Górski et al. (1996) explicitly construct orthonormal functions on the cut sphere and decompose the anisotropy data with respect to these modes. They form and evaluate an exact Gaussian likelihood directly in terms of this mode decomposition. The results of these analyses are summarized in Table 2. Further details, including results from other data combinations are given in the respective papers.

Hinshaw et al. (1996a) evaluate a Gaussian likelihood directly in terms of a full pixel-

pixel covariance matrix, a technique applied to the 2-year data by Tegmark & Bunn (1995). The results of the power-law spectrum fits are summarized in Table 2. Hinshaw et al. (1996a) also analyze the quadrupole anisotropy separately from the higher-order modes, to complement the analysis of Kogut et al. (1996b). They compute a likelihood for the quadrupole mode C_2, nearly independent of higher-order power, and show that it peaks between 6 and 10 μK, depending on Galactic model, but that its distribution is so wide that it is easily consistent with $15.3^{+3.8}_{-2.8}$ μK, the value derived using the full power spectrum.

An important lesson from fitting with different cuts and configurations and from our Monte Carlo simulations is that the best fitted parameters depend both upon the random statistics of CMB fluctuations and on the choice of cuts and fitting parameters. Table 1 in Górski et al. 1996 is very indicative of the range of results obtainable using a robust and stable approach.

Tests for Gaussian Statistics: It is important to determine whether the primordial fluctuations are Gaussian. The probability distribution of temperature residuals should be close to Gaussian if the sky variance is Gaussian and the receiver noise is Gaussian. The receiver noise varies somewhat from pixel to pixel because the observation times are not all the same, but when this is taken into account the data appear Gaussian (Smoot et al. 1994). There is no evidence for an excess of large deviations, as would be expected if there were an unknown population of point sources. A search for point sources in the 2-year maps found none (Kogut et al. 1994). Given the large beam of the instrument and the variance of both cosmic signals and receiver noise, it is still possible for interesting signals to be hidden in the data.

Kogut et al. (1996c) compare the 4-year DMR maps to Monte Carlo simulations of Gaussian power-law CMB anisotropy. The 3-point correlation function, the 2-point correlation of temperature extrema, and the topological genus are all in excellent agreement with the hypothesis that the CMB anisotropy on angular scales of 7° or larger represents a random-phase Gaussian field. A likelihood comparison of the DMR maps against non-Gaussian χ^2_N toy models tests the alternate hypothesis that the CMB is a random realization of a field whose spherical harmonic coefficients $a_{\ell m}$ are drawn from a χ^2 distribution with N degrees of freedom. Not only do Gaussian power-law models provide an adequate description of the large-scale CMB anisotropy, but non-Gaussian models with $1<N<60$ are five times less likely to describe the true statistical distribution than the exact Gaussian model.

5. Summary of 4-year *COBE* DMR CMB Measurements

(1) The full 4-year set of *COBE* DMR observations is analyzed and full sky maps are presented. The typical signal-to-noise ratio in a 10° smoothed map is $\sim$ 2 in the frequency-averaged map, enough to provide a visual impression of the anisotropy.

(2) We derive a CMB monopole temperature from DMR (despite its being a differential instrument) of $T_0 = 2.725 \pm 0.020$ K (Kogut et al. 1996a). This is in excellent agreement with the *COBE* FIRAS precision measurement of the spectrum of the CMB, $T_0 = 2.728 \pm 0.002$ K (Fixsen et al. 1996).

(3) The CMB dipole from DMR has amplitude 3.353 ± 0.024 mK toward Galactic coordinates $(l, b) = (264^\circ\!.26 \pm 0^\circ\!.33, 48^\circ\!.22 \pm 0^\circ\!.13)$, or equatorial coordinates $(\alpha, \delta) = (11^h 12^m 2 \pm 0^m 8, -7^\circ\!.06 \pm 0^\circ\!.16)$ epoch J2000. This is consistent with the dipole amplitude and direction derived by *COBE* FIRAS (Fixsen et al. 1996).

(4) The 95% confidence interval for the observed $\ell = 2$ quadrupole amplitude is 4 μK

Technique	n [a]	Q_{rms-PS} [b] (μK)	$Q_{rms-PS\|n=1}$ [c] (μK)
	No Galaxy Correction [d]		
2-point function [e]	—	—	$17.5^{+1.4}_{-1.4}$
Orthog. functions [f]	$1.21^{+0.24}_{-0.28}$	$15.2^{+3.7}_{-2.6}$	$17.7^{+1.3}_{-1.2}$
Pixel temps [g]	$1.23^{+0.26}_{-0.27}$	$15.2^{+3.6}_{-2.8}$	$17.8^{+1.3}_{-1.3}$
Hauser-Peebles [h]	—	—	—
	Internal Combination Galaxy Correction [j]		
2-point function [e]	—	—	$16.7^{+2.0}_{-2.0}$
Orthog. functions [f]	$1.11^{+0.38}_{-0.42}$	$16.3^{+5.2}_{-3.7}$	$17.4^{+1.8}_{-1.7}$
Pixel temps [g]	$1.00^{+0.40}_{-0.43}$	$17.2^{+5.6}_{-4.0}$	$17.2^{+1.9}_{-1.7}$
Hauser-Peebles [h]	$1.62^{+0.44}_{-0.50}$	—	$19.6^{+2.5}_{-2.5}$

[a] Mode & 68% confidence range of the projection of the 2-d likelihood, $L(Q, n)$, on n
[b] Mode & 68% confidence range of the projection of the 2-d likelihood, $L(Q, n)$, on Q
[c] Mode & 68% confidence range of the slice of the 2-d likelihood, $L(Q, n)$, at $n = 1$
[d] Formed from the weighted average of all 6 channels
[e] Hinshaw et al. 1996b
[f] Górski et al. 1996
[g] Hinshaw et al. 1996a
[h] Wright et al. 1996
[i] weighted average of 6 channels with the Galactic maps subtracted (Kogut et al. 1996b)
[j] from a linear combination of 6 channel maps canceling free-free emission (Kogut et al. 1996)

TABLE 2. Summary of DMR 4-Year Power Spectrum Fitting Results.

$\leq Q_{rms} \leq 28$ μK. This is consistent with the value predicted by a power-law fit to the power spectrum: $Q_{rms-PS} = 15.3^{+3.8}_{-2.8}$ μK (Kogut et al. 1996b; Hinshaw et al. 1996a).

(5) The power spectrum of large angular scale CMB measurements are consistent with an $n = 1$ power-law (Górski et al. 1996; Hinshaw et al. 1996a; Wright et al. 1996). If the effects of a standard cold dark matter model are included, *COBE* DMR should find $n_{eff} \approx 1.1$ for a $n = 1$ universe. With full use of the multi-frequency 4-year DMR data, including our estimate of the effects of Galactic emission, we find a power-law spectral index of $n = 1.2 \pm 0.3$ and a quadrupole normalization $Q_{rms-PS} = 15.3^{+3.8}_{-2.8}$ μK. For $n = 1$ the best-fit normalization is $Q_{rms-PS|n=1} = 18 \pm 1.6$ μK. Differences in the derived values of Q and n between various analyses of DMR data are much more dependent on the detailed data selection effects than on the analysis technique.

(6) The DMR anisotropy data are consistent with Gaussian statistics. Statistical tests prefer Gaussian over other toy statistical models by a factor of ~ 5 (Kogut et al. 1996c).

6. Conclusions

The COBE-discovered (Smoot et al. 1992) higher-order ($\ell \geq 2$) anisotropy is interpreted as being the result of perturbations in the energy density of the early Universe, manifesting themselves at the epoch of the CMB's last scattering. In the standard scenario the last scattering takes place at a redshift of approximately 1100, at which epoch the large number of photons was no longer able to keep the hydrogen sufficiently ionized. The optical thickness of the cosmic photosphere is roughly $\Delta z \sim 100$ or about 5 arcminutes, so that features smaller than this size are damped.

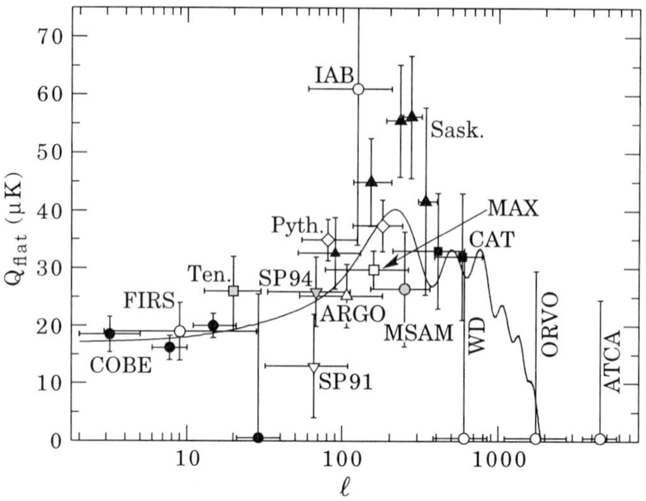

FIGURE 3. Current status of CMB anisotropy observations, adapted from Scott, Silk, and White (1995). Plotted are the multipole amplitudes for a flat (unprocessed scale-invariant spectrum of primordial perturbations, i.e., a horizontal line) anisotropy spectrum that would give the observed results for the experiment. The vertical error bars represent estimates of 68% CL, while the upper limits are at 95% CL. Horizontal bars indicate the range of ℓ values sampled. The curve indicates the expected spectrum for a standard CDM model ($\Omega_0 = 1$, $\Omega_B = 0.05$, $h = 0.5$), although true comparison with models should involve convolution of this curve with each experimental filter function.

At this time on the order of ten experiments have observed CMB anisotropies. Anisotropies are observed on angular scales larger than this damping scale (see Figure 3) and are consistent with those expected from an initially scale-invariant power spectrum of potential and thus metric fluctuations. It is believed that the large scale structure in the Universe developed through the process of gravitational instability where small primordial perturbations in energy density were amplified by gravity over the course of time. The initial spectrum of density perturbations can evolve significantly in the epoch $z > 1100$ for causally connected regions (angles $\lesssim 1^\circ\ \Omega_{tot}^{1/2}$). The primary mode of evolution is through adiabatic (acoustic) oscillations, leading to a series of peaks that encode information about the perturbations and geometry of the universe, as well as information on Ω_0, Ω_B, Ω_Λ (cosmological constant), and H_0 (Scott, Silk, & White 1995). The location of the first acoustic peak is predicted to be at $\ell \sim 220\ \Omega_{tot}^{-1/2}$ or $\theta \sim 1^\circ\ \Omega_{tot}^{1/2}$ and its amplitude increases with increasing Ω_B.

Figure 4 shows the theoretically predicted power spectrum for a sample of models, plotted as $\ell(\ell+1)C_\ell$ versus ℓ which is the power per logarithmic interval in ℓ or, equivalently, the two-dimensional power spectrum. If the initial power spectrum of perturbations is the result of quantum mechanical fluctuations produced and amplified during inflation, then the anisotropy spectrum and the fractional contribution from density (scalar) and gravity wave (tensor) perturbations are coupled. If the energy scale of inflation at the appropriate epoch is at the level of $\simeq 10^{16}$GeV, then detection of gravitons is possible, as well as partial reconstruction of the inflaton potential. If the energy scale is $\lesssim 10^{14}$GeV, then density fluctuations dominate and less constraint is possible.

Fits to data over smaller angular scales are often quoted as the expected value of the quadrupole $\langle Q \rangle$ for some specific theory, e.g. a model with power-law initial conditions (primordial density perturbation power spectrum $P(k) \propto k^n$). The full 4-year COBE

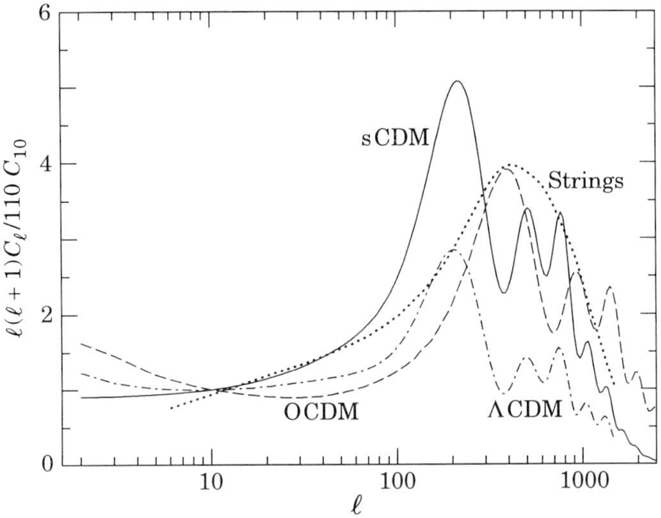

FIGURE 4. Examples of theoretically predicted $\ell(\ell+1)C_\ell$ or CMB anisotropy power spectra. **sCDM** is the standard cold dark matter model with $h = 0.5$ and $\Omega_B = 0.05$. **ΛCDM** is a model with $\Omega_{tot} = \Omega_\Lambda + \Omega_0 = 1$ where $\Omega_\Lambda = 0.3$ and $h = 0.8$. **OCDM** is an open model with $\Omega_0 = 0.3$ and $h = 0.75$. [See (White 1996) for models.] **Strings** is a model where cosmic strings are the primary source of large scale structure (Albrecht et al. 1995). The plot indicates that precise measurements of the CMB anisotropy power spectrum could distinguish between current models.

DMR data give $\langle Q \rangle = 15.3^{+3.7}_{-2.8}\,\mu$K, after projecting out the slope dependence, while the best-fit slope is $n = 1.2 \pm 0.3$, and for an $n = 1$ (scale-invariant potential perturbation) spectrum $\langle Q \rangle\,(n = 1) = 18 \pm 1.6\,\mu$K (Bennett et al. 1996; Górski et al. 1996). The conventional notation is such that $Q^2_{\rm rms}/T^2_\gamma = 5C_2/4\pi$.

Only somewhat weak conclusions can be drawn based on the current smaller angular scale data (see Figure 3). However, new data are being acquired at an increasing rate. With future experiments and the prospect of next generation satellite missions: MAP and COBRAS/SAMBA, a precise measurement of the CMB anisotropy power spectrum is possible and likely, allowing us to decode the information that it contains (Knox 1995; Jungman et al. 1996).

The DMR time-ordered data, map data, and ancillary data sets (including the revised Galactic cut) are publicly available through the National Space Science Data Center (NSSDC) at http://www.gsfc.nasa.gov/astro/cobe/cobe_home.html. *COBE* support was provided by the NASA Office of Space Sciences (OSS). This work was supported in part by the Director, Office of Energy Research, Office of High Energy and Nuclear Physics, Division of High Energy Physics of the U.S. Department of Energy under contract No. DE-AC03-76SF00098.

REFERENCES

ALBRECHT, A. ET AL. 1995 *Phys. Rev. Lett.*, in press, astro-ph/9505030.

ALPHER, R. A. & HERMAN, R. C. 1948 *Physics Today*, **41**, No. 8, p. 24.

BANDAY, A. ET AL. 1996a *Astrophys. J.*, submitted.

BANDAY, A. ET AL. 1996b *Astrophys. J.*, submitted.

BENNETT, C. L. ET AL. 1996 *Astrophys. J.*, accepted.

BENNETT, C. L. ET AL. 1994 *Astrophys. J.* **436**, 423.

BENNETT, C. L. ET AL. 1993 *Astrophys. J.* **414**, L77.

BENNETT, C. L. ET AL. 1992 *Astrophys. J.* **396**, L7.

BENNETT, C. L. ET AL. 1991 *Astrophys. J.* **391**, 466.

DICKE, R. H., PEEBLES, P. J. E., ROLL, P. G. & WILKINSON, D. T. 1965 *Astrophys. J.* **142**, 414.

DOROSHKEVICH, A. G., LUKASH V. N. & NOVIKOV I. D. 1973 The Isotropization of Homogeneous Cosmological Models. *Zh. Eksper. Teor. Fiz* **64**, 739–746.

DOROSHKEVICH, A. G., LUKASH V. N. & NOVIKOV I. D. 1974 Primordial Radiation in a Homogeneous but Anisotropic Universe. *Astron. Zh.* **51**, 554–560.

FIXSEN, D. J., ET AL. 1996 *Astrophys. J.*, submitted.

GANGA, K., ET AL. 1993 *Astrophys. J.* **410**, L57.

GÓRSKI, K. M. ET AL. 1996 *Astrophys. J.*, submitted.

HARRISON, E. R. 1970 *Phys. Rev. D* **1**, 2726.

HAUSER, M. G. & PEEBLES, P. J. E. 1973 *Astrophys. J.* **185**, 757.

HINSHAW, G. ET AL. 1996a *Astrophys. J. Lett.*, submitted.

HINSHAW, G. ET AL. 1996b, *Astrophys. J. Lett.*, submitted.

JUNGMAN, G., KAMIONKOWSKI, M., KOSOWSKY, A., & SPERGEL, D. N. 1995 astro-ph/9512139.

KNOX, L. 1995 *Phys. Rev. D*, in press, astro-ph/9504054.

KOGUT, A. ET AL. 1992 *Astrophys. J.* **401**, 1.

KOGUT, A., BANDAY, A. J., BENNETT, C. L., HINSHAW, G., LOEWENSTEIN, K., LUBIN, P., SMOOT, G. F., & WRIGHT, E. L. 1994 *Astrophys. J.* **433**, 435.

KOGUT, A. ET AL. 1996a *Astrophys. J.*, submitted.

KOGUT, A. ET AL. 1996b *Astrophys. J.*, submitted.

KOGUT, A. ET AL. 1996c *Astrophys. J.*, submitted.

LINEWEAVER, C. ET AL. 1995 *Astrophys. J.* **448**, 482-487.

LINEWEAVER, C. ET AL. 1996 *Astrophys. J.*, accepted, astro-ph/9601151.

MATHER, J. C. ET AL. 1994 *Astrophys. J.* **420**, 439.

PEEBLES, P. J. E. 1973 *Astrophys. J.* **185**, 413.

PEEBLES, P. J. E. 1993 *Principles of Physical Cosmology*, p. 168. Princeton U. Press.

PEEBLES, P. J. E. & YU, J. T. 1970 *Astrophys. J.* **162**, 815.

PENZIAS, A. A. & WILSON, R. 1965 *Astrophys. J.* **142**, 419.

SACHS R. K. & WOLFE A. M. 1967 *Astrophys. J.* **147**, 73.

SCOTT, D., SILK, J. & WHITE, M. 1995 *Science* **268**, 829.

SILK, J. 1967 *Nature* **215**, 1155-1156.

SMOOT, G. F. ET AL. 1990 *Astrophys. J.* **360**, 685.

SMOOT, G. F. ET AL. 1992 *Astrophys. J.* **396**, L1.

SMOOT, G. F., TENORIO, L., BANDAY, A. J., KOGUT, A., WRIGHT, E. L., HINSHAW, G., & BENNETT, C. L. 1994 *Astrophys. J.* **437**, 1.

TEGMARK, M. & BUNN, E. F. 1995 *Astrophys. J.* **455**, 1.

WHITE, M. 1996 *Phys. Rev. D*, in press, astro-ph/9601158.

WHITE, M., SCOTT, D. & SILK, J. 1994 *Ann. Rev. Astron. & Astrophys.* 1994 **32**, 329.

WRIGHT, E. L. ET AL. 1996 *Astrophys. J. Lett.*, submitted.

WRIGHT, E. L. ET AL. 1992 *Astrophys. J.* **396**, L13.

ZELDOVICH, YA. B. 1972 *Mon. Not. Roy. Astr. Soc.* **160**, 1.

Future prospects in measuring the CMB power spectrum

By P. M. LUBIN†

Department of Physics, University of California, Santa Barbara, CA 93106-9530, USA
and Center for Particle Astrophysics, University of California, Berkeley, CA 94720, USA

Measurements of the CMB power spectrum promise to revolutionize our understanding of the early universe and the parameters involved. We already have a tentative understanding of the power spectrum from COBE at large angular scales and from ground based and balloon borne experiments at small angular scales. The currently observed spectrum shows a strong rise as we approach sub-degree scales and even a possible fall at smaller scales. The capability now exists to make a precision measurement of the power spectrum from both sub-orbital and orbital platforms in the next 5–10 years. The results of such a measurement will profoundly effect our knowledge of the fundamental parameters that characterize our universe.

1. Introduction

Studies of the Cosmic Microwave Background (CMB) have provided some of the most challenging and fruitful measurements of the fundamental properties of the universe. The fundamental assumptions made over half a century ago about a homogeneous and isotropic universe are still consistent with our current knowledge. In particular, we know from studies of the CMB that the assumption of isotropy is valid to an astonishing level, one which has profoundly affected the theoretical concepts of how we view the very early universe. Indeed, during the long march from the discovery of the CMB in 1965 to our first hints of fundamental structure in the CMB in 1991–3, the very fact that we placed increasingly stringent limits on CMB anisotropies was strongly influencing the theoretical framework. This was particularly true regarding inflationary scenarios of the very early universe. We now have a rudimentary knowledge of the CMB power spectrum sufficient to constrain and even make preliminary measurements of the fundamental parameters such as total density, baryon fraction, etc. We also have at present sufficiently sensitive, degree-scale measurements in selected regions of the sky to give confidence that we will be able to map large portions of the sky with enough accuracy to reduce the errors on many fundamental cosmological parameters to the few percent level or better. Combining these maps with maps from other surveys, such as LSS optical and IR surveys, will provide a database we could only dream about even a decade ago. While there are still many potential pitfalls that remain to be crossed towards this goal, from our recent degree scale measurements we now understand many of the issues of foreground contamination and systematic errors that we worried about in the past. We are now very optimistic about the science that will result from the next generation of experiments.

2. Early Measurements of the Power Spectrum — Prior to COBE

As a historical footnote, we now know we had detected degree scale structure and hence had a preliminary measurement of power spectrum prior to 1992 but needed confirmation

† Parts of this paper are adapted from my other papers.

of what we had measured. Our (and others') subsequent measurements have now shown that these early measurements were correct. Prior to the COBE launch in 1989, our Advanced Cosmic Microwave Explorer (ACME) payload had made two balloon flights and one South Pole expedition performing measurements from 0.3 to 3 degrees with SIS and bolometric detectors. Prior to the April 1992 COBE announcement, ACME had flown four times and made two South Pole trips with a total of seven measurements. Our 1988 South Pole trip with ACME outfitted with a sensitive SIS (Superconductor-Insulator-Superconductor) receiver resulted in an upper limit of $\Delta T/T \leq 3.5 \times 10^{-5}$ at $0.5°$ for a Gaussian sky. This was tantalizingly close to the "minimal predictions" of anisotropy at the time and — as we were to subsequently measure — just barely above the level of detectability.

In the fall of 1989, we had our second ACME flight. This flight used a bolometric detector in place of the SIS detector and was the first in a series (the MAX flights) using bolometric detectors. The third ACME (second MAX) flight was the next summer (1990). This MAX-II flight resulted in structure being detected that was consistent with a cosmological spectrum. The data were taken in a low dust region and showed no evidence for galactic contamination. These data in the Gamma Ursa Minoris region ("GUM data") were first published in Alsop et al. (1992) *prior* to the COBE detections.

At the time, our most serious concern was of atmospheric stability, so we revisited this region in the next ACME flight in June 1991. ACME was next shipped to the South Pole for the second time in October 1991, with both SIS and HEMT (High Electron Mobility Transistor) based detectors for additional ground based measurements. This South Pole trip — the so-called SP91 data — produced a number of interesting results and still represents the most sensitive per point measurement of the CMB ever made. There were seven strips measured, the most sensitive two of which are referred to as the "9" and "15" point scans, respectively published in Gaier et al. (1992) and Schuster et al. (1993). These data were taken using a broad-band HEMT between 25-35 GHz, split into four bands. The beam size was about 1.5 degrees, but varied from 1.4 to 2.0 degrees over the lowest to the highest frequency bands. Both the 9 and 15 point scans showed significant structure at the level of $\Delta T/T = 1 \times 10^{-5}$. Since the beam size varied with band and since the data were taken in a "step scan" mode, a falling (negative spectral index) spectrum could result if the primary structure was smaller than the beam size and not centered in the beam. The 4 channel added data in both the 9 and 15 point scans show a very similar level of structure at the 10 ppm level, with a significant negative spectral index in the 9 point data and a less negative (consistent with flat at 2 sigma) in the 15 point. As mentioned, one cannot conclude unique spectral information about the source spectrum from these "spectral" indexes due to the beam size variation and scan mode. The fact that they both yielded similar amplitude was intriguing. A conservative upper limit was published in Gaier et al. (1992) of the 9 point scan of $\Delta T/T \leq 1.4 \times 10^{-5}$ at $1.2°$. This upper limit for a Gaussian autocorrelation function was computed from the highest frequency channel. The four channel co-add (assuming it is of cosmological origin) of this same data set yields a strong detection at a level of $\Delta T/T = 1 \times 10^{-5}$. The 15 point data are even more sensitive and, as mentioned, yield a strong detection at a similar level of $\Delta T/T = 1 \times 10^{-5}$ (Schuster et al. 1993).

Analyses of dust and synchrotron maps from the area of the sky surveyed, with reasonable assumptions about the spectral indices, predict that the signal level we observed is not consistent with expected dust or synchrotron radiation. Indeed, the low frequency synchrotron maps do not show similar morphology and would predict an amplitude that is much smaller ($< 7\ \mu$K). No evidence for point source contamination was found either and a recent intensive measurement of the field yielded a negative result for point source contamination (Gundersen et al. 1997). The 1σ error measured per point in the 15 point scan is 14 μK or $\Delta T/T = 5 \times 10^{-6}$. Per pixel, this is the most sensitive CMB measurement to date at any angular scale. Perhaps the most intriguing aspect — and the most confusing at the time — was the strong rise in the power spectrum from 10 ppm at 1.5 degrees from the SP 91 data to 40 ppm in the ACME-MAX-II 0.5 degree data. This was the first indication of a rise in the power spectrum as we approached the (possibly) sub-horizon scale. The fact that these two SP 91 data sets yielded similar amplitudes prompted us to return to the South Pole in 1993 with two HEMT detectors with a larger frequency range, 25-35 and 38-45 GHz in seven bands. We also used a "smooth scan" instead of the "step scan" for these data. These so-called SP93 (or sometimes SP94) data also yielded strong detections. Since the beam size was now ranging from 1 to 2 degrees we could compare with the SP 91 results as well as the ACME-MAX results at 0.5 degrees. The SP 93/94 results measure an amplitude consistent with the SP 91 data but somewhat larger at the smaller beam size of 1 degree.

The relevant measurements just prior to the COBE announcement are summarized in Figure 1. While these early data were quite confusing at the time, particularly the rise in the power spectrum from 1.5 to 0.5 degrees, it is now seen to be consistent with our (and others') additional measurements. Without the large-scale normalization of the COBE data, however, it seemed difficult to reconcile the apparently discordant data. This normalization at large angular scales was provided by the COBE data in 1992, and as shown in Fig. 5, our earlier degree-scale measurements are consistent with the COBE detections.

3. Foreground Issues

To make measurements of the CMB at the part per million level we have to contend with radiation from a number of other sources. Included in this are emission from our own galaxy in the form of interstellar dust, Bremsstrahlung and synchrotron radiation. Additional significant radiation from the earth, sun and moon and the atmosphere can be a problem that needs to be considered. The approximate relevant level of the foregrounds is shown in Figure 2. This needs to be interpreted carefully as the real picture is quite complicated depending on the spatial morphology being studied and — for the atmosphere — on time-dependent effects. In any case it is a good guide to discuss the relevant issues.

At high frequencies, above 100 GHz the primary galactic contaminant is thermal radiation from interstellar dust grains. The composition and structure of these dusts grains is not well understood and the generic properties are lumped into a description of the effective emissivity and temperature. The emissivity is frequency dependent while the temperature is spatially dependent. Since we are not generally interested in making measurements close to the galactic plane we do not have to deal with the extreme complexities in this region. With the ACME-MAX experiments we have made measurements both in and off the galactic plane of the interstellar dust in a few selected regions. While we cannot claim to have made an exhaustive survey, the results are consistent morphologically with the shorter 100 micron IRAS maps suitably scaled to our wavelengths. This allows us to at least predict the approximate level of expected contamination at shorter wavelengths using the IRAS and DIRBE maps as templates. The good news is that at the 10 ppm level where we have strong detections, there are large regions of the sky that should be relatively uncontaminated as we show in Figure 3.

At lower frequencies below 50 GHz, the primary contaminant is Bremsstrahlung and synchrotron radiation. Here we rely on a similar comparison using maps at much lower

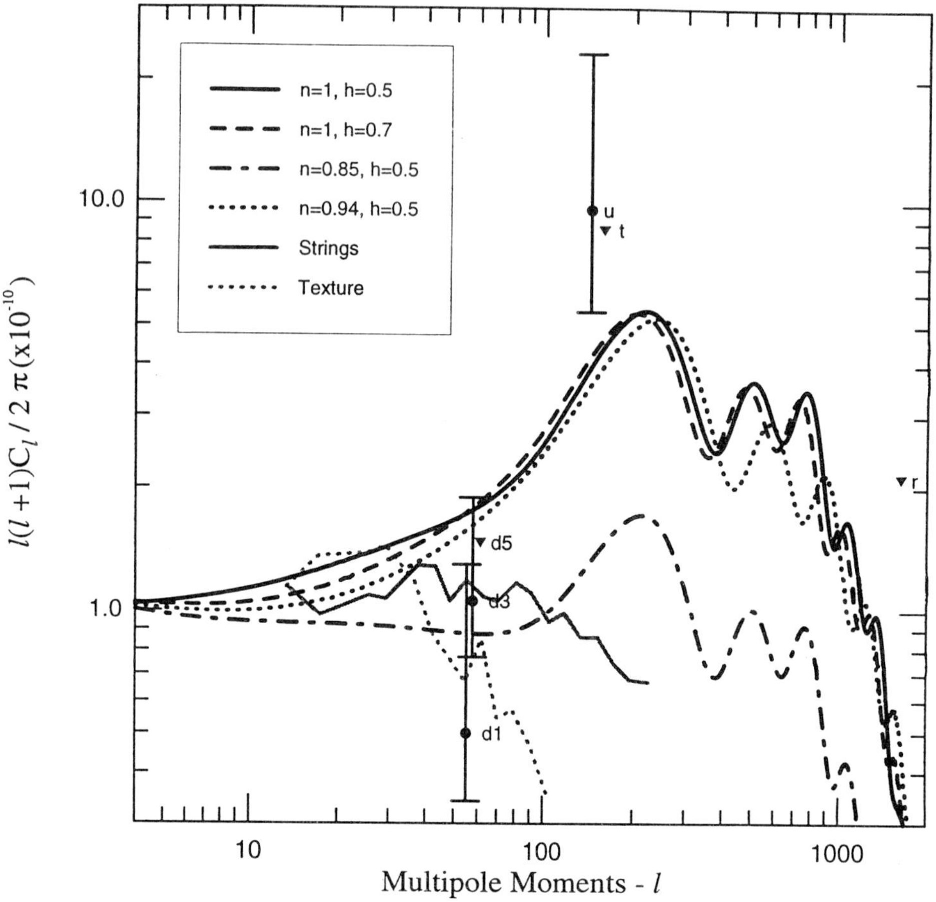

FIGURE 1. ACME CMB power spectrum data prior to the COBE detection. Theoretical curves are from Steinhardt and Bond (private communication). See KEY in Fig. 5 caption.

frequencies, notably the 0.4 and 1.4 GHz maps. The synchrotron radiation is generally not expected to be much of a problem due to its steep spectral index (typically -2.9 off the plane). The Bremsstrahlung is more complicated. First, we do not know much about the diffuse Bremsstrahlung emission and secondly, it does not decrease nearly as rapidly as the synchrotron emission at higher frequencies. We can, however, measure the H-alpha and related optical emission to constrain or measure it indirectly. Reynolds and others have done selected regions and are planning much more ambitious surveys that will help us in the future. In the end, it is the data themselves that will provide the best indicator of contamination or lack thereof. By making measurements over a wide range of frequencies we can distinguish between cosmological and galactic emission.

At all observation frequencies extragalactic radio sources are a potential concern. Extragalactic radio sources have the disadvantage that there is no well-known spectrum which describes the whole class of sources. For this reason, measurements over a very large range of frequencies and angular scales are required for CMB anisotropy measurements in order to achieve a sensitivity of $\Delta T/T \approx 1 \times 10^{-6}$. As we proceed towards smaller beams sizes the problem becomes progressively more difficult for point source

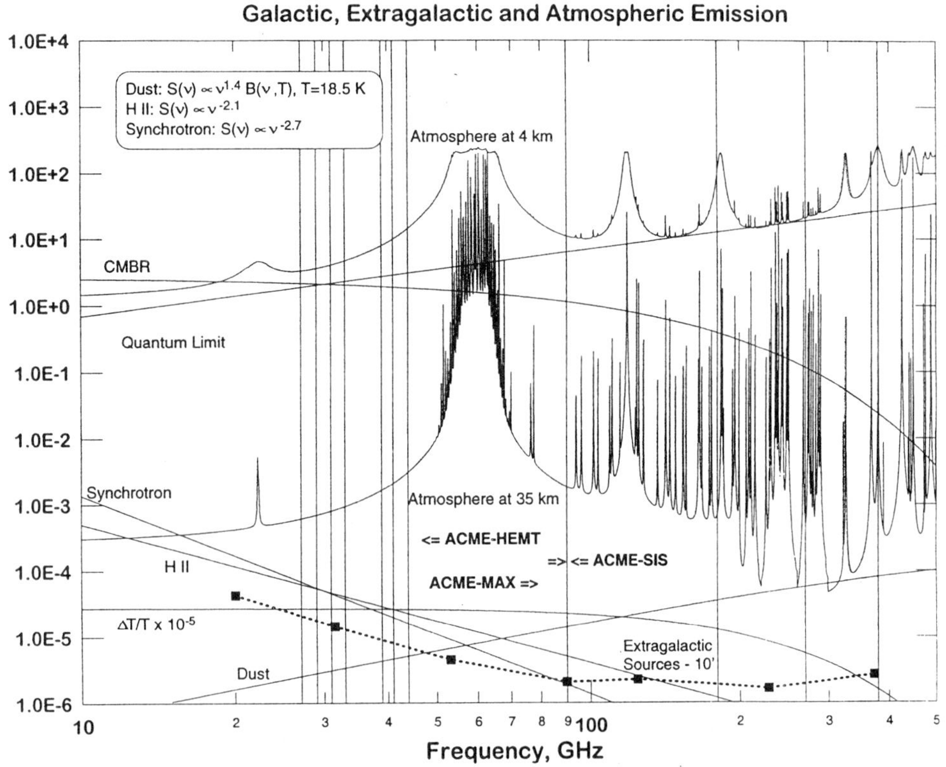

FIGURE 2. Relevant backgrounds for terrestrial measurements at the South Pole and at balloon altitudes (35 km) where ACME observes. Representative galactic backgrounds are shown for synchrotron, bremsstrahlung, and interstellar dust emission as well as the various ACME (center) wavelength bands.

contamination. However, as we show in Figure 2, the estimated contribution for 10' beam surveys is still encouraging. We can make an estimate of the portions of the sky that are likely to be contaminated by galactic emission by combining the low and high frequency galaxy maps with our "knowledge" of the spectral indices of the various species. In Fig. 2, we show such an estimate for the fraction of sky uncontaminated by the direct emission from our galaxy. Keep in mind that this is only an estimate based on our present data. While additional galactic surveys would help to determine the galactic component better, it is assumed at the current time that the best data on galactic contamination will come from the future CMB maps themselves.

Additional ground-based surveys of point sources and H-II emission will be a great aid in CMB studies, however. As an example, for our SP '94 data set we have made a point source survey at a number of frequencies to determine the spectra of the brightest sources in our field (Gundersen et al. 1997). No evidence of significant point source contamination was found for this data set. Ideally, such surveys could be carried out for the entire sky. Such a task is daunting, however, considering the number of known, let alone unknown, point sources. For example, in ACME, the sensitivity to point sources is about 47 μK per Jansky at 40 GHz. This is for a 1 degree beam. A 10' beam size experiment will face a point source sensitivity of about 1 mK per Jansky. A 1 ppm per pixel survey of the CMB at 10' resolution will require milli-Jansky point source rejection ability. One of the objectives of the next generation of CMB measurements will indeed be greatly refined galactic and extra-galactic source models and measurements.

Date	Site	Detector System	Beam FWHM (deg.)	Instrument Sensitivity
1988 Sept	Balloon^P	90 GHz SIS receiver	0.5	4 mK $s^{1/2}$
1988 Nov-89 Jan	South Pole	90 GHz SIS receiver	0.5	3.2
1989 Nov	Balloon^{FS}	MAX photometer (3, 6, 9, 12 cm^{-1}) ^3He	0.5	12, 2, 5.7, 7.1
1990 Jul	Balloon^P	MAX photometer (6, 9, 12 cm^{-1}) ^3He	0.5	0.7, 0.7, 5.4
1990 Nov-90 Dec	South Pole	90 GHz SIS receiver	0.5	3.2
1990 Dec-91 Jan	South Pole	4 Channel HEMT amp (25-35 GHz)	1.5	0.8
1991 Jun	Balloon^P	MAX photometer (6, 9, 12 cm^{-1}) ^3He	0.5	0.6, 0.6, 4.6
1993 Jun	Balloon	MAX photometer (3, 6, 9, 12 cm^{-1}) ADR	0.55-0.75	0.6, 0.5, 0.8, 3.0
1993 Nov-94 Jan	South Pole	HEMT 25-35 GHz	1.5	0.8
1993 Nov-94 Jan	South Pole	HEMT 38-45 GHz	1.0	0.5
1994 Jun	Balloon	MAX photometer (3, 6, 9, 14 cm^{-1}) ADR	0.55-0.75	0.4, 0.4, 0.8, 3.0
1996 Feb	Balloon	HACME HEMT 38-45 GHz	0.7	0.4, 0.5, 0.8
1996 Jun	Balloon	HACME HEMT 38-45 GHz	0.7	0.4, 0.5, 0.8

TABLE 1. CMB measurements made with ACME.

4. Atmosphere

For sub orbital missions, balloon borne and ground based, atmospheric emission is a dominant concern. The atmosphere is a complex source being frequency, spatially and temporally dependent. The frequency dependence is actually useful since it allows us to monitor it and allows narrow band instruments to achieve extremely low atmospheric residuals. The emission consists primarily of the molecular transitions of oxygen, water and ozone. Except for the effect of pressure waves, the oxygen is thought to be very uniformly distributed. Water is known to be very inhomogeneous and ozone appears to have some inhomogeneities as well. For virtually all CMB experiments these are the only relevant species. Below 60 GHz there are no ozone lines of significance and hence at longer wavelengths we only need to contend with oxygen and water. As seen from Figure 2 the atmospheric emission is vastly different depending on the frequency and bandwidth of the observations. This is particularly true at high altitudes (balloons), where the pressure broadening of the molecular lines is less prominent. Here the ratio of peak to valley emission can be as large as a factor of 1000 or more. Hence, judicious choice of wavelength can drastically reduce the atmospheric emission. As an example, we can reduce the atmospheric emission for a 40 GHz HEMT based detector from 7 Kelvin at the South Pole to less than 1 milliKelvin at balloon altitudes. At some wavelengths this allows us to operate without worrying about any significant atmosphere.

5. ACME Results

There have been a total of 13 ACME observations/flights from 1988 to 1996. Over 20 papers and proceedings articles have resulted from these measurements as well as nine Ph.D. theses. ACME-SIS and ACME-HEMT articles by Meinhold and Lubin (1991), Meinhold et al. (1992), Gaier et al. (1992), Schuster et al. (1993), Gundersen et al. (1995) and ACME-MAX articles by Fischer et al. (1992), Alsop et al. (1992), Meinhold et al. (1993), Gundersen et al. (1993), Devlin et al. (1994), Clapp et al. (1994), Tanaka et al. (1996) and Lim et al. (1996) summarize the results to date.

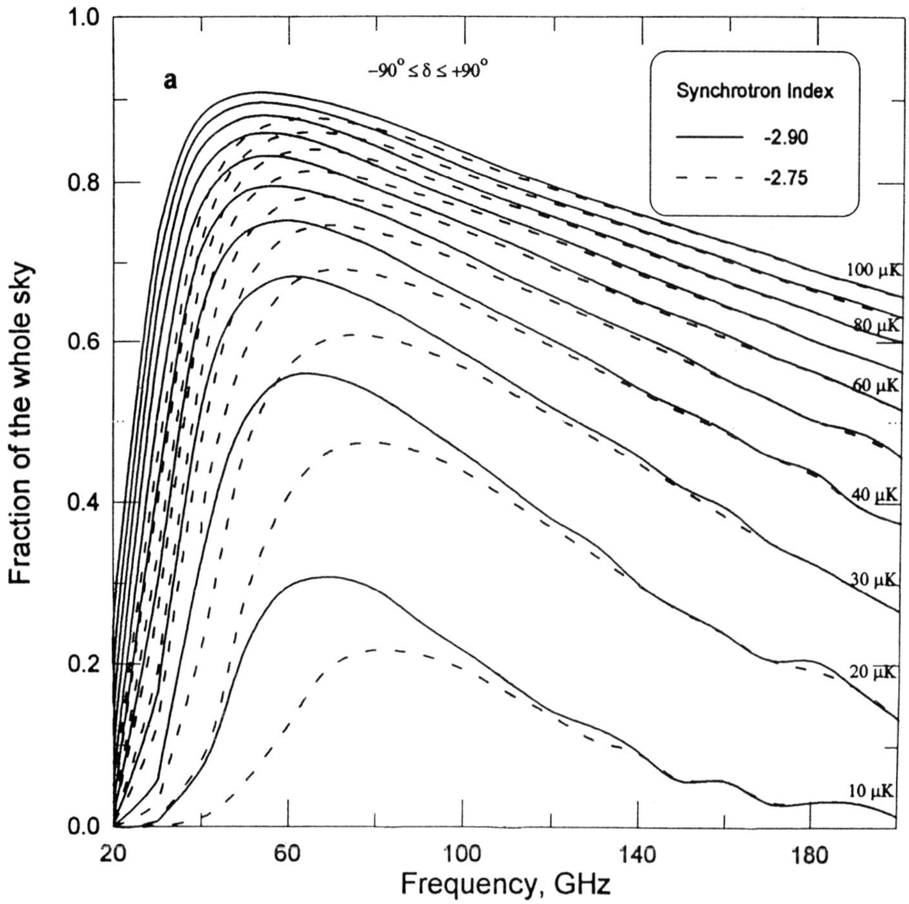

FIGURE 3. Galactic model estimates for the fraction of sky uncontaminated below a given level. The model includes synchrotron, bremsstrahlung, and dust emission. The synchrotron model is given for two different spectral indices. Far off the galactic plane (where we are most interested in measuring) the steeper spectral index is more appropriate. The dust model is based on the IRAS 100 micron map combined with our dust data from the mu-Pegasus region.

6. Future

As exciting and rewarding as the past five years have been in CMB studies, the next decade truly promises to be profound in its results. We have now made measurements sensitive enough to extract many of the cosmological parameters but lack the sky coverage to reduce the sample and cosmic variance to a reasonable level. We are now at a point, both technologically and in our understanding of the relevant CMB signals and non cosmological backgrounds, that we can seriously contemplate making a precision measurement of the CMB power spectrum. What is needed is not necessarily more sensitivity (though more is always welcome) but more samples of the sky. Some of the current and past confusion in measurements of the power spectrum are undoubtably due to small statistics. For example, in Figure 4 we show the estimated cosmological parameter extraction precision versus beam size and pixel sensitivity for an experiment that covers 50% of the sky. While the extraction procedures and error estimates are still being

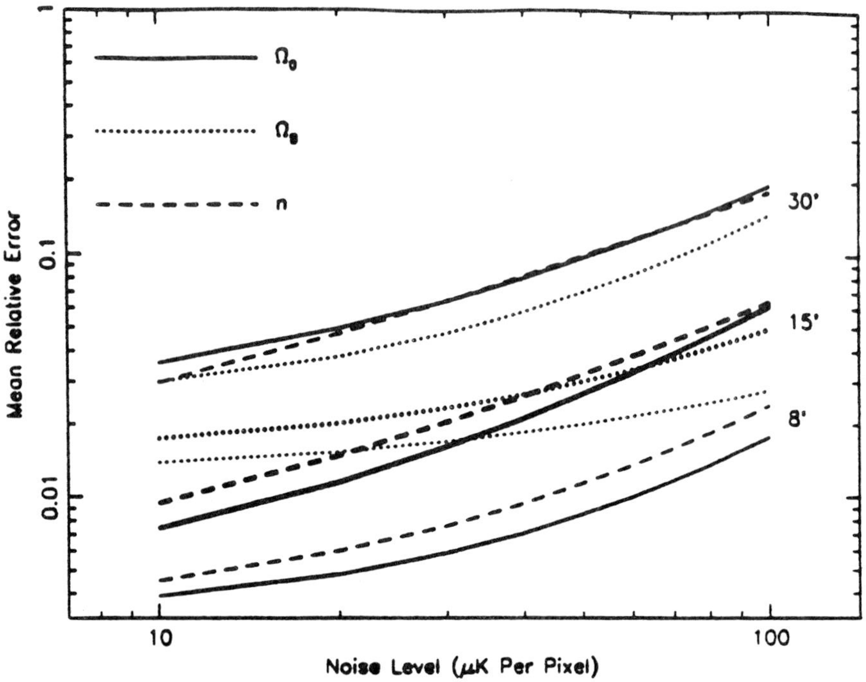

FIGURE 4. Estimated parameter extraction versus sensitivity, assuming 50% sky coverage.

debated, it appears feasible to get a precision in measuring the various parameters to the percent level within the next decade. This implicitly assumes a cosmological model that is "well behaved" and assumes no additional pathological foregrounds.

7. Mapping Techniques

To date essentially all the sub degree experiments have been done in a one-dimensional scan mode and have not been able to make a "true map". This has been done because of the atmospheric emission restricting us to make one dimensional, constant elevation angle chops. All of the ACME measurements are of this type. An orbital mission does not suffer from this problem, of course. However even then one is making 1-D strips and then "sewing" them together to make a map. The COBE mission was no exception. The only "real maps" so far have been made by interferometers and then a "map" must be carefully scrutinized since the data weighting is far from uniform. With the exception of interferometers, experiments have been either differential or "total" power (this means a direct measure of the flux is made). Virtually all experiments (including our own) have opted for the differential technique as it is less demanding on various system stabilities. Recently experiments have shown that a direct "total" power measurement is possible and the era of sub-degree resolution maps is about to begin. It is easy to see why this is difficult. The typical equivalent noise of a state-of-the-art detector is 10–30K. This is to be compared to the desired sensitivity of 3–30 micro Kelvin. This requires a system stability of the order of 1–10 ppm.

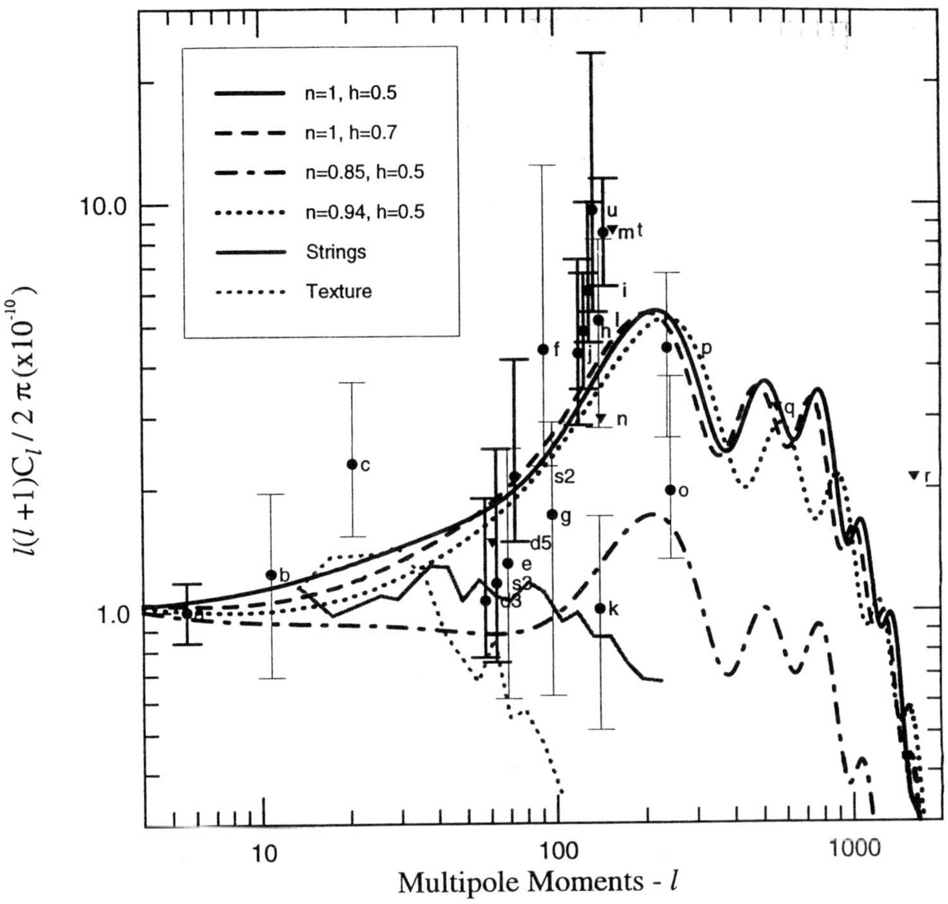

FIGURE 5. Recent ACME results (in BOLD) along with results from other groups. Key: a-COBE, b-FIRS, c-Tenerife, d1-SP91 9 pt. 4 channel analysis-Bond '93, d3-SP91 9+13 pt. 4 channel analysis-Bond '93, d5-SP91 9 pt. Gaier et al. '92, e-Big Plate, f-PYTHON, g-ARGO, h-MAX4-Iota Dra, i-MAX4-GUM, j-MAX4-Sig Herc, k-MSAM2, l-MSAM2, m-MAX3-GUM, n-MAX3-mu Peg, o-MSAM3, p-MSAM3, q-Wh. Dish, r-OVRO7, s2-SP94-Q, s3-SP94-Ka, t-SP89, u-MAX2-GUM, many from Steinhardt and Bond by private communication.

8. Polarization

The CMB is characterized by the four Stokes parameters I, Q, U and V in each direction and at each frequency observed. I characterizes the overall spectrum (flux) while Q and U give the linear polarization and V the circular polarization. Very little effort has been directed towards the measurement of the polarization of the CMB compared to the effort in anisotropy detection. In part, this is due to the low level of linear polarization expected. Typically, the polarization is only 1–30% of the anisotropy and depends strongly on the model parameters. This is an area which, in theory, can give information about the reionization history, scalar and tensor gravity wave modes, and large-scale geometry effects. It is now possible to measure CMB polarization to a sensitivity of better than 1 ppm on limited portions of the sky. In the future, this will be a very fruitful area of inquiry, particularly when combined with overlapping anisotropy measurements. In particular, the polarization-anisotropy cross correlation is a very powerful technique for

Publication	Configuration	Beam FWHM (deg.)	$\Delta T/T \times 10^{-6}$
Meinhold & Lubin '91	ACME-SIS SP '89	0.5	<35
Alsop et al., '92	ACME-MAX-II (GUM)	0.5	45^{+57}_{-26}
Gaier et al., '92	ACME-HEMT SP '91	1.5	<14
Meinhold et al., '92	ACME-MAX-III (μ Peg - upper limit)	0.5	<25
Meinhold et al., '92	ACME-MAX-III (μ Peg - detection)	0.5	15^{+11}_{-7}
Schuster et al., '93	ACME-HEMT SP '91	1.5	9^{+4}_{-2}
Gundersen et al., '93	ACME-MAX-III (GUM)	0.5	42^{+17}_{-11}
Devlin et al., '94	ACME-MAX-IV (GUM)	0.55-0.75	37^{+19}_{-11}
Clapp et al., '94	ACME-MAX-IV (Iota Draconis)	0.55-0.75	33^{+11}_{-11}
Clapp et al., '94	ACME-MAX-IV (Sigma Hercules)	0.55-0.75	31^{+17}_{-13}
Gundersen et al., '95	ACME-HEMT SP '94	1	$15^{+5.7}_{-2.5}$
Lim et al., '96	ACME-MAX-V (μ Peg)	0.5	<13
Tanaka et al., '96	ACME-MAX-V (HR5127)	0.5	12^{+4}_{-3}
Tanaka et al., '96	ACME-MAX-V (Phi Herculis)	0.5	19^{+7}_{-4}

TABLE 2. Recent ACME degree-scale results.

understanding and breaking degeneracies in the parameter extraction. At mm and cm wavelengths dedicated polarization measurements designed to reach a level of sensitivity of 1 ppm or better are feasible to do from the ground since the atmosphere is essentially unpolarized at these wavelengths.

9. Orbital vs Sub Orbital Missions

There are currently several major proposed (and now just approved) satellite missions. The European COBRAS/SAMBA proposal is a combined HEMT and bolometer mission, and would cover from about 1 to 10 mm with resolution varying from about 4 to 40 arc minutes depending on the frequency. In the U.S., the Mid-Explorer class MAP mission is a HEMT based mission that will cover from 3 to 15 mm with a beam size from 20 to 80 arc minutes. There are significant differences in technical and programmatic approaches being taken with the European being a more ambitious, and hence, more costly experiment. The U.S. Mid-Explorer mission is designed with more limited objectives, but at a significantly lower price and at a possibly shorter time scale to launch. Either of these missions would provide invaluable data that could revolutionize our understanding about early universe physics. Currently, it can be assumed that these missions will not produce data before 2002 at the earliest for the US mission, and 2006 for the European one, and hence it is to be anticipated that continued vigorous ground-based and suborbital experiments will continue to produce valuable data.

Indeed, per pixel sensitivities with suborbital missions in the μK region are now achievable with current and new technologies, HEMTs and bolometers, over hundreds to hundreds of thousands of pixels and possibly over large portions of the sky. With the HEMT and bolometer array receivers that are being planned now for sub orbital measurements, it is quite possible that the integrated sensitivity and resolution of these measurements will exceed that of the US mission. The major issue will be atmospheric stability, control of sidelobes and getting a uniform dataset. Ideally, to reduce systematic effects, full-sky coverage would be best and this is one area where a space-based measurement will excel. In a number of ways the sub-orbital and orbital missions will be complimentary and allow for a very powerful combined dataset.

9.1. *The ACE Experiment*

The main objective at this time is not necessarily more sensitivity but rather increased sky coverage. We now have detections at a S/N of 1 to 10. While more sensitivity is always welcome, the primary goal is to increase the statistics of the sky sample. For sub-orbital approaches the primary balance is between increased exposure time from the ground but much larger atmosphere versus much less exposure time with balloons but with much lower atmosphere and increased detector sensitivity. Long duration balloons circumnavigating the arctic for periods of 10-15 days will allow us to significantly extend the conventional balloon exposure. A new approach using ultra long duration balloons with 100 day exposure may allow an excellent low cost alternative to orbital missions in this and other areas. As an example of this approach, the ACE experiment is designed to achieve beam sizes of less than 10 arc minutes at 100 GHz and exposure times of 100 days or more. This system will be the equivalent of a small satellite with closed cycle refrigerators to cool the HEMTs to 20K and autonomous control. Figure 6 shows a schematic drawing of this experiment.

10. Acknowledgements

This work was supported by the National Science Foundation Center for Particle Astrophysics, the National Aeronautics and Space Administration, the NASA Graduate Student Research Program, the National Science Foundation Polar Program, the California Space Institute and the University of California. Its success is the result of the work of a number of individuals, particularly the graduate students involved with the experiment and our collaborators, most notably Paul Richards and Andrew Lange on ACME-MAX. The exceptional HEMT amplifier was provided by NRAO. Robert Wilson, Anthony Stark, and Corrado Dragone, all of AT&T Bell Laboratories, provided critical support and discussion regarding the early design of the telescope and receiver system. We would like to thank all of the South Pole support staff for highly successful 1988–1989, 1990–1991 and 1993–1994 polar summers. In addition, we want to acknowledge the crucial contributions of the entire team of the National Scientific Balloon Facility in Palestine, Texas for their continued excellent support.

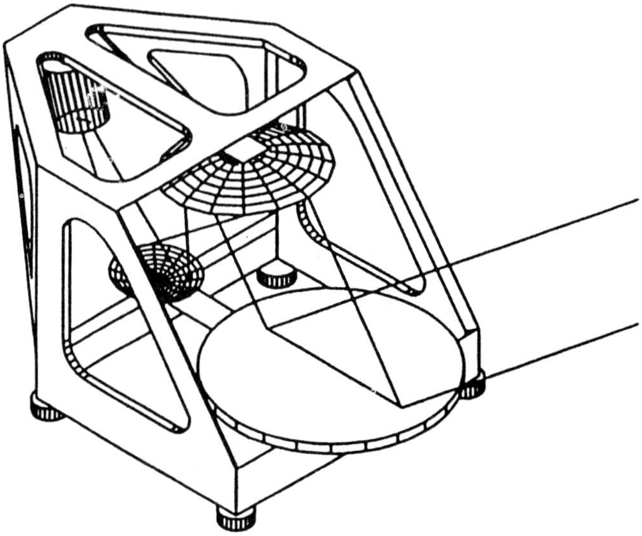

FIGURE 6. The ACE Experiment - An ultra long duration balloon experiment.

REFERENCES

ALSOP, D. C *et al.* 1992 *Astrophys. J.* **317**, 146.
CLAPP, A. *et al.* 1994 *Astrophys. J.* **433**, L57.
DEVLIN. M. *et al.* 1994 *Astrophys. J.* **430**, L1.
FISCHER, M. *et al.* 1992 *Astrophys. J.* **388**, 242.
GAIER, T. *et al.* 1992 *Astrophys. J.* **398**, L1.
GUNDERSEN, J. O. *et al.* 1993 *Astrophys. J.* **413**, L1.
GUNDERSEN, J. O. *et al.* 1995 *Astrophys. J.* **443**, L57.
GUNDERSEN, J. O. *et al.* 1997 *to be submitted to Astrophys. J.*
LIM, M. *et al.* 1996 in press.
LUBIN, P. *et al.* 1985 *Astrophys. J.* **298**, L1.
MEINHOLD, P.R. & LUBIN, P. M. 1991 *Astrophys. J.* **370**, L11.
MEINHOLD, P. *et al.* 1992 *Astrophys. J.* **406**, 12.
MEINHOLD, P. *et al.* 1993 *Astrophys. J.* **409**, L1.
POSPIESZALSKI, M. W. *et al.* 1990 *IEEE MTT-S Digest*, 1253.
SCHUSTER, J. *et al.* 1993 *Astrophys. J.* **412**, L47.
SMOOT, G. F. *et al.* 1992 *Astrophys. J.* **396**, L1.
TANAKA, S. *et al.* 1996 in press.
WOLLACK, E. *et al.* 1994 *Astrophys. J.* **419**, L49.

Inflationary Cosmology

By MICHAEL S. TURNER[1,2] †,

[1]Departments of Physics and of Astronomy and Astrophysics, Enrico Fermi Institute, The University of Chicago, Chicago, IL 60637-1433, USA

[2]NASA/Fermilab Astrophysics Center, Fermilab, Batavia, IL 60510-0500, USA

Inflation is a bold and expansive extension of the Standard Cosmology. It holds the promise to extend our understanding of the Universe to within 10^{-32} sec of the big bang and answer most of the pressing questions in cosmology. Its boldest assertion is that all the structure observed in the Universe today arose from quantum-mechanical fluctuations on subatomic scales. A flood of cosmological observations and laboratory experiments are testing inflation and within five years we should know whether inflation is to become part of a new standard cosmology.

---

1. From Hubble to Zel'dovich & Novikov

The value of the Hubble constant has changed by about a factor of ten since Edwin Hubble's pioneering measurements. The context in which we view the Universe has changed even more profoundly. Until 1964 cosmology was mostly concerned with cosmography; the spirit of this period was perhaps best captured by Sandage, "the quest for two numbers (H_0 and q_0)." The discovery of the Cosmic Background Radiation (CBR) led to the establishment of a physical foundation for the expanding Universe – the hot big-bang cosmology. The 1970s saw this model become firmly established as the Standard Cosmology. Igor Novikov played an important role in this enterprise, and with Zel'dovich wrote one of the first reviews of the Standard Cosmology (Novikov & Zel'dovich 1967) as well as a comprehensive and authoritative summary of it in Volume 2 of *Relativistic Astrophysics (Structure and Evolution of the Universe)* (Zel'dovich & Novikov 1983).

The hot big-bang cosmology is a remarkable achievement. It provides a reliable account of the Universe from about 10^{-2} sec to the present (Peebles et al. 1991; Turner 1993a). Further, the hot big-bang model together with modern ideas in particle physics — the Standard Model, supersymmetry, grand unification, and superstring theory — provides a framework for sensible speculation all the way back to the Planck epoch, and perhaps even earlier. Here too, Novikov was one of the pioneering figures in the study of the earliest moments (Novikov & Zel'dovich 1973; Zel'dovich & Novikov 1983).

Building upon the Standard Cosmology, in the 1980s cosmologists and particle physicists began an ambitious program to root the hot big-bang model in fundamental physics and to address a deeper set of questions. What is the nature of the ubiquitous dark matter that is the dominant component of the mass density? Why does the Universe contain only matter? What is the origin of the tiny inhomogeneities that seeded the formation of structure, and how did that structure evolve? Why is the portion of the Universe that we can see so flat and smooth? What is the value of the cosmological constant? What is the origin of the entropy (photons in the CBR)? How did the expansion begin—or was there a beginning? What was the big bang?

In the past fifteen years much progress has been made, and many believe that the answers to all these questions involve events that took place during the earliest moments and involved physics beyond the Standard Model (Kolb & Turner 1990). For example, the matter-antimatter asymmetry, quantified as a net baryon number of about 10^{-10}

† This work was supported by the US DoE (at Chicago and Fermilab) and the US NASA (at Fermilab).

per photon, is believed to have developed through interactions that do not conserve baryon number or C, CP and occurred out of thermal equilibrium. If "baryogenesis" involved unification-scale physics the baryon asymmetry developed around 10^{-34} sec (Kolb & Turner 1983); on the other hand, baryogenesis may have occurred at the weak scale ($T \sim 300\,\mathrm{GeV}$ and $t \sim 10^{-11}\,\mathrm{sec}$) through baryon-number violation within the Standard Model and C, CP violation from physics beyond the Standard Model (Cohen et al. 1992; Dolgov 1992).

The most optimistic early-Universe cosmologists (of which I am one) believe that we are on the verge of solving almost all of the above problems and extending our knowledge of the Universe back to around 10^{-32} sec after "the bang." The key is inflation. Among other things, inflation has led to the cold dark matter model of structure formation, whose basic tenets are scale-invariant density perturbations and dark matter whose primary composition is slowly moving elementary particles (e.g., axions or neutralinos). Cold dark matter provides a crucial test of inflation and a possible window to physics at unification-scale energies. If cold dark matter is correct, it would complete the Standard Cosmology by connecting the theorist's smooth early Universe to the astronomer's Universe which abounds with structure, galaxies, clusters of galaxies, superclusters, voids and great walls (Huchra & Geller 1989).

2. Inflation

As successful as the big-bang cosmology is, it suffers from a dilemma involving initial data. Extrapolating back, one finds that the Universe apparently began from a very special state: A slightly inhomogeneous and very flat Robertson-Walker spacetime. Collins and Hawking showed that the set of initial data that evolve to a spacetime that is as smooth and flat as ours is today is of measure zero (Collins & Hawking 1973). [In the context of simple grand unified theories, the hot big-bang model suffers from another serious problem: the extreme overproduction of superheavy magnetic monopoless (Preskill 1984); in fact, attempting to solve the monopole problem led Guth to inflation.]

The cosmological appeal of inflation is the lessening of the dependence of the present state of the Universe upon its initial state. Two elements are essential: (1) accelerated ("superluminal") expansion and the concomitant growth of the scale factor; and (2) massive entropy production (Hu, Turner & Weinberg 1994). Through inflation, a small, smooth subhorizon-sized patch of the early Universe grows to a large enough size and comes to contain enough heat (entropy in excess of 10^{88}) to encompass our present Hubble volume. In addition, superluminal expansion guarantees that the Universe today appears flat (just as any small portion of the surface of a large sphere appears flat).

While there is presently no standard model of inflation—just as there is no standard model for physics at these energies (typically 10^{15} GeV or so)—viable models have much in common. They are based upon well posed, albeit highly speculative, microphysics involving the classical evolution of a scalar field. A nearly exponential expansion is driven by the potential energy ("vacuum energy") that arises when the scalar field is displaced from its potential-energy minimum. Provided the potential is flat, during the time it takes for the field to roll to the minimum of its potential the Universe undergoes many e-foldings of expansion (more than around 60 or so are required to realize the beneficial features of inflation). As the scalar field nears the minimum, the vacuum energy has been converted to coherent oscillations of the scalar field, which correspond to nonrelativistic scalar-field particles. The eventual decay of these particles into lighter particles and their thermalization results in the "reheating" of the Universe and accounts for all the heat in the Universe today (the entropy production event).

The growth of the cosmic scale factor (by a factor greater than that since the end of inflation) allows quantum fluctuations excited on very small scales ($\lesssim 10^{-23}$ cm) to be stretched to astrophysical scales ($\gtrsim 10^{25}$ cm). Quantum fluctuations in the scalar field responsible for inflation ultimately lead to an almost scale-invariant spectrum of density perturbations (Guth & Pi 1982; Hawking 1982; Starobinsky 1982; Bardin, Steinhardt & Turner 1983), and quantum fluctuations in the metric itself lead to an almost scale-invariant spectrum of gravity-waves (Rubakov, Sazhin & Veryaskin 1984; Fabbri & Pollock 1983; Starobinsky 1983a; Abbott & Wise 1984). Scale invariance for density perturbations means scale-independent fluctuations in the gravitational potential (equivalently, density perturbations of different wavelength cross the horizon with the same amplitude); scale invariance for gravity waves means that gravity waves of all wavelengths cross the horizon with the same amplitude. Because of subsequent evolution, neither the scalar nor the tensor perturbations are scale invariant today.

2.1. *Grander Implications*

Inflation lessens greatly the "specialness" problem, but does not eliminate it (Turner & Widrow 1986; Jensen & Schabes 1987; Starobinsky 1983b). All open FRW models will inflate and become flat; however, many closed FRW models will recollapse before they can inflate. If one imagines the most general initial spacetime as being comprised of negatively and positively curved FRW (or Bianchi) models that are stitched together, the failure of the positively curved regions to inflate is of little consequence: because of exponential expansion during inflation the negatively curved regions will occupy most of the space today. Inflation does not solve the smoothness problem forever; it just postpones the problem into the exponentially distant future: We will be able to see outside our smooth inflationary patch at a time $t \sim t_0 \exp[3(N - N_{\rm min}]$, where N is the actual number of e-foldings of inflation and $N_{\rm min} \sim 60$ is the minimum required to solve the horizon/flatness problems.

Linde has emphasized that inflation has changed our view of the Universe in a very fundamental way (Linde 1990). While cosmologists have long used the Copernician principle to argue that the Universe must be smooth because of the smoothness of our Hubble volume, in the post-inflation view, our Hubble volume is smooth because it is a small part of a region that underwent inflation. On the largest scales the structure of the Universe is likely to be very rich: Different regions may have undergone different amounts of inflation, may have different realizations of the laws of physics because they evolved into different vacuum states (of equivalent energy), and may even have different numbers of spatial dimensions. Since it is likely that most of the volume of the Universe is still undergoing inflation and that inflationary patches are being constantly produced (eternal inflation), the age of the Universe is a meaningless concept and our expansion age merely measures the time back to our big bang – the nucleation of our inflationary bubble.

2.2. *Specifics*

Guth's seminal paper (Guth 1981) both introduced the idea of inflation and showed that the model that he based the idea upon did not work! Thanks to very important contributions by Linde (Linde 1982) and Albrecht and Steinhardt (Albrecht & Steinhardt 1982) that was quickly remedied, and today there are many viable models of inflation. That of course is good news and bad news – since it means that there is no standard model of inflation. The absence of a standard model of inflation should of course be viewed in the light of our general ignorance about physics at unification-scale energies.

Many different approaches have taken in constructing particle-physics models for inflation. Some have focussed on very simple scalar potentials, e.g., $V(\phi) = \lambda\phi^4$ or

$= m^2\phi^2/2$, without regard to connecting the model to any underlying theory (Linde 1983; Steinhardt & Turner 1984). Others have proposed more complicated models that attempt to make contact with speculations about physics at very high energies, e.g., grand unification (Shafi & Vilenkin 1984; Pi 1984) supersymmetry (Holman, Ramond & Ross 1984; Olive 1990; Murayama 1994; Randall, Soljacic & Guth 1996) preonic physics (Cvetic et al. 1990), or supergravity (Copeland et al. 1994). Several authors have attempted to link inflation with superstring theory (Gasperini & Veneziano 1994; Brustein & Veneziano 1994; Banks et al. 1995) or "generic predictions" of superstring theory such as pseudo-Nambu-Goldstone boson fields (Freese, Frieman, & and Olinto 1990). While the scale of the vacuum energy that drives inflation is typically of order $(10^{15}\,\mathrm{GeV})^4$, a model of inflation at the electroweak scale, vacuum energy $\approx (1\,\mathrm{TeV})^4$, has been proposed (Knox & Turner 1993). There are also models in which there are multiple epochs of inflation (Silk & Turner 1986; Kofman, Linde & Einsato 1987).

In all of the models above gravity is described by general relativity. A qualitatively different approach is to consider inflation in the context of alternative theories of gravity.† The most successful of these models is first-order inflation (La & Steinhardt 1991; Kolb 1991). First-order inflation returns to Guth's original idea of a strongly first-order phase transition; in the context of general relativity Guth's model failed because the phase transition, if inflationary, never completed. In theories where the effective strength of gravity evolves, like Brans-Dicke theory, the weakening of gravity during inflation allows the transition to complete. In other models based upon nonstandard gravitation theory, the scalar field responsible for inflation is itself related to the size of additional spatial dimensions, and inflation then also explains why our three spatial dimensions are so big, while the other spatial dimensions are so small.

All models of inflation have one feature in common: the scalar field responsible for inflation has a very flat potential-energy curve and is very weakly coupled. Invariably, this leads to a very small dimensionless number, usually a coupling constant of the order of 10^{-14}. Such a small number, like other small numbers in physics (e.g., the ratio of the weak to Planck scales $\approx 10^{-17}$ or the ratio of the mass of the electron to the W/Z boson masses $\approx 10^{-5}$), runs counter to one's belief that a truly fundamental theory should have no tiny parameters, and cries out for an explanation. At the very least, this small number must be stabilized against quantum corrections—which it is in all of the previously mentioned models.‡ In some models, the small number in the inflationary potential is related to other small numbers in particle physics: for example, the ratio of the electron mass to the weak scale or the ratio of the unification scale to the Planck scale. Explaining the origin of the small number that seems to be associated with inflation is both a challenge and an opportunity.

Because of the growing base of observations that bear on inflation, another approach to model building is emerging: the use of observations to constrain the underlying inflationary model. In Section 4 I will discuss the possibilities for reconstructing the inflationary potential.

† After all, inflation probably involves physics at energy scales not too different from the Planck scale and the effective theory of gravity at these energies could well be very different from general relativity; in fact, there are some indications from superstring theory that gravity in these circumstances might be described by a Brans-Dicke-like theory.

‡ It is sometimes stated that inflation is unnatural because of the small coupling of the scalar field responsible for inflation; while the small coupling certainly begs explanation, these inflationary models are not unnatural in the technical sense as the small number is stable against quantum fluctuations.

2.3. *Three Robust Predictions*

While there are many varieties of inflation, there are three robust predictions which are crucial to sharply testing inflation.¶

(*a*) **Flat universe.** Because solving the "horizon" problem (large-scale smoothness in spite of small particle horizons at early times) and solving the "flatness" problem (maintaining Ω very close to unity until the present epoch) are linked geometrically (Kolb & Turner 1990; Hu, Turner & Weinberg 1994), this is the most robust prediction of inflation. Said another way, it is the prediction that most inflationists would be least willing to give up. [Even so, models of inflation have been constructed where the amount of inflation is tuned just to give Ω_0 less than one today (Bucher, Goldhaber & Turok 1995; Bucher & Turok 1995).] Through the Friedmann equation for the scale factor, flatness implies that the total energy density (matter, radiation, vacuum energy, and anything else) is equal to the critical density.

(*b*) **Nearly scale-invariant spectrum of gaussian density perturbations.** Essentially all inflation models predict a nearly scale-invariant spectrum of gaussian density perturbations. Described in terms of a power spectrum, $P(k) \equiv \langle|\delta_k|^2\rangle = Ak^n$, where δ_k is the Fourier transform of the primeval density perturbations, and the spectral index $n = 1$ in the scale-invariant limit. The overall amplitude A is model dependent. Density perturbations give rise to CBR anisotropy as well as seeding structure formation. Requiring that the density perturbations are consistent with the observed level of anisotropy of the CBR (and large enough to produce the observed structure formation) is the most severe constraint on inflationary models and leads to the small dimensionless number that all inflationary models have.

(*c*) **Nearly scale-invariant spectrum of gravitational waves.** These gravitational waves have wavelengths from around 1 km to the size of the present Hubble radius and beyond. Described in terms of a power spectrum for the dimensionless gravity-wave amplitude at early times, $P_T(k) \equiv \langle|h_k|^2\rangle = A_T k^{n_T-3}$, where the spectral index $n_T = 0$ in the scale-invariant limit. Once again, thc ovcrall amplitude A_T is model dependent (varying as the value of the inflationary vacuum energy). Unlike density perturbations, which are required to initiate structure formation, there is no cosmological lower bound to the amplitude of the gravity-wave perturbations. Tensor perturbations also give rise to CBR anisotropy; requiring that they do not lead to excessive anisotropy implies that the energy density that drove inflation must be less than about $(10^{16}\,\mathrm{GeV})^4$. This indicates that if inflation took place, it did so at an energy well below the Planck scale.†

There are other interesting consequences of inflation that are less generic. For example, in first-order inflation, where reheating occurs through the nucleation and collision of vacuum bubbles, there is an additional, larger amplitude, but narrow-band, spectrum of gravitational waves ($\Omega_{\mathrm{GW}}h^2 \sim 10^{-6}$) (Turner & Wilczek 1990; Kosowsky, Turner & Watkins 1992). In other models large-scale primeval magnetic fields of interesting size are seeded during inflation (Turner & Widrow 1988; Ratra 1992).

I want to emphasize the importance of the tensor perturbations. The attractiveness of a flat Universe with scale-invariant density perturbations was appreciated long before inflation. Verifying these two predictions of inflation, while important, will not provide a "smoking gun." A spectrum of nearly scale-invariant tensor perturbations is a defining

¶ Because theorists are so clever, it is not possible nor prudent to use the word immutable. Models that violate any or all of these "robust predications" can and have been constructed.

† To be more precise, the part of inflation that led to perturbations on scales within the present horizon involved subPlanckian energy densities. In some models of inflation, the earliest stage of inflation, which only influences scales much larger than the present horizon, involve energies as large as the Planck energy density.

feature of inflation, and further, is crucial to obtaining information about the underlying scalar potential.

CBR anisotropy probably provides the best possibility of detecting the tensor perturbations, but their contribution to CBR anisotropy has to be separated that of the scalar perturbations. Because the sky is finite, sampling variance sets a fundamental limit: the tensor contribution to CBR anisotropy can only be separated from that of the scalar if T/S is greater than about 0.14 (Knox, L. & Turner 1994) (T is the contribution of tensor perturbations to the variance of the CBR quadrupole and S is the same for scalar perturbations).

It is possible that the stochastic background of gravitational waves itself can be directly detected, though it appears that the LIGO facilities being built will lack the sensitivity and even space-based interferometery (e.g., LISA) is not a sure bet (Turner, Lidsey & White 1993; Turner 1996).

3. Cold Dark Matter

Cold dark matter actually draws from three important ideas – inflation, big-bang nucleosynthesis, and the quest to better understand the fundamental forces and particles. As discussed above, inflation predicts a flat Universe (total energy density equal to the critical density) and nearly scale-invariant density perturbations. Big-bang nucleosynthesis provides the most precise determination of the density of ordinary matter, present density between $1.5\times10^{-31}\,\mathrm{g\,cm^{-3}}$ and $4.5\times10^{-31}\,\mathrm{g\,cm^{-3}}$, or fraction of critical density $\Omega_B = 0.008h^{-2} - 0.024h^{-2}$, where $H_0 = 100h\,\mathrm{km\,s^{-1}\,Mpc^{-1}}$ (Copi, Schramm & Turner 1995). Allowing $h = 0.4 - 0.8$, consistent with modern measurements (Fukugita, Hogan & Peebles 1993; Jacoby G. et al. 1992), implies that ordinary matter can contribute at most 10% of the critical density. If the inflationary prediction is correct, then most of the matter in the Universe must be nonbaryonic.

This idea has received indirect support from particle physics. Attempts to further our understanding of the particles and forces have led to the prediction of new, stable or long-lived particles that interact very feebly with ordinary matter. These particles, if they exist, should have been present in great numbers during the earliest moments and remain today in numbers sufficient to contribute the critical density (Turner 1991). Two of the most attractive possibilities behave like cold dark matter: a neutralino of mass 10 GeV to 1000 GeV, predicted in supersymmetric theories, and an axion of mass 10^{-6} eV to 10^{-4} eV, needed to solve a subtle problem of the standard model of particle physics (strong-CP problem). The third interesting possibility is that one of the three neutrino species has a mass between 5 eV and 30 eV; neutrinos move very fast and are referred to as hot dark matter.†

According to cold dark matter theory CDM particles provide the cosmic infrastructure: It is their gravitational attraction that forms and holds cosmic structures together. Structure forms in a hierarchical manner, with galaxies forming first and successively larger objects forming thereafter (Blumenthal et al. 1984). Quasars and other rare objects form at redshifts of up to five, with ordinary galaxies forming a short time later. Today, superclusters, objects made of several clusters of galaxies, are just becoming bound by the gravity of their CDM constituents. The formation of larger and larger

† The possibility that most of the exotic particles are fast-moving neutrinos – hot dark matter – was explored first and found to be inconsistent with observations (White, Frenk & Davis 1983). The problem is that large structures form first and must fragment into smaller structures, which conflicts with the fact that large structures are just forming today.

objects continues. In the clustering process regions of space are left devoid of matter – and galaxies – leading to voids.

3.1. *Standard Cold Dark Matter*

When the cold dark matter scenario emerged more than a decade ago many referred to it as a no parameter theory because it was so specific compared to previous models for the formation of structure. This was an overstatement as there are cosmological quantities that must be known to determine the development of structure in detail. However, the data available did not require precise knowledge of these quantities to test the model.

Broadly speaking these parameters can be organized into two groups. First are the cosmological parameters: the Hubble constant, specified by h; the density of ordinary matter, specified by $\Omega_B h^2$; the power-law index n and normalization A that quantify the density perturbations; and the amplitude and spectral index n_T that quantify the gravitational waves. The inflationary parameters fall into this category because there is no standard model of inflation. On the other hand, once determined they can be used to discriminate between models of inflation.

The other quantities specify the composition of invisible matter in the Universe: radiation, dark matter, and a possible cosmological constant. Radiation refers to relativistic particles: the photons in the CBR, three massless neutrino species (assuming none of the neutrino species has a mass), and possibly other undetected relativistic particles (some particle-physics theories predict the existence of additional massless particle species). At present relativistic particles contribute almost nothing to the energy density in the Universe, $\Omega_R \simeq 4.2 \times 10^{-5} h^{-2}$; early on — when the Universe was smaller than about 10^{-5} of its present size — they dominated the energy content.

In addition to CDM particles, the dark matter could include other particle relics. For example, each neutrino species has a number density of $113\,\text{cm}^{-3}$, and a neutrino species of mass 5 eV would account for about 20% of the critical density ($\Omega_\nu = m_\nu / 90h^2\,\text{eV}$). Predictions for neutrino masses range from 10^{-12} eV to several MeV, and there is some experimental evidence that at least one of the neutrino species has a small mass (Parke 1995; Athanassopoulos et al. 1995; Hill 1995; Hirata et al. 1992; Fukuda et al. 1994; Becker-Szendy 1992).

Finally, there is the cosmological constant. Both introduced and abandoned by Einstein, it is still with us. In the modern context it corresponds to an energy density associated with the quantum vacuum. At present, there is no reliable calculation of the value that the cosmological constant should take, and so its existence must be regarded as a logical possibility (Weinberg 1989).

The original no parameter cold dark matter model, referred to as standard CDM, is characterized by: $h = 0.5$, $\Omega_B = 0.05$, $\Omega_{\rm CDM} = 0.95$, $n = 1$, no gravitational waves and standard radiation content. The overall normalization of the density perturbations was fixed by comparing the predicted level of inhomogeneity with that seen today in the distribution of bright galaxies. Specifically, the amplitude A was determined by comparing the expected mass fluctuations in spheres of radius $8h^{-1}$ Mpc (denoted by σ_8) to the galaxy-number fluctuations in spheres of the same size. The galaxy-number fluctuations on the scale $8h^{-1}$ Mpc are unity; adjusting A to achieve $\sigma_8 = 1$ corresponds to the assumption that light, in the form of bright galaxies, traces mass. Choosing σ_8 to be less than one means that light is more clustered than mass and is a biased tracer of mass. There is some evidence that bright galaxies are somewhat more clumped than mass with biasing factor $b \equiv 1/\sigma_8 \simeq 1 - 2$ (Lin 1996).

A dramatic change occurred with the detection of CBR anisotropy by COBE in 1992 (Smoot et al. 1992). The COBE measurement permitted a precise normalization of the

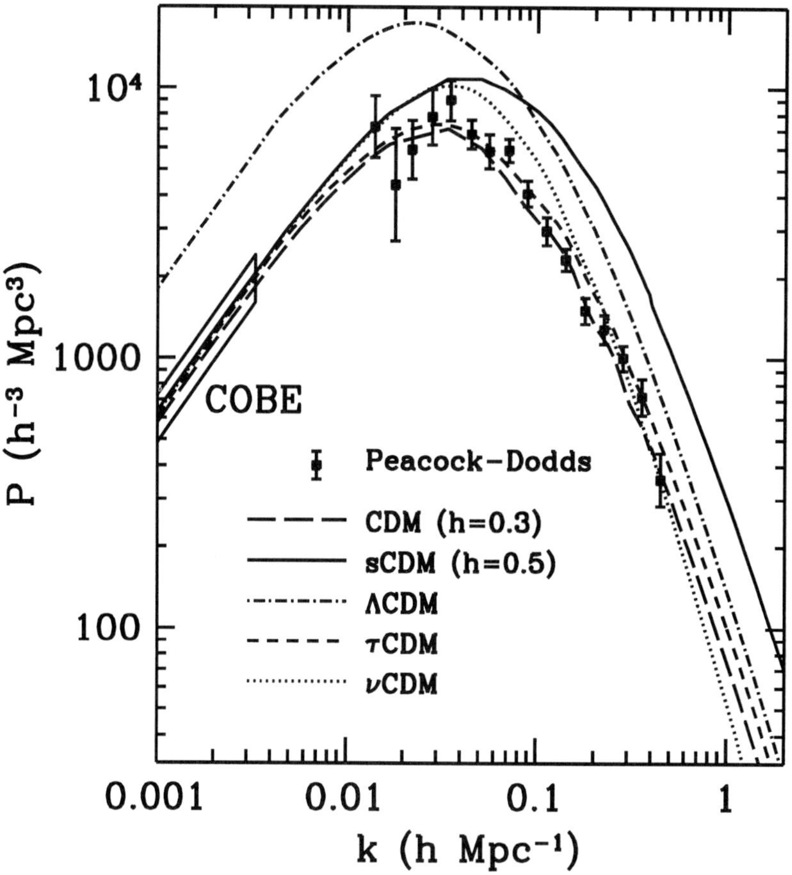

FIGURE 1. Measurements of the power spectrum, $P(k) \equiv |\delta_k|^2$, and the predictions of different COBE-normalized CDM models. The points are from several redshift surveys as analyzed in Ref. (Peacock & Dodds 1994); the models are: ΛCDM with $\Omega_\Lambda = 0.6$ and $h = 0.65$; standard CDM (sCDM), CDM with $h = 0.35$; τCDM (with the energy equivalent of 12 massless neutrino species) and νCDM with $\Omega_\nu = 0.2$ (unspecified parameters have their standard CDM values).

amplitude of density perturbations on very large scales ($\lambda \sim 10^4 h^{-1}$ Mpc) without regard to the issue of biasing. [CBR anisotropy on the angular scale θ arises primarily due to inhomogeneity on length scales $\lambda \sim 100h^{-1}\,\mathrm{Mpc}(\theta/\mathrm{deg})$.] For standard CDM, the COBE normalization leads to: $\sigma_8 = 1.2 \pm 0.1$ or anti-bias since $b = 1/\sigma_8 \simeq 0.7$. The pre-COBE normalization ($\sigma_8 = 0.5$) led to too little power on scales of $30h^{-1}$ Mpc to $300h^{-1}$ Mpc, as compared to what was indicated in redshift surveys, the angular correlations of galaxies on the sky and the peculiar velocities of galaxies. The COBE normalization leads to about the right amount of power on these scales, but predicts too much power on small scales ($\lesssim 8h^{-1}$ Mpc); see Fig. 1.

While standard CDM is in general agreement with the observations, a consensus has developed that the conflict just mentioned is probably significant (Ostriker 1993; Liddle & Lyth 1993). This has led to a new look at the cosmological and invisible-matter parameters and to the realization that the problems of standard CDM are simply a poor choice for the standard parameters.

4. A Flood of Data

Standard CDM has served well to focus everyone's attention on the same problem. However, the quality and quantity of data have improved and knowledge of the cosmological and invisible-matter parameters has become important for serious testing of CDM and inflation. There are a variety of combinations of the parameters that lead to good agreement with the existing data on both large and small length scales – and thus can make a claim to being the new standard CDM model. Figure 2 shows the allowed values of the cosmological parameters for several COBE-normalized CDM models.†

More precisely, for a given CDM model — specified by the cosmological and invisible-matter parameters — the expected CBR anisotropy is computed and required to be consistent with the four-year COBE data set at the two-sigma level (Bennett et al. 1996; Gorski et al. 1996; White & Bunn 1995). The expected level of inhomogeneity in the Universe today is compared to three robust measurements of inhomogeneity: the shape of the power spectrum as inferred from surveys of the distribution of galaxies today (Peacock & Dodds 1994); a determination of σ_8 based upon the abundance of rich, x-ray emitting clusters (White, Efstathiou & Frenk 1993); and the abundance of hydrogen clouds at high redshift (which probes early structure formation) (Lanzetta, Wolfe & Turnshek 1995; Storrie-Lombardi et al. 1995).

Figure 2 summarizes the overall picture. The simplest CDM models – those with standard invisible-matter content – lie in a region that runs diagonally from smaller Hubble constant and larger n to larger Hubble constant and smaller n. That is, higher values of the Hubble constant require more tilt (tilt referring to deviation from scale invariance). Note too that standard CDM is well outside of the allowed range. Current measurements of CBR anisotropy on the degree scale, as well as the COBE four-year anisotropy data, preclude n less than about 0.7 (see Fig. 3). This implies that the largest Hubble constant consistent with the simplest CDM models is slightly less than $60\,\mathrm{km\,s^{-1}\,Mpc^{-1}}$. If the invisible-matter content is nonstandard, higher values of the Hubble constant can be accommodated. In Fig. 2, Ω_ν is taken to be 0.2; in fact, this is essentially the largest value allowed by measurements of the power spectrum (Dodelson, Gates & Turner 1996; Ma & Bertschinger 1994; Liddle et al. 1995). On the other hand, even $\Omega_\nu = 0.05$ (around 1 eV worth of neutrinos) can have important consequences (e.g., accommodating a higher value of the Hubble constant or more nearly scale-invariant density perturbations).

Changes in the different parameters from their standard CDM values alleviate the excess power on small scales in different ways. Tilt has the effect of reducing power on small scales when power on very large scales is fixed by COBE. A small admixture of hot dark matter works because fast moving neutrinos suppress the growth of inhomogeneity on small scales by streaming from regions of higher density and to regions of lower density. (It was in fact this feature of hot dark matter that led to the demise of the hot dark matter model for structure formation.)

A low value of the Hubble constant, additional radiation or a cosmological constant all reduce power on small scales by lowering the ratio of matter to radiation. Since the critical density depends upon the square of the Hubble constant, $\rho_{\mathrm{Critical}} = 3H_0^2/8\pi G$, a smaller value corresponds to a lower matter density since $\rho_{\mathrm{Matter}} = \rho_{\mathrm{Critical}}$ for a

† Computation of both the CBR anisotropy and the level of inhomogeneity today depends upon the invisible-matter content and the cosmological parameters and requires that the distribution of matter and radiation be evolved numerically; for details see Refs. (Peebles & Yu 1970; Wilson & Silk 1981; Bond & Efstathiou, 1984; Holtzman 1989; van Dalen & Schaefer 1992; Sugiyama & Gouda 1992; Dodelson & Jubas 1993). The discussion of viable models is a summary of collaborative work (Dodelson, Gates & Turner 1996).

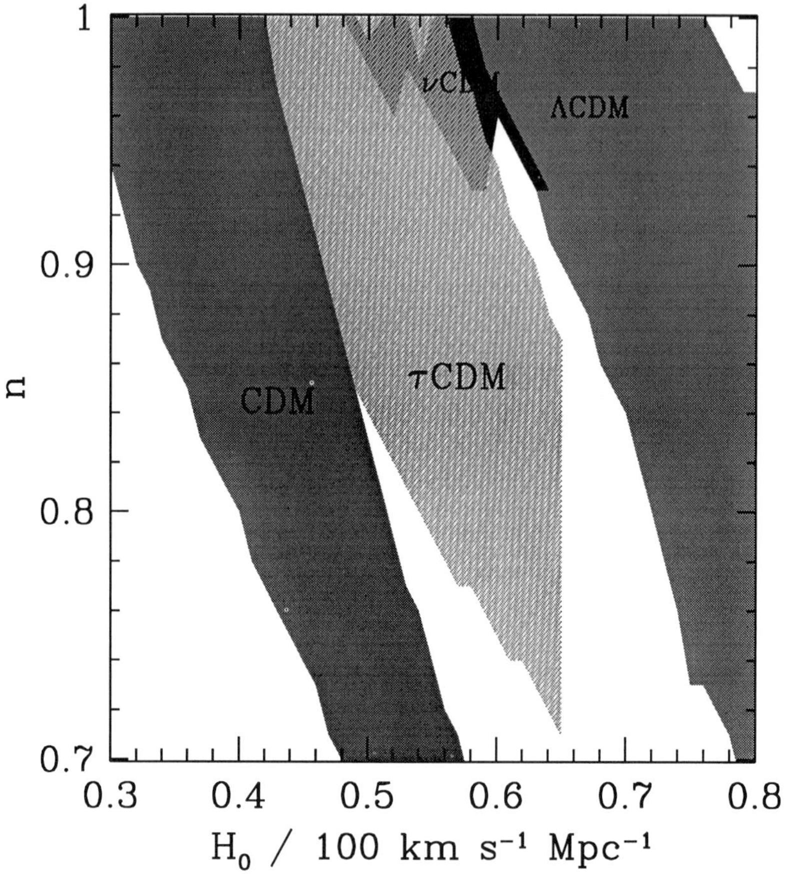

FIGURE 2. Acceptable values of the cosmological parameters n and h for CDM models with standard invisible-matter content (CDM), with 20% hot dark matter (νCDM), with additional relativistic particles (the energy equivalent of 12 massless neutrino species, denoted τCDM), and with a cosmological constant that accounts for 60% of the critical density (ΛCDM). Note that standard CDM ($n = 1$ and $h = 0.5$) is not viable.

flat Universe without a cosmological constant. Shifting some of the critical density to vacuum energy also reduces the matter density since $\Omega_{\rm Matter} = 1 - \Omega_\Lambda$. Lowering the ratio of matter to radiation reduces the power on small scales in a subtle way. While the primeval fluctuations in the gravitational potential are nearly scale-invariant, density perturbations today are not because the Universe made a transition from an early radiation-dominated phase ($t \lesssim 1000\,$yrs), where the growth of density perturbations is inhibited, to the matter-dominated phase, where growth proceeds unimpeded. This introduces a feature in the power spectrum today (see Fig. 1), whose location depends upon the relative amounts of matter and radiation. Lowering the ratio of matter to radiation shifts the feature to larger scales and with power on large scales fixed by COBE this leads to less power on small scales.

Some of the viable models have been discussed as singular solutions — cosmological constant (Turner, Steigman & Krauss 1984; Turner 1991; Peebles 1984; Efstathiou et al. 1990; Kofman & Starobinskii 1985), very low Hubble constant (Bartlett et al. 1995), tilt (Cen et al. 1992; Davis et al. 1992; Lucchin, Mattarese & Mollerach 1992; Salopek 1992; Liddle & Lyth 1992; Lidsey & Coles 1992; Souradeep & Sahni 1992; Adams et al. 1993), tilt + low Hubble constant (White et al. 1995), extra radiation (Dodelson, Gyuk &

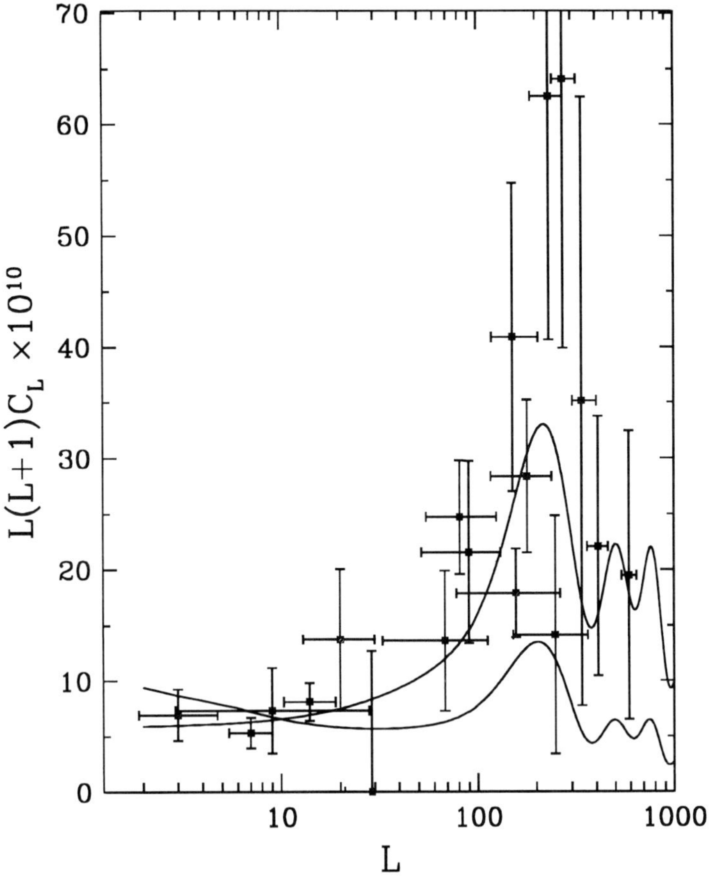

FIGURE 3. Summary of CBR anisotropy measurements and predictions for three CDM models [adapted from Ref. (White, Scott, & and Silk 1995)]. Plotted are the squares of the measured multipole amplitudes ($C_l = \langle|a_{lm}|^2\rangle$) in units of 7×10^{-10} vs. multipole number l. The temperature difference on angular scale θ is given roughly by $\sqrt{l(l+1)C_l}$ with $l \sim 200°/\theta$. The theoretical curves are standard CDM and CDM with $n = 0.7$ and $h = 0.5$.

Turner 1994; Bond & Efstathiou 1991), an admixture of hot dark matter (Shafi & Stecker 1984; Davis, Summers & Schlegel 1992; Primack et al. 1995; Pogosyan & Starobinsky 1995). There is actually a continuum of viable models, as can be seen in Fig. 2, which arises because of imprecise knowledge of cosmological parameters and the invisible-matter sector and not the inventiveness of theorists.

4.1. *Other and Future Considerations*

There are many other observations that bear on structure formation. However, with cosmological data systematic error and interpretational issues are important considerations. In fact, if all extant observations were taken at face value, there is no viable model for structure formation, cold dark matter or otherwise! With this as a preface, I now discuss some of the other existing data as well as future measurements that will more sharply test cold dark matter.

There is tension between measures of the age of the Universe and determinations of the Hubble constant. It arises because determinations of the ages of the oldest stars lie between 13 Gyr and 19 Gyr (Chaboyer et al. 1996; Bolte & Hogan 1995) and recent measurements of the Hubble constant favor values between 60 km s^{-1} Mpc^{-1} and

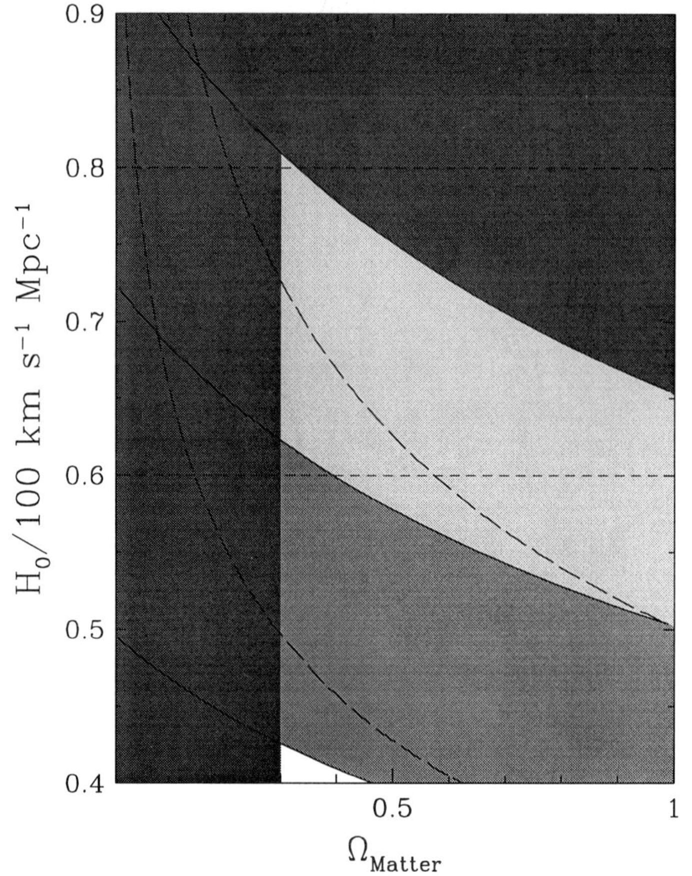

FIGURE 4. Isochrones in the H_0 - $\Omega_{\rm Matter}$ plane. The bottom band corresponds to time back to the bang of between 13 Gyr and 19 Gyr; the lightest band to between 10 Gyr and 13 Gyr; the darkest region is disallowed: $\Omega_{\rm Matter} < 0.3$ or expansion time less than 10 Gyr. Broken horizontal lines indicate the range favored for the Hubble constant, $80\,{\rm km\,s^{-1}\,Mpc^{-1}} > H_0 > 60\,{\rm km\,s^{-1}\,Mpc^{-1}}$. The age – Hubble constant tension is clear, especially for the inflationary prediction of $\Omega_{\rm Matter} = 1$. The broken curves denote the 13 Gyr–19 Gyr isochrone for ΛCDM; a cosmological constant greatly lessens the tension.

$80\,{\rm km\,s^{-1}\,Mpc^{-1}}$ (Reiss, Kirshner & Press 1995; Hamuy 1995; Freedman et al. 1994), which, for $\Omega_{\rm Matter} = 1$, leads to a time back to the bang of 11 Gyr or less (see Fig. 4).† These age determinations receive additional support from estimates of the age of the galaxy based upon the decay of long-lived radioactive isotopes and the cooling of white-dwarf stars, and all methods taken together make a strong case for an absolute minimum age of 10 Gyr (Cowan, Thieleman & Truran 1991). It should be noted that within the uncertainties there is no inconsistency, even for $\Omega_{\rm Matter} = 1$.

While age is not a major issue for cold dark matter — large-scale structure favors an older Universe by virtue of a lower Hubble constant or cosmological constant (see Fig. 5) — the Hubble constant still has great leverage. If it is determined to be greater than about $60\,{\rm km\,s^{-1}\,Mpc^{-1}}$, then only CDM models with nonstandard invisible-matter content – a cosmological constant or additional radiation – can be consistent with large-

† The time back to the bang depends upon H_0, $\Omega_{\rm Matter}$ and Ω_Λ; for $\Omega_{\rm Matter} = 1$ and $\Omega_\Lambda = 0$, $t_{\rm BB} = \frac{2}{3}H_0^{-1}$, or 13 Gyr for $h = 0.5$ and 10 Gyr for $h = 0.65$. For a flat Universe with a cosmological constant the numerical factor is larger than 2/3 (see Fig. 4).

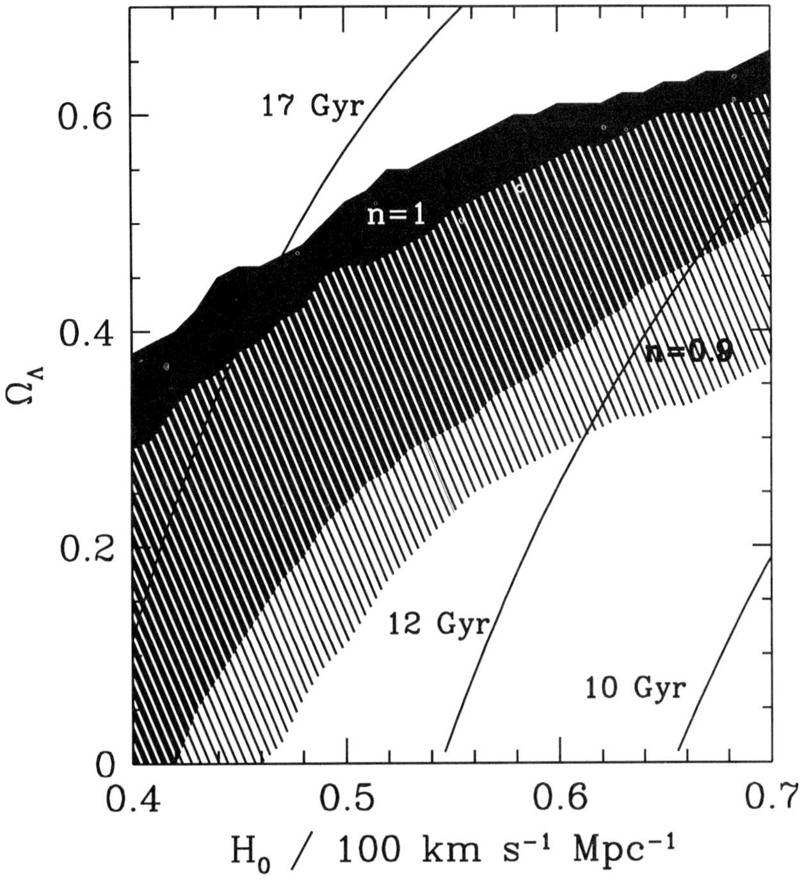

FIGURE 5. Acceptable values of Ω_Λ and h for $n = 0.9, 1.0$. Note that large-scale structure considerations generally favor a more aged Universe – smaller h or larger Ω_Λ.

scale structure. If H_0 is greater than $65\,\mathrm{km\,s^{-1}\,Mpc^{-1}}$, consideration of the age of the Universe leaves ΛCDM as the lone possibility. The issue of H_0 is not settled, but the use of Type Ia supernovae as standard candles, the study of Cepheid variable stars in nearby galaxies using the Hubble Space Telescope, and other methods make it likely that it will be soon.

If CDM is correct, baryons make up a small fraction of matter in the Universe. Most of the baryons in galaxy clusters are in the hot, x-ray emitting intracluster gas and not the luminous galaxies. The measured x-ray flux fixes the mass in baryons, while the measured x-ray temperature fixes the total mass (through the virial theorem). The baryon-to-total-mass has been determined from x-ray measurements for more than ten clusters and is found to be $M_{\rm B}/M_{\rm TOT} \simeq (0.04 - 0.1)h^{-3/2}$ (White et al. 1993; Briel et al. 1992; White & Fabian 1995). Because of their size, clusters should represent a fair sample of the cosmos and thus the baryon-to-total mass ratio should reflect its universal value, $\Omega_B/\Omega_{\rm Matter} \simeq (0.01 - 0.02)h^{-2}/\Omega_{\rm Matter}$. These two ratios are consistent for models with a very low Hubble constant, $h \sim 0.4$ and $\Omega_{\rm Matter} = 1$, or with a cosmological constant and $\Omega_{\rm Matter} \sim 0.3$. However, important assumptions are made in this analysis – that the hot gas is unclumped and in virial equilibrium and that magnetic fields do not provide significant pressure support for the gas – if any one of them is not valid the actual baryon fraction would be smaller (Babul & Katz 1993), allowing for consistency with a larger value of H_0 without recourse to a cosmological constant.

The halos of individual spiral galaxies like our own are not large enough to provide a fair sample of matter in the Universe – for example, much of the baryonic matter has undergone dissipation and condensed into the disk of the galaxy. Nonetheless, the content of halos is expected to be primarily CDM particles. This is consistent with the fact that visible stars, hot gas, dust, and even dark stars acting as microlenses (known as MACHOs) account for only a fraction of the mass of our own halo (Alcock et al. 1995; Gates, Gyuk & Turner 1995; Bahcall et al. 1994; Flynn, Gould & Bahcall 1996).

Determining the mean mass density of the Universe would discriminate between models with and without a cosmological constant, as well as test the inflationary prediction of a flat Universe. A definitive determination is still lacking. The measurement that averages over the largest volume – and thus is potentially most useful – uses the peculiar velocities of galaxies. Peculiar velocities arise due to the inhomogeneous distribution of matter, and the mean matter density can be determined by relating the peculiar velocities to the observed distribution of galaxies. The results of this technique indicate that $\Omega_{\rm Matter}$ is at least 0.3 and perhaps as large as unity (Strauss & Willick 1995; Dekel 1994). Though not definitive, this provides strong evidence for the existence of nonbaryonic dark matter, a key aspect of cold dark matter.

A different approach to the mean density is through the deceleration parameter q_0, which quantifies the slowing of the expansion due to the gravitational attraction of matter in the Universe. Its value is given by $q_0 = \frac{1}{2}\Omega_{\rm Matter} - \Omega_\Lambda$ (vacuum energy actually leads to accelerated expansion) and can be determined by relating the distances and redshifts of distant objects. In all but the ΛCDM scenario, $q_0 = 0.5$; for ΛCDM, $q_0 \sim -0.5$. Two groups are trying to measure q_0 by using high redshift ($z \sim 0.4 - 0.7$) Type Ia supernovae as standard candles; the preliminary results of one group suggest that q_0 is positive (Perlmutter et al. 1996). More than a dozen distant Type Ia supernovae were discovered this year and both groups should soon have enough to measure q_0 with a precision of ± 0.2.

Gravitational lensing of distant QSOs by intervening galaxies is another way to measure q_0, and the frequency of QSO lensing suggests that $q_0 > -0.6$ (Kochanek 1996). The distance to a QSO of given redshift is larger for smaller q_0, and thus the probability for its being lensed by an intervening galaxy is greater.

The 10 m Keck Telescope and the Hubble Space Telescope are providing the deepest images of the Universe ever and are revealing details of galaxy formation as well as the formation and evolution of clusters of galaxies (Fukugita, Hogan & Peebles 1996). The Keck has made the first detection of deuterium in high redshift hydrogen clouds (Songaila et al. 1994; Carswell et al. 1994; Rugers & Hogan 1996a; Tytler, Fan & Burles 1996; Burles & Tytler 1996; Wampler et al. 1996; Carswell, R. F. et al. 1996; Rugers & Hogan 1996b; Cowie & Songaila 1996). This is a new confirmation of big-bang nucleosynthesis and has the potential of pinning down the density of ordinary matter to a precision of 10%.

The level of inhomogeneity in the Universe today is determined largely from redshift surveys, the largest of which contain of order 10^4 galaxies. Two larger surveys are in progress. The Sloan Digital Sky Survey will obtain positions for two-hundred million galaxies and redshifts for one million over a quarter of the sky; the Anglo - Australian Two-Degree Field will gather 250,000 redshifts in hundreds of two-degree fields. They will allow the power spectrum to be measured more precisely and out to large enough scales ($500h^{-1}$ Mpc) to connect with measurements from CBR anisotropy on angular scales of up to five degrees.

The most fundamental element of cold dark matter – the existence of the CDM particles themselves – is being tested. While the interaction of CDM particles with ordinary matter occurs through very feeble forces and makes their existence difficult to test, experiments with sufficient sensitivity to detect the CDM particles that hold our own galaxy together if they are in the form of axions of mass $10^{-6}\,\mathrm{eV} - 10^{-4}\,\mathrm{eV}$ (Rosenberg 1995) or neutralinos of mass tens of GeV (Caldwell 1995) are now underway. Evidence for the existence of the neutralino could also come from particle accelerators searching for other supersymmetric particles. In addition, several experiments sensitive to neutrino masses are operating or are planned, ranging from accelerator-based neutrino oscillation experiments to the detection of solar neutrinos to the study of the tau neutrino at e^+e^- colliders.

CBR anisotropy probes the power spectrum most cleanly as it is related directly to the distribution of matter when density perturbations were very small (Hu & Sugiyama 1995). Current measurements are beginning to test CDM and differentiate between the variants (see Fig. 3); e.g., a spectral index $n < 0.7$ is strongly disfavored. More than ten groups are making measurements with instruments in space, on balloons and at the South Pole. Two new satellite-borne experiments have been approved: NASA will launch MAP in 2001 and ESA will launch COBRAS/SAMBA in 2004. They will map the anisotropy of the CBR with better than 0.2° resolution (more than 30 times better than COBE). The results from these maps should easily discriminate between the different variants of CDM (see Fig. 6).

The first and most powerful test to emerge from these measurements will be the location of the first (Doppler) peak in the angular power spectrum (see Fig. 6) (Kamionkowski, Spergel & Sugiyama 1994; Jungman et al. 1996a; Hu & White 1996). All variants of CDM predict the location of the first peak to lie in roughly the same place. On the other hand, in an open Universe (total energy density less than critical) the first peak occurs at a larger value of l (much smaller angular scale). This will provide an important test of inflation. In addition, theoretical studies (Knox 1995; Jungman et al. 1996b) indicate that n could be determined to a precision of a few percent, Ω_Λ to ten percent, and perhaps even Ω_ν to enough precision to test νCDM (Dodelson, Gates & Stebbins 1996).

If *all* the current observations – from recent Hubble constant determinations to the cluster baryon fraction – are taken at face value, the cosmological constant + cold dark matter model is probably the best fit (Krauss & Turner 1995; Ostriker & Steinhardt 1995), though there may be a conflict with the measurement of q_0 with Type Ia supernovae. ΛCDM raises a fundamental question – the origin of the implied vacuum energy, about $(10^{-2}\,\mathrm{eV})^4$ – since there is no known principle or mechanism that explains why it is less than $(300\,\mathrm{GeV})^4$, let alone $(10^{-2}\,\mathrm{eV})^4$ (Weinberg 1989). In any case, it would be imprudent to take all the observational data at face value because of important systematic and interpretational uncertainties. To paraphrase the biologist Francis Crick, a theory that fits all the data at any given time is probably wrong as some of the data are probably not correct.

4.2. *Reconstruction*

If inflation and the cold dark matter theory are shown to be correct, a window to the very early Universe ($t \sim 10^{-32}$ sec) will have been opened. While it is certainly premature to jump to this conclusion, I would like to illustrate one example of what one could hope to learn. The spectra and amplitudes of the the tensor and scalar metric perturbations predicted by inflation depend upon the underlying model, to be specific, the shape of the inflationary scalar-field potential. If one can measure the power-law index of the scalar spectrum and the amplitudes of the scalar and tensor spectra, one can recover the value of the potential and its first two derivatives around the point on the potential where

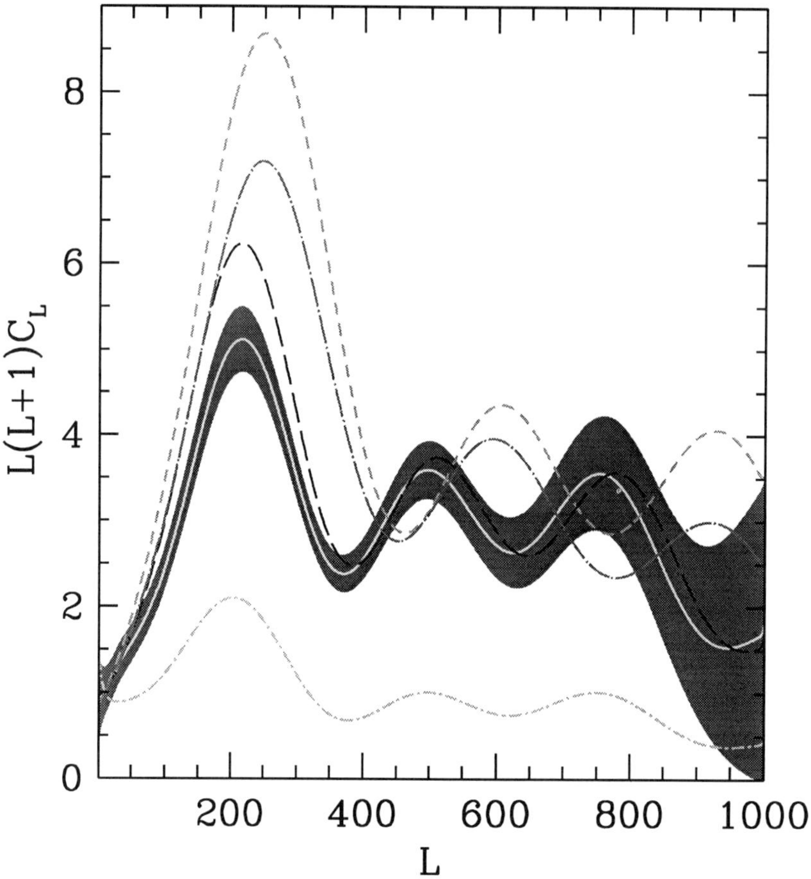

FIGURE 6. Predicted angular power spectra of CBR anisotropy for several viable CDM models and the anticipated uncertainty from a CBR satellite experiment with angular resolution of $0.3°$. From top to bottom the models are: CDM with $h = 0.35$, τCDM with the energy equivalent of 12 massless neutrino species, ΛCDM with $h = 0.65$ and $\Omega_\Lambda = 0.6$, νCDM with $\Omega_\nu = 0.2$, and CDM with $n = 0.7$ (unspecified parameters have their standard CDM values).

inflation took place (Turner 1993b, Lidsey et al. 1995):

$$V = 1.65T\, {m_{\rm Pl}}^4, \tag{4.1}$$

$$V' = \pm\sqrt{\frac{8\pi r}{7}}\, V/m_{\rm Pl}, \tag{4.2}$$

$$V'' = 4\pi\left[(n-1) + \frac{3}{7}r\right]\, V/{m_{\rm Pl}}^2, \tag{4.3}$$

where $r \equiv T/S$ (T is the contribution of tensor perturbations to the variance of the CBR quadrupole and S is the same for scalar perturbations), prime indicates derivative with respect to ϕ, $m_{\rm Pl} = 1.22 \times 10^{19}\,$GeV is the Planck energy, and the sign of V' is indeterminate. In addition, if the tensor spectral index n_T can be measured a consistency relation, $n_T = -r/7$, can be used to test inflation. Reconstruction of the inflationary scalar potential would shed light on the underlying physics of inflation as well as physics at energies of the order of $10^{15}\,$GeV.

5. Concluding Remarks

Thanks to the work of Igor Novikov and others, the fifteen years following the discovery of the CBR saw the establishment of a physical basis for the expanding Universe, the hot big-bang cosmology. With this as a firm foundation, the decade of the 1980s has produced many bold and interesting speculations about the earliest history of the Universe which address the most fundamental problems in cosmology. Two of these ideas, inflation and cold dark matter, are so attractive that experimenters and observers took them seriously. The decade of the 1990s is producing a flood of data that are testing inflation and cold dark matter. The stakes for both cosmology and fundamental physics are high: if correct, inflation and cold dark matter represent a major extension of the big bang and our understanding of the Universe, which would certainly shed light on fundamental physics at energies beyond the reach of terrestrial accelerators.

REFERENCES

ABBOTT, L. & WISE, M. 1984 *Nucl. Phys. B* **244**, 541.

ADAMS, F. C. ET AL. 1993 *Phys. Rev.* **D47**, 426.

ALBRECHT, A. & STEINHARDT, P. J. 1982 *Phys. Rev. Lett.* **48**, 1220.

ALCOCK, C. ET AL. 1995 *Phys. Rev. Lett.* **74**, 2867.

ATHANASSOPOULOS, C. ET AL. 1995 *Phys. Rev. Lett.* **75**, 2650.

BABUL, A. & KATZ, N. 1993 *Astrophys. J.* **406**, L51.

BAHCALL, J. ET AL. 1994 *Astrophys. J. Lett.* **435**, L51.

BANKS, T. ET AL. 1995 hep-th/9503114.

BARDEEN, J. M., STEINHARDT, P. J. & TURNER, M. S. 1983 *Phys. Rev. D* **28**, 697.

BARTLETT, J. ET AL. 1995 *Science* **267**, 980.

BECKER-SZENDY, R. ET AL. 1992 *Phys. Rev. D* **46**, 3720.

BENNETT, C. L. ET AL. 1996 *Astrophys. J. (Lett.)* in press, astro-ph/9601067.

BLUMENTHAL, G. ET AL. 1984 *Nature* **311**, 517.

BOLTE, M. & HOGAN, C. J. 1995 *Nature* **376**, 399.

BOND, J. R. & EFSTATHIOU, G. 1984 *Astrophys. J.* **285**, L45.

BOND, J. R. & EFSTATHIOU, G. 1991 *Phys. Lett. B* **265**, 245.

BRIEL, U. G. ET AL. 1992 *Astron. Astrophys.* **259**, L31.

BRUSTEIN, R. & VENEZIANO, G. 1994 *Phys. Lett. B* **329**, 429.

BUCHER, M., GOLDHABER, A. S. & TUROK, N. 1995 *Phys. Rev. D* **52**, 3314.

BUCHER, M. & TUROK, N. 1995 hep-ph/9503393.

BURLES, S. & TYTLER, D. 1996 astro-ph/9603070.

CALDWELL, D. O. 1995 in *Particle and Nuclear Astrophysics in the Next Millennium* (eds. E.W. Kolb and R.D. Peccei), p. 38. World Scientific, Singapore.

CARSWELL, R. F. ET AL. 1994 *Mon. Not. Roy. Astr. Soc.* **268**, L1.

CARSWELL, R. F. ET AL. 1996 *Mon. Not. Roy. Astron. Soc.* **278**, 506.

CEN, R., GNEDIN, N., KOFMAN, L., & OSTRIKER, J. P. 1992 *Astrophys. J.* **399**, L11.

CHABOYER, B., DEMARQUE, P., KERNAN, P. J. & KRAUSS, L. M. 1996 *Science* **271**, 957.

COHEN, A., KAPLAN, D. & NELSON, A. 1992 *Annu. Rev. Nucl. Part. Sci.* **43**, 27.

COLLINS, C. B. & HAWKING, S. W. 1973 *Astrophys. J.* **180**, 317.

COPELAND, E. J. ET AL. 1994 *Phys. Rev. D* **49**, 6410.

COPI, C., SCHRAMM, D. N. & TURNER, M. S. 1995 *Science* **267**, 192.

COWAN, J., THIELEMAN, F. & TRURAN, J. 1991 *Ann. Rev. Astron. Astrophys.* **29**, 447.

Cowie, L. & Songaila, A. 1996 in preparation.

Cvetic, M., Hubsch, T., Pati, J. & Stremnitzer, H. 1990 *Phys. Rev. D* **40**, 1311.

Dalen, A. van & Schaefer, R. K. 1992 *Astrophys. J.* **398**, 33.

Davis, R. et al. 1992 *Phys. Rev. Lett.* **69**, 1856.

Davis, M., Summers, F. & Schlegel, D. 1992 *Nature* **359**.

Dekel, A. 1994 *Ann. Rev. Astron. Astrophys.* **32**, 319.

Dodelson, S., Gates, E. & Stebbins, A. S. 1996 *Astrophys. J.*, in press, astro-ph/9509147.

Dodelson, S., Gates, E. I. & Turner, M. S. 1996 astro-ph/9603081.

Dodelson, S., Gyuk, G. & Turner, M. S. 1994 *Phys. Rev. Lett.* **72**, 3578.

Dodelson, S. & Jubas, J. M. 1993 *Phys. Rev. Lett.* **70**, 2224.

Dolgov, A. 1992 *Phys. Repts.* **222**, 309.

Efstathiou, G. et al. 1990 *Nature* **348**, 705.

Fabbri, R. & Pollock, M. 1983 *Phys. Lett. B* **125**, 445.

Flynn, C., Gould, A. & Bahcall, J. 1996 astro-ph/9603035.

Freedman, W. et al. 1994 *Nature* **371**, 757.

Freese, K., Frieman, J. A. & and Olinto, A. 1990 *Phys. Rev. Lett.* **65**, 3233.

Fukuda, Y. et al. 1994 *Phys. Lett. B* **335**, 237.

Fukugita, M., Hogan, C. J. & Peebles, P. J. E. 1993 *Nature* **366**, 309.

Fukugita, M., Hogan, C. J. & Peebles, P. J. E. 1996 *Nature* **381**, 489.

Gasperini, M. & Veneziano, G. 1994 *Phys. Rev. D* **50**, 2519.

Gates, E., Gyuk, G. & Turner, M. S. 1995 *Phys. Rev. Lett.* **74**, 3724.

Gorski, K. M. et al. 1996 *Astrophys. J. (Lett.)* in press.

Guth, A. H. 1981 *Phys. Rev. D* **23**, 347.

Guth, A. H. & Pi, S.-Y. 1982 *Phys. Rev. Lett.* **49**, 1110.

Hamuy, M. et al. 1995 *Astron. J.* **109**, 1.

Hawking, S. W. 1982 *Phys. Lett. B* **115**, 295.

Hill, J. E. 1995 *Phys. Rev. Lett.*, 2654.

Hirata, K. S. et al. 1992 *Phys. Lett. B* **280**, 146.

Holman, R., Ramond, P. & Ross, G. G. 1984 *Phys. Lett. B* **137**, 343.

Holtzman, J. A. 1989 *Astrophys. J. Suppl.* **71**, 1.

Hu, W. & Sugiyama, N. 1995 *Phys. Rev. D* **51**, 2599.

Hu, W. & White, M. 1996 astro-ph/9602020.

Hu, Y., Turner, M. S. & Weinberg, E. J. 1994 *Phys. Rev. D* **49**, 3830.

Huchra, J. & Geller, M. 1989 *Science* **246**, 897.

Jacoby, G. et al. 1992 *Proc. Astron. Soc. Pac.* **104**, 599.

Jensen, L. & Stein-Schabes, J. 1987 *Phys. Rev. D* **35**, 1146.

Jungman, G., Kamionkowski, M., Kosowsky, A. & Spergel, D. 1996a *Phys. Rev. Lett.* **76**, 1007.

Jungman, G., Kamionkowski, M., Kosowsky, A. & Spergel, D. 1996b *Phys. Rev. D*, in press, astro-ph/9512139.

Kamionkowski, M., Spergel, D. N. & Sugiyama, N. 1994 *Astrophys. J.* **426**, L57.

Knox, L. 1995 *Phys. Rev. D* **52**, 4307.

Knox, L. & Turner, M. S. 1993 *Phys. Rev. Lett.* **70**, 371.

Knox, L. & Turner, M. S. 1994 *Phys. Rev. Lett.* **73**, 3347.

Kochanek, C. S. 1996 *Astrophys. J.*, in press, astro-ph/9510077.

Kofman, L. A., Linde, A. D. & Einsato, J. 1987 *Nature* **326**, 48.

KOFMAN, L. & STAROBINSKII, A. A. 1985 *Sov. Astron. Lett.* **11**, 271.

KOLB, E. W. 1991 *Physica Scripta* **T36**, 199.

KOLB, E. W. & TURNER, M. S. 1983 *Ann. Rev. Nucl. Part. Sci.* **33**, 645.

KOLB, E. W. & TURNER, M. S. 1990 *The Early Universe*. Addison-Wesley, Redwood City.

KOSOWSKY, A., TURNER, M. S. & WATKINS, R. 1992 *Phys. Rev. Lett.* **69**, 2026.

KRAUSS, L. & TURNER, M. S. 1995 *Gen. Rel. Grav.* **27**, 1137.

LA, D. & STEINHARDT, P. J. 1991 *Phys. Rev. Lett.* **62**, 376.

LANZETTA, K., WOLFE, A. M. AND TURNSHEK, D. A. 1995 *Astrophys. J.* **440**, 435.

LIDDLE, A. ET AL. 1995 astro-ph/9511957.

LIDDLE, A. & LYTH, D. 1992 *Phys. Lett. B* **291**, 391.

LIDDLE, A. & LYTH, D. 1993 *Phys. Repts.* **231**, 1.

LIDSEY, J. E. & COLES, P. 1992 *Mon. Not. Roy. Astron. Soc.* **258**, 57.

LIDSEY, J. ET AL. 1995 astro-ph/9508078.

LIN, H. ET AL. 1996 *Astrophys. J.*, in press, astro-ph/9602064.

LINDE, A. D. 1982 *Phys. Lett. B* **108**, 389.

LINDE, A. D. 1983 *Phys. Lett. B* **129**, 177.

LINDE, A. D., 1990 *Inflation and Quantum Cosmology*. Academic Press.

LUCCHIN, F., MATTARESE, S., & MOLLERACH, S. 1992 *Astrophys. J.* **401**, L49.

MA, C. P. & BERTSCHINGER, E. 1994 *Astrophys. J.* **434**, L5.

MURAYAMA, H. ET AL. 1994 *Phys. Rev. D* **50**, R2356.

NOVIKOV, I. D. & ZEL'DOVICH, YA. B. 1967 *Ann. Rev. Astron. Astrophys.* **5**, 627.

NOVIKOV, I. D. & ZEL'DOVICH, YA. B. 1973 *Ann. Rev. Astron. Astrophys.* **11**, 387.

OLIVE, K. 1990 *Phys. Repts.* **190**, 309.

OSTRIKER, J. P. 1993 *Ann. Rev. Astron. Astrophys.* **31**, 689.

OSTRIKER, J. P. & STEINHARDT, P. J. 1995 *Nature* **377**, 600.

PARKE, S. 1995 *Phys. Rev. Lett.* **74**, 839.

PEACOCK, J. & DODDS, S. 1994 *Mon. Not. Roy. Astron. Soc.* **267**, 1020.

PEEBLES, P. J. E. 1984 *Astrophys. J.* **284**, 439.

PEEBLES, P. J. E., SCHRAMM, D. N., TURNER, E., & KRON, R. 1991 *Nature* **352**, 769.

PEEBLES P. J. E. & YU, J. T. 1970 *Astrophys. J.* **162**, 815.

PERLMUTTER, S. ET AL. 1996 astro-ph/9602122.

PI, S.-Y. 1984 *Phys. Rev. Lett.* **52**, 1725.

POGOSYAN, D. & STAROBINSKII, A. A. 1995 astro-ph/9502019.

PRESKILL, J. 1984 *Ann. Rev. Nucl. Part. Sci.* **34**, 461.

PRIMACK, J. ET AL. 1995 *Phys. Rev. Lett.* **74**, 2160.

RANDALL, L., SOLJACIC, M. & GUTH, A. H. 1996 hep-ph/9601296.

RATRA, B. 1992 *Astrophys. J.* **391**, L1.

REISS, A., KIRSHNER, R. P. & PRESS, W. 1995 *Astrophys. J.* **438**, L17.

ROSENBERG, L. J. 1995 *Particle World* **4**, 3.

RUBAKOV, V. A., SAZHIN, M. & VERYASKIN, A. 1982 *Phys. Lett. B* **115**, 189.

RUGERS, M. & HOGAN, C. J. 1996a *Astrophys. J. (Lett.)*, in press, astro-ph/9512004.

RUGERS, M. & HOGAN, C. J. 1996b astro-ph/9603084.

SALOPEK, D. 1992 *Phys. Rev. Lett.* **69**, 3602.

SHAFI, Q. & STECKER, F. 1984 *Phys. Rev. Lett.* **53**, 1292.

SHAFI, Q. & VILENKIN, A. 1984 *Phys. Rev. Lett.* **52**, 691.

SILK, J. & TURNER, M. S. 1986 *Phys. Rev. D* **35**, 419.

SMOOT, G. F. ET AL. 1992 *Astrophys. J.* **396**, L1.

SONGAILA, A. ET AL. 1994 *Nature* **368**, 599.

SOURADEEP, T. & SAHNI, V. 1992 *Mod. Phys. Lett. A* **7**, 3541.

STAROBINSKY, A. A. 1982 *Phys. Lett. B* **117**, 175.

STAROBINSKY, A. A. 1983a *Sov. Astron. Lett.* **9**, 302.

STAROBINSKY, A. A. 1983b *JETP Lett.* **37**, 66.

STEINHARDT, P. J. & TURNER, M. S. 1984 *Phys. Rev. D* **29**, 2162.

STORRIE-LOMBARDI, L. J., MCMAHON, R. G., IRWIN, M. J. & HAZARD, C. 1995 astro-ph/9503089.

STRAUSS, M. & WILLICK, J. 1995 *Phys. Repts.* **261**, 271.

SUGIYAMA, N. & GOUDA, N. 1992 *Prog. Theor. Phys.* **88**, 803.

TRIMBLE, V. 1987 *Ann. Rev. Astron. Astrophys.* **25**, 425.

TURNER, M. S. 1993 *Proc. Nat. Acad. Sci. (USA)* **90**, 4827.

TURNER, M. S. 1991 *Physica Scripta* **T36**, 167.

TURNER, M. S. 1993a *Science* **262**, 861.

TURNER, M. S. 1993b *Phys. Rev. D* **48**, 3502.

TURNER, M. S. 1996 astro-ph/9607066.

TURNER, M. S., LIDSEY, J. & WHITE, M. 1993 *Phys. Rev. D* **48**, 4613.

TURNER, M. S., STEIGMAN, G. & KRAUSS, L. 1984 *Phys. Rev. Lett.* **52**, 2090.

TURNER, M. S. & WIDROW, L. M. 1986 *Phys. Rev. Lett.* **57**, 2237.

TURNER, M. S. & WIDROW, L. M. 1988 *Phys. Rev. D* **37**, 2743.

TURNER, M. S. & WILCZEK, F. 1990 *Phys. Rev. Lett.* **65**, 3080.

TYTLER, D., FAN, K.-M. & BURLES, S. 1996 astro-ph/9603069.

WAMPLER, E. J. ET AL. 1996 *Astron. Astrophys.*, in press, astro-ph/9512084.

WEINBERG, S. 1989 *Rev. Mod. Phys.* **61**, 1.

WHITE, D. A. & FABIAN, A. C. 1995 *Mon. Not. Roy. Astron. Soc.* **273**, 72.

WHITE, M. & BUNN, E. F. 1995 *Astrophys. J. (Lett.)* **450**, 477.

WHITE, M., SCOTT, D. & SILK, J. 1995 *Science* **268**, 829.

WHITE, M., SCOTT, D., SILK, J. & DAVIS, M. 1995 *Mon. Not. Roy. Astron. Soc.* **276**, L69.

WHITE, S. D. M. ET AL. 1993 *Nature* **366**, 429.

WHITE, S. D. M., EFSTATHIOU, G. & FRENK, C. S. 1993 *Mon. Not. Roy. Astron. Soc.* **262**, 1023.

WHITE, S. D. M., FRENK, C. & DAVIS, M. 1983 *Astrophys. J.* **274**, L1.

WILSON, M. L. & SILK, J. 1981 *Astrophys. J.* **243**, 14.

ZEL'DOVICH, YA. B. & NOVIKOV, I. D. 1983 *Relativistic Astrophysics (Structure and Evolution of the Universe)*. Univ. Chicago Press.

The Signature of the Universe

By BERNARD J.T. JONES

Imperial College, London, UK

Discovering the specific cosmological model that best describes our Universe is one of the central tasks of modern observational cosmology. In particular, we would like to know the densities of the various types of matter that make up the Universe relative to the total mass density. The ability to measure and characterise fluctuations in the angular distribution of the cosmic microwave background radiation temperature provides a very powerful tool for measuring these quantities.

The cosmic microwave background radiation fluctuations directly reflect the distribution of the cosmic material at the time the cosmic plasma recombined. In this article I review the physical processes that are relevant to the evolution of the matter distribution and that take place just before and during the period of recombination, concentrating particularly on the dissipative processes. The approach is gasdynamic, using flux limited transport coefficients to describe processes such as viscosity, heat conduction and the role of pressure gradients in the transition between optically thick and thin plasma. The discussion should be of relevance to other branches of astrophysical plasmas of unit optical depth.

---

1. Introduction

The discovery by Penzias and Wilson (1965) of a Microwave Background Radiation and its interpretation by Dicke et al. (1965) as the "Relict Cosmic Radiation" left over from the Big Bang rates as one of the most fundamental discoveries of our century. In 1929 Hubble had discovered a redshift - distance relationship for some of the external galaxies for which Slipher (1915) had measured radial velocities. It was Robertson (1928) who, on the basis of preliminary data, first offered the explanation that this was due to cosmic expansion. The inference that the Universe started with a Big Bang was easy to draw, but did not go uncontested. Hoyle (1948) and Bondi and Gold (1948) put forward an alternative cosmological model in which the universe expanded without having had a singularity in its past: the "Steady State Theory". The debate was only settled once and for all by the discovery of Penzias and Wilson, though confirmation of the Planckian nature of the radiation spectrum had to await the measurements made by the FIRAS experiment on the COBE satellite.

1965 was the dawn of a new age in Cosmology: development of the cosmological theory proceeded rapidly with a series of papers on galaxy formation and the cosmic background radiation. It is interesting, in the present context, that the Zel'dovich school in Moscow were writing papers on the Hot Big Bang theory long before the Penzias-Wilson discovery. They were apparently impressed by the nucleosynthesis arguments. Smirnov (1964) had written a definitive paper on cosmic Helium production and put the cosmic radiation temperature in the range 1 - 10 K. Doroshkevich and Novikov then wrote a paper in 1964 (Doroshkevich & Novikov 1964) discussing how consistent this was with data available at the time. Igor recounts that the Moscow group felt certain that the cosmic background radiation could be discovered and that they asked many of their colleagues about this. Generally they were told by the experts that the experiment was difficult or impossible with the technology available at the time. The experts were unaware that in the Soviet Union in 1957 Shmaonov had obtained a marginal detection of an isotropic relict radiation with a temperature of $4 \pm 3K$. Observations of the relict radiation prior to 1965 are well recounted by Wilson (1985).

The paper of Dicke et al. (1965) recognised that at a redshift of around 1000 the matter in the Universe underwent a rapid transition from being totally ionised to almost neutral, and that the radiation detected by Penzias and Wilson almost certainly came straight to us from that time. The central role played in the formation of cosmic structure by this "recombination" event was quickly recognised. We saw papers by Peebles (1965), Doroshkevich, Zeldovich and Novikov (1965), Dicke and Peebles (1968), Silk (1967), Michie (unpublished) laying down the foundation of the subject of galaxy formation as we know it today.

One of the outstanding papers of those early years was that of Sachs and Wolfe (1967) who predicted that the distribution of the Cosmic Microwave Background Radiation on the sky should show angular structure due to the existence of primordial density perturbations. This has become known as the "Sachs-Wolfe" effect. The detection of angular fluctuations of the Comsic Microwave Background temperature by the COBE satellite (see Smoot's contribution in this volume) was a dramatic confirmation of the prediction of Sachs and Wolfe.

Igor Novikov has worked throughout his career on various aspects of the microwave background radiation anisotropy. Most recently he has written papers about the so-called Sakharov Oscillations that might be detected in the angular scale dependence of the cosmic microwave background temperature fluctuations (Naselsky and Novikov, 1993; Jørgensen et al., 1993, 1995). In this article I will review the origin of these oscillations, concentrating on what is happening to the baryonic component of the cosmic plasma (there are numerous fine articles discussing the radiation field). Since much of the present discussion will focus around the time when the cosmic plasma makes the transition from the optically thick to optically thin regime, this also impinges on one of Igor's other interests, accretion disks (see Artemova et al., 1996).

Indeed, the approach adopted here should not be seen as a replacement for the conventional method of handling the cosmic recombination process. The virtue of the approach is that it falls within the framework of gasdynamics using the same macroscopic variables as are used when there are no finite optical depth effects. The technique may therefore have wider applicability than dissipation in the cosmic plasma.

2. Cosmic Constituents

The Universe is known to contain baryons and radiation - we see these directly. The photons making up the microwave background radiation field are far more numerous than the baryons, but today they have almost no role to play since the Universe is optically thin at the relevant wavelengths. They simply carry important information from the distant past, providing us with a direct glimpse of the time when the Universe was barely a million years old.

We know the temperature of the cosmic microwave background radiation with high precision and so we know its mass density with high precision. Our knowledge of the baryonic density is by comparison highly uncertain. We have upper limits on this density from cosmic nucleosynthesis, and we have lower limits by virtue of those baryons that we see directly: the "luminous baryons". Two questions arise naturally. Is there any baryonic matter that we do not see and if so how much? Is there any substantial quantity of matter that is neither baryonic nor photons, and again, if so - how much?

The ultimate fate of our Universe is determined by the total density. Given the present expansion rate (the "Hubble Constant"), H_0, if the cosmic mass density exceeds a critical value (the "Einstein de Sitter critical density")

$$\rho_c = \frac{3H_0^2}{8\pi G}. \tag{2.1}$$

the expansion will eventually stop and be followed by a collapse to a future singularity. If the cosmic mass density is less than this value, the expansion will continue forever. By convention, densities are frequently measured in units of the critical density and denoted by the symbol Ω_X, where the subscript denotes the type of material the density value refers to:

$$\Omega_X = \frac{\rho_X}{\rho_c} \tag{2.2}$$

Note that since ρ_c depends on the Hubble constant, the value of Ω_X for a particular constituent might also depend on the Hubble constant. However, the determined value of ρ_X usually depends on the Hubble constant in the same way as does ρ_c, and then Ω_X is independent of H_0.

The upper limit on the baryonic density set by the considerations of cosmic nucleosynthesis are well below this critical density, and so it is important to know whether there is a substantial quantity of unseen (dark) non-baryonic material. There are numerous suggestions as to what form this hypothetical dark non-baryonic matter might take. This is an important question if we are to attempt to detect it directly.

Cosmology affords us with several ways of inferring the densities, or at least, of putting limits on them. The cosmic microwave background radiation density has been measured directly. There are limits on the baryonic component. But we have only indirect evidence that there is any dark non-baryonic matter at all. This evidence comes from looking at the dynamics of the material in the Universe. Because the Universe is inhomogeneous, consisting of galaxies and clusters and superclusters of galaxies, the galaxies are in motion relative to one another. This relative motion is over and above the relative motion they would have from the simple Hubble flow expected in a perfectly homogeneous and isotropic expanding Universe. This "peculiar" motion of the galaxies is driven by inhomogeneities in the gravitational field due to the irregular distribution of the galaxies (and perhaps the unseen non-baryonic material). The amount of peculiar motion might provide a measure of the total cosmic density.

Indications are that these peculiar motions are more substantial than could be accounted for in terms of the observed distribution of luminous matter. However, the interpretation of the data is not simple since it is predicated on assumptions as to the extent to which the observed light distribution traces the mass distribution.

It turns out that the character of the angular distribution on the sky of the cosmic microwave background radiation temperature is sensitive to the densities of baryonic and non-baryonic matter. Indeed it is even sensitive to the type on non-baryonic matter and other things like the cosmological constant. So simply measuring the amount of temperature variation on different angular scales provides the required information. The variation of temperature with scale is a signature of the cosmological model.

The key issue then is what the expected variation in temperature is on various angular scales for various cosmological models containing different varieties of matter. To get at this we need to understand how inhomogeneities of different kinds behave just before and during the recombination of the primeval plasma.

3. The Primeval Plasma

3.1. *The pre-recombination plasma*

As the Universe expands, the baryonic density declines like $\rho_B \propto R^{-3}$, where $R(t)$ is the expansion factor. This equation is no more than a statement about the conservation of baryons. The radiation temperature declines as $T_\gamma \propto R^{-1}$ and so the radiation energy density declines as $\rho_\gamma \propto R^{-4}$. So even though today the baryonic mass density certainly exceeds the photon density by many orders of magnitude, this was not always the case. In the past the Universe was "radiation dominated": $\rho_\gamma > \rho_B$. The time when these densities were equal is denoted by t_{eq} and corresponded to a redshift denoted by z_{eq}:

$$z_{eq} = 2 \times 10^4 \Omega_B h^2 \tag{3.3}$$

where the value of the Hubble constant has been parametrised as $100h$ km.s^{-1}.Mpc^{-1}. At this epoch the temperature was on the order of 10^4K, and the cosmic plasma was fully ionised. The Universe in fact did not become neutral until its temperature fell to around 3000K (this happens at a redshift of about 1000).

Prior to the time when the Universe became neutral, the cosmic plasma was dominated by the interaction between the photons and the free electrons. The photon mean free path is given by $l_\gamma \sim \sigma_T n_e$ where $\sigma_T = 6.65 \times 10^{-25}\mathrm{cm}^2$ is the Thomson scattering cross-section. The photons vastly outnumber the electrons (by a factor $aT^4/nkT \sim 10^9$ (independent of redshift) and so the plasma behaves like a gas with $\gamma = 4/3$ and the electron and photon temperatures are essentially equal.

This scattering makes the radiation pressure the dominant contributor to the total pressure of the plasma (again by a factor $\sim 10^9$):

$$p = \frac{1}{3}aT^4 + n_e kT \tag{3.4}$$

where T is the plasma temperature and n is the electron number density. The adiabatic sound speed in the cosmic medium is therefore dominated by radiation pressure and is easily shown to be

$$c_s = \frac{c}{\sqrt{3}} \frac{1}{\left[1 + \dfrac{3\rho_m}{4\rho_r}\right]^{1/2}}. \tag{3.5}$$

Here ρ_m is the density of the baryonic component and ρ_r is the mass density of the radiation field. In the situation where the radiation mass density is dominant, the sound speed is close to the speed of light: $c_s \to c/\sqrt{3}, \quad \rho_r \gg \rho_m$.

The sound speed c_S determines the scale on which pressure forces can balance the effect of gravity:

$$\lambda_J \simeq c_S t_{exp} \tag{3.6}$$

On scales $l < \lambda_J$, pressure waves can inhibit the tendency for perturbations in the density to be amplified gravitationally, while on scales $l > \lambda_J$ gravity will enhance the amplitude. The equations for this are presented below.

As photons diffuse across density and velocity gradients they transport energy and momentum from one part of the fluid to another. Photon diffusion is the major dissipative mechanism and while the Universe is ionised is responsible for dissipative processes due to viscosity and thermal conduction. The shear viscosity, for example, depends directly on the photon mean free path:

$$\nu = \frac{4}{15} c l_\gamma \tag{3.7}$$

Velocity, density and temperature gradients on small scales are damped. By "small" we mean roughly small enough that the photons can diffuse across that scale distance within the cosmic expansion timescale: $l < l_{damp} \sim \sqrt{\nu t}$. This is discussed in much more detail below.

3.2. *What happens during recombination?*

The recombination of the primeval plasma is a little different from the recombination of a simple gas of ionised hydrogen. This is because there are enough photons of high energy in the microwave background radiation spectrum at that time to prevent recombination to the ground state until the temperature falls below $\sim 3500K$. The rate of recombination is then essentially fixed by the rate at which cosmic microwave background photons are redshifted out of the Ly-α line. The recombination process is not instantaneous, it takes some 10% of the expansion time.

During the recombination process the number of free electrons decreases rapidly and the photon mean free path becomes very long compared with the scale ct of the Universe at that time. The photons cease to interact with electrons and the baryonic and radiation components are dynamically decoupled (and hence the alternative name "decoupling epoch" for "recombination"). One of the questions one might ask is "What happens to the pressure gradient terms in the equations of motion as the Universe becomes optically thin and the photons stop colliding with the electrons?". Before recombination, the pressure is the radiation pressure $\frac{1}{3}aT^4$ and afterwards it is nkT. How are we to write down the equations of motion under these circumstances?

The physics at this time is quite complex because the photon mean free path is changing more rapidly than the photon diffusion timescale, or indeed any other dynamical timescale in the problem. This is not a classical gas and some care is needed in calculating the appropriate transport coefficients. Technically we are making a transition from a regime where the optical depth to Thomson scattering is very large to a regime where is very small.

The finite duration of the recombination means that the interaction between electrons and photons is still able to damp velocity fields. It is inappropriate to regard this as a viscosity since the fluctuations in density and velocity are of roughly unit optical depth, and equation (3.7) doesn't hold. When the fluctuation scale is optically thin we can use a "radiation drag" force on an electron moving through the radiation due to Thomson scattering:

$$F = -\frac{4}{3}\sigma_T a T^4 \frac{v}{c} \tag{3.8}$$

where v is the electron velocity relative to the frame in which the radiation field is isotropic. We can consider the force on a small volume δV within an optically thin cloud in the case that $\rho_B c^2 \gg aT^4$ and deduce the deceleration of the volume:

$$\dot{v} = -\frac{4}{3}\frac{aT^4}{\rho_B c^2}(\sigma_T n_e c)\, v \tag{3.9}$$

where n_e is the free electron density. The timescale on which the electrons can transfer their momentum to the radiation field is thus

$$t_T = \frac{3}{4}\frac{\rho_B c^2}{aT^4}\frac{1}{\sigma_T n_e c} \tag{3.10}$$

The timescale $t_{coll} = \sigma_T n_e$ is the photon collision time and this becomes longer as the ionisation falls and the density of free electrons falls towards zero.

We can make a crude estimate of the scale on which velocities are damped during the recombination period (Jones, 1977) by noting that the time available for damping is the time it takes the Universe to recombine - roughly 1/10th. of the cosmic expansion time at that epoch - and that most of the damping will take place while a scale is of unit optical depth. So, from (3.9) we estimate

$$v = v_0 \exp(-\Delta_l), \quad \Delta_l = \frac{4}{3}\frac{aT^4}{\rho_m c^2}\left(\frac{1}{10}\frac{ct_{rec}}{l}\right) \tag{3.11}$$

for the velocity attenuation on scale l during recombination. We have put $(\sigma_T n_e)^{-1} \simeq l$ since the damping takes place mostly around the time when the optical depth of the scale

l is unity. If the recombination were for some reason slower, the damping on a given scale would be greater (we would change the 1/10 to something larger).

The important scale is that for which $\Delta_l = 1$. This defines a scale l_* which is the largest scale on which there is any attenuation of the velocity. Scales $l > l_*$ are not damped, while scales $l < l_*$ are. We have

$$l_* = \frac{4aT^4}{3\rho_m c^2}\left(\frac{ct_{rec}}{10}\right) \sim 10^{21}(\Omega_B h^2)^{-7/2}\ cm. \tag{3.12}$$

where all quantities are evaluated at recombination (or more precisely when the scale l_* is of optical depth unity).

Calculating this effect accurately through the period of recombination is nontrivial and is one of the subjects of this article.

3.3. *After recombination*

Prior to recombination the photon gas was the prime contributor to the pressure of the cosmic medium. It was also the agent through which energy is transported and hence the dominant contributor to the shear viscosity and thermal conductivity of the medium. After recombination the pressure acting on the gas is due to collisions among baryons and is simply nkT. This results in a post-recombination sound speed of only a few kilometers per second and a post-recombination Jeans mass of around $10^5\ M_\odot$: everything that survives the pre-recombination damping can be gravitationally enhanced.

It was originally thought that the transition from a radiation pressure dominated state to an almost zero-pressure state would enable large shock waves to be generated immediately after recombination. Motions that were subsonic prior to recombination would be supersonic after recombination, resulting in density fluctuations on the order of the square of the post recombination Mach number. The paper of Jones (1977) integrated the velocity field through recombination using an interpolation formula for the viscosity and showed that there would in fact be no substantial amplification of the density fluctuations. We shall return to the subject of this interpolation procedure later on.

In the case of primeval adiabatic perturbations there is some residual velocity left in perturbations on scales $l > l_*$ (equation 3.12). Whether this velocity acts as to expand or contract the perturbation depends on the phase of the oscillation at the time of damping. This phase is correlated with scale for reasons we shall discuss below. In any case, the magnitude of the velocity field is equivalent to a fractional change $(c_s/c)_{rec}$ in the perturbation amplitude and so has only a very small effect on the time that a perturbation will collapse to form a bound object.

4. A Specific Model

The focus here is on the physics of the cosmic plasma and so we can for didactic reasons cast the discussion in terms of a very simple model for the Universe at around the time it became neutral.

4.1. *The background cosmology*

In general, the expansion of the background Universe is described by the expansion factor $R(t)$, given by the appropriate Friedmann-Lemaitre equations:

$$\frac{1}{R}\frac{d^2R}{dt^2} = -\frac{4}{3}\pi G\left[\rho(t) + 3\frac{p(t)}{c^2}\right] - \frac{\Lambda}{3} \tag{4.13}$$

where $\rho(t)$ is the total matter density, $p(t)$ is the pressure and Λ is the cosmological constant. A decision is required as to what types of matter should be included in $\rho(t)$

and an equation of state is required to give us the contributions to the total pressure from each of these species. Doing the first integral of this equation requires a constant of integration which is set by the boundary conditions for the origin of the Universe. The choice of this constant determines whether the universe expands forever or recollapses at some time in its future.

The simplest models set $\Lambda = 0$ and use baryons plus radiation as the main constituents. In that case the pressure before recombination is the radiation pressure $p_r = \frac{1}{3}\rho_r c^2 = \frac{1}{3}aT_r^4$ and after recombination it is $p = 0$. A more "realistic" model may additionally contain dark nonbaryonic matter. The contribution of the dark matter to the total pressure depends on whether or not the matter is relativistic.

4.2. *Linear Perturbation Modes*

It has long been recognised that today's cosmic structures grew out of density irregularities that must have existed at the time the Universe was born. There are no general inhomogeneous solutions to the Einstein equations that enable us to deal with this and so the practise has been to use linear perturbation theory to discuss the deviation from homogeneity. Linear perturbation theory assumes that the deviations are of small amplitude. We now know that this is a good approximation since the COBE satellite has measured the amplitudes (see Smoot's cahpter in this volume).

Linear perturbation theory is relatively simple since the irregularities in the universe can be simply viewed as a superposition of independent modes. We can thus tackle the general problem by identifying the possible modes and dealing with them individually. The catalogue of modes even in the simple case of a Universe consisting only of baryons and photons is quite large.

There are modes in which the density is left unchanged and only the velocity field is perturbed. These are shear modes and vorticity modes and they are a key element of the cosmic turbulence theory for the origin of structure that was popular in the late 1960's. Subsonic turbulence prior to recombination could be considered incompressible. The idea behind the theory was that such velocity perturbations might generate density fluctuations when the turbulence becomes subsonic after recombination. For a variety of good reasons, this theory is no longer considered seriously.

There is also a class of compressive density mode in which only the divergence of the velocity field is affected. Among these density modes we can identify the important "adiabatic modes" in which the matter and radiation field are perturbed together in such a way as to maintain a spatially constant entropy per baryon:

$$\delta = \frac{\delta\rho_B}{\rho_B}, \qquad \tau = \frac{\delta T}{T}, \qquad \delta = \frac{1}{3}\tau \tag{4.14}$$

(The entropy per baryon is proportional to T^3/ρ_B.) It is however possible to consider radiation modes in which the radiation density is perturbed without affecting the baryons.

Both these density modes perturb the gravitational potential (through the Poisson equation). It is possible to consider a mode that consists of perturbations of both the radiation and matter fields such as to leave the gravitational potential homogeneous.

Adding dark (nonbaryonic) matter to the soup introduces modes that are fluctuations in the gravitational potential: these drive their own density perturbations and are independent of the fluctuations in the gravitational potential due to the previously discussed density modes. The cosmic plasma gravitates towards these potential wells under the influence of pressure and dissipative forces.

We do not know which modes were present, but we may be able to tell when we have mapped the microwave background radiation temperature over a wide range of angular scales. For the moment we have to rely on theories for the origin of the initial inhomogeneities.

4.3. *Simple linear perturbations*

Linear perturbation theory is discussed in many textbooks on cosmology, the most exhaustive treatments being those of Padmanabhan (1993) and of Coles and Lucchin (1995). (The former deals with all possible cases on a blow-by-blow basis. The latter has a particularly thorough discussion on dissipation of perturbations).

The goal here is merely to illustrate important features of the evolution of perturbations and so we might as well fix on a specific (and simple) case and follow the Newtonian discussion of Peebles (1980) for the matter dominated case when $\rho_m \gg \rho_\gamma$. We shall also assume that the fluctuations in the gravitational potential arise from the perturbations in density themselves. Peebles' equations (10.2) and (10.3) describe the evolution of the density contrast $\delta = \delta\rho/\rho$ under these circumstances. In a Universe where the scale factor is $R(t)$:

$$\frac{1}{R^2}\frac{d}{dt}\left(R^2\frac{d\delta}{dt}\right) = \frac{1}{R^2}\frac{\nabla^2 p}{\rho}\delta + 4\pi G\rho_b\delta \tag{4.15}$$

$R(t)$ is a known function of time given the cosmological model† The gravitational potential has been eliminated using the perturbed Poisson equation, this results in the $4\pi G\rho_B$ term. The pressure is eliminated in favour of the density via the equation of state; this brings the sound speed of the medium into the equation.

It is usual to consider the evolution of a single spatial Fourier component of the density field in co-expanding co-ordinates (a plane wave):

$$\delta(\mathbf{x}, t) = \delta(t) \exp\left(i\frac{\mathbf{k}.\mathbf{x}}{R(t)}\right). \tag{4.16}$$

The wave-vector $\mathbf{k}$ is considered to be fixed and the wavelength is

$$\lambda = \frac{2\pi R(t)}{k}. \tag{4.17}$$

Under the given circumstances the behaviour of the Fourier amplitude $\delta(t)$ of the perturbation in the density is described by the following set of equations.

$$\frac{1}{R^2}\frac{d}{dt}\left(R^2\frac{d\delta}{dt}\right) = -\left(c_s^2\frac{k^2}{R^2} - 4\pi G\rho\right)\delta \tag{4.18}$$

The time derivative operator reflects the effect of the cosmic expansion, and the sign of the coefficient of δ determines the nature of the solution - whether it is oscillatory or not. It is this latter coefficient that determines the Jeans Length at a given instant:

$$\lambda_J = \left(\frac{\pi c_s^2}{G\rho}\right)^{1/2} \tag{4.19}$$

(The exact coefficient depends on the case being considered, but the form is generically correct).

4.4. *Growing and Oscillatory solutions*

Equation (4.18) has simple limiting solutions in the cases $\lambda \gg \lambda_J$ and $\lambda \ll \lambda_J$. In the former case gravitational forces dominate. The equation is simply the homogeneous equation

$$\ddot{\delta} + \frac{2}{3t}\dot{\delta} - \frac{2}{3t^2}\delta = 0 \tag{4.20}$$

† Padmanabhan (1993) gives a generic version of this equation. See his equation (4.88).

from which we get the famous solution

$$\delta(t) = At^{2/3} + Bt^{-1} \tag{4.21}$$

describing a linear combination of a growing mode and a decaying mode. The decaying mode corresponds to a non-linear initial condition in the past and since we assume that the universe in the past was more homogeneous than it is now is is usual to set this decaying term equal to zero.

The equation for the other case in which the pressure gradients stabilise the perturbation against gravitational amplification can be written as

$$\ddot{\delta} + \frac{2}{3t}\dot{\delta} + \frac{1}{t^2}\alpha^2 = 0, \qquad \alpha = \left.\frac{8\pi}{3}\frac{ct}{\lambda}\right|_{eq} \tag{4.22}$$

This is again a homogeneous equation and for large α has solution

$$\delta = t^{-1/6}\left[C\cos\left(\alpha\log\frac{t}{t_{eq}}\right) + D\sin\left(\alpha\log\frac{t}{t_{eq}}\right)\right] \tag{4.23}$$

This is an oscillatory solution of constant period when plotted in terms of $\log t$. Viewed as a function of t, the oscillations slow down as time goes on. The frequency of the oscillations depends on the wavelength (inversely proportional to α). Thus small wavelengths oscillate with higher frequency. The factor $t^{-1/6}$ is interesting: this represents a decrease in the amplitude due to the fact that the wave is oscillating in an expanding medium.

The initial conditions for specifying the full solution to equation (4.18) consist of specifying the constant A and B in the asymptotic region described by the solution (4.21). As remarked earlier, we generally pick a pure growing mode, $B = 0$, for the initial state. We wish to follow that solution into the oscillatory regime, but it is not immediately clear how to calculate the values of the constants C and D of equation given the initial conditions.† This is where the WKB method of solving equation (4.18) comes in.

4.5. *Phase Coherence in Adiabatic Oscillations*

If we start with the linear equation for the perturbation amplitude in the form (4.18), we can make a transformation of the time variable

$$dx = R(t)^{-2}dt \tag{4.24}$$

In the case $R \propto t^{3/3}$ this is just

$$x = \left(\frac{t_\dagger}{t}\right)^{1/3} \tag{4.25}$$

where we have chose to normalise x to unity at the time $t_\dagger$ when, for a given scale k the pressure gradient and gravitational forces balance. For $t < t_\dagger$ gravitational forces dominate the perturbation, while for $t > t_\dagger$ pressure forces dominate.

The equation for the perturbations then takes the form

$$\frac{d^2\delta}{dx^2} + Q(x)\delta = 0 \tag{4.26}$$

where

$$Q(x) = R^4\left(c_s^2\frac{k^2}{R^2} - 4\pi G\rho\right) \tag{4.27}$$

† In fact, because in this rather special model the term on the right hand side of (4.18) is always proportional to t^{-2} whatever the value of k, the exact solution can easily be written down.

has to be expressed as a function of the new variable x. In this rather trivial case, $Q(x)$ is just a constant depending on the scale k of the perturbation. Our normalisation is such that $Q(x) = 0$ at $x = 1$, in the general case this is a different physical time for different scales.

Given the purely growing initial solution, the WKB method tells us which is the corresponding oscillatory solution for $Q(x) > 0$:

$$\delta(x) \propto Q(x)^{-1/4} \cos\left[\int_1^x \sqrt{Q(x)}dx - \frac{\pi}{4}\right] \tag{4.28}$$

The first thing to note is that, from (4.27), $Q \propto t^{2/3}$ and so the term $Q^{-1/4} \propto t^{-1/6}$. The amplitude of the sound waves decreases as the Universe expands (compare equation (4.23)).

The next thing to note is that, at a given instant of time, the phase of the wave depends only on its wavenumber k. The situation at the epoch of recombination is that waves of particular frequencies have attained there maximum compression. For our simple model these are wavelengths

$$\lambda \sim \frac{\lambda_J}{n} \tag{4.29}$$

for larger values of n. The mean square amplitude of fluctuations (the Power Spectrum) is thus a function with well separated peaks. These peaks are occasionally referred to as Sakharov Oscillations after an early paper of Sakharov (1965). In that paper, Sakharov discusses what happens in a cold Universe, but the principle is nevertheless the same. The first paper to display the relationship between phase and wavelength in the current cosmological context was that of Peebles and Yu (1970).

4.6. *Comments*

The fact that the initial conditions are specified uniquely by the growing mode is of utmost importance in getting the phase-correlated (Sakharov) oscillations in the sub-Jeans Length regime. If we imagine that perturbations are produced by some other means, for example if there were a forcing term on the right hand side of equation (4.18), then there may be no phase correlations - the asymptotic WKB equation (4.28) need not be valid. In particular, models of the early Universe in which inhomogeneities are created by topological defects may not show these oscillations in the power spectrum (Albrecht et al. 1996).

Since these oscillations would be reflected in the angular distribution of the cosmic microwave background fluctuations, seeing or not seeing peaks is a matter of fundamental importance in the understanding of our Universe.

5. Dissipative effects

The only physics contained in equation (4.18) is amplification due to gravity. There is no description of dissipative effects due to viscosity or thermal conduction. As first pointed out by Silk (1967), the dissipation is important on scales even larger than galaxies, and in the more open cosmological models it could be important on scales larger even than galaxy clusters. Whereas dissipative effects are not important for large scale cosmic structure, nor for perturbations in the gravitational potential, they will dominate other aspects on galaxy scales. As we probe the angular structure of the cosmic microwave background radiation on smaller scales we may see these effects.

The question of dissipation is a complex one, and so I will first review the situation as it appears in the viscous flow of a perfect compressible gas. Then I shall show how this picture is changed for the cosmic fluid, and finally comment on what happens when the fluctuations become optically transparent.

5.1. *Radiative Viscosity and Thermal Conduction*

In standard gasdynamic/fluid problems, the collision mean free path of the particles, l, is much smaller than the characteristic lengthscale, L, of the problem. (L might represent the scale of density variations). The equations of motion, including expressions for the transport coefficients, can then be derived from the Boltzmann equation by expanding the distribution function in a power series in l/L. This is the Chapman-Enskog method.

The momentum equation for the fluid motion is then written as

$$\rho \frac{d\mathbf{v}}{dt} = -\nabla p + \hat{\mathbf{e}}_i \frac{\partial}{\partial x_j}\left[\mu\left(\frac{\partial v_i}{\partial x_j} + \frac{\partial v_j}{\partial x_i} - \frac{2}{3}\delta_{ij}\nabla.\mathbf{v}\right)\right] + \rho\mathbf{g} \tag{5.30}$$

where the summation convention over dummy vector indices is implied, and $\hat{\mathbf{e}}_i$ denotes three unit vectors in a rectangular coordinate system. $\mathbf{g}$ is the acceleration due to external forces (like gravity), and μ is the coefficient of viscosity, given by the Chapman Enskog method.

The energy equation is

$$\rho \frac{dE}{dt} = -p\nabla.\mathbf{v} + \mu\left[\frac{1}{2}\left(\frac{\partial v_i}{\partial x_j} + \frac{\partial v_j}{\partial x_i}\right)^2 - \frac{2}{3}(\nabla.\mathbf{v})^2\right] + \nabla.(\kappa\nabla kT) \tag{5.31}$$

κ is the coefficient of thermal conductivity, determined by the Chapman Enskog method. The heat flux is $\mathbf{Q} = -\kappa\nabla kT$, which is known as Fouriers Law for heat conduction.

The orders of magnitude of these transport coefficients are

$$\kappa \simeq n(l\bar{v}) \qquad \mu \simeq \rho(l\bar{v}) \tag{5.32}$$

where l denotes the particle collision mean free path and $\bar{v}$ demotes the rms thermal velocity. The Chapman Enskog method can of course give more accurate values.

In these classical equations we note that the thermal conductivity only appears in the energy equation. In a relativistic gas this is not the case, it also appears in the momentum equation since heat flux can carry momentum. The neglect of the momentum transport due to heat flux was a common error of the earlier papers on damping.

5.2. *Damping of Adiabatic Perturbations*

In the pre-recombination cosmic plasma the viscosity and thermal conduction are determined in terms of the photon mean free path to Thomson scattering, $l_\gamma = (\sigma_T n_e)^{-1}$:

$$\lambda = \frac{8}{27} c l_\gamma a T^4, \qquad \chi = \frac{4}{3} c l_\gamma a T^3 \tag{5.33}$$

The equation for the propagation of a small amplitude density wave are then as follows. First define the relative fluctuations in temperature and baryonic density in the wave as

$$\tau = \frac{T_1}{T_0} \qquad \delta = \frac{\rho_1}{\rho_0} \tag{5.34}$$

Here $T_1(\mathbf{x}, t)$ and $\rho_1(\mathbf{x}, t)$ are the small variations in temperature and density about their mean values $T_0(t)$ and $\rho_0(t)$. With this the pressure fluctuation is

$$p_1 = \frac{4}{3} a T_0^4 \tau \tag{5.35}$$

and the associated fluctuation on the gravitational potential, given by Poissons equation, is

$$\nabla^2 \phi_1 == \frac{4\pi G \rho_0}{c^2}\left[8\frac{aT_0^4}{\rho_0 c^2}\tau + \delta\right] \tag{5.36}$$

The factor 8 comes from the fact that the fluctuation in T^4 is four times the fluctuation in T, and that the source of the gravitational field for the radiation is $\rho_r + 3p_r = 2\rho_r$. Conservation of baryons tells us that the fluctuation in the divergence of the baryonic velocity field is

$$\theta_1 = -\delta \tag{5.37}$$

The heat flux is

$$\mathbf{q} = \psi^{-1}\chi\nabla\tau, \qquad \psi = 1 + \frac{4aT_0^4}{3\rho_0 c^2} \tag{5.38}$$

The pre-factor ψ is a relativistic correction factor, and is such as to decrease the heat flux when the mass density of the radiation field becomes dominant.

Now we are set up to write down the Navier Stokes equations for the perturbed quantities. In order to keep the equations simple (ie: short!) we can suppress terms relating to the cosmic expansion. These terms contain coefficients involving the expansion rate $\dot{R}/R$. Neglecting the expansion terms in this way amounts to using the approximation that the damping timescales are shorter than the cosmic expansion timescale: we are in the dissipative regime. We shall also drop the terms arising from fluctuations in the gravitational potential for the same reason. This means that our equations apply to scales on which gravity is not important over the dissipation timescale. Putting these terms back again is not a serious problem - it is easier to see the essential physics without them.

The energy equation is the simpler:

$$\dot{\tau} - \frac{1}{3}\dot{\delta} = -\frac{1}{\psi}\frac{\chi}{4aT^4}\nabla^2\tau = -\psi^{-1}cl_\gamma\,\nabla^2\tau \tag{5.39}$$

In the dissipationless limit ($l_\gamma \to 0$) we have the standard result for adiabatic perturbations that the fluctuation in the temperature is one third the baryonic density fluctuation (in other words aT^3/ρ is held constant). The heat flux determines the departure from adiabaticity.

We use the divergence of the momentum equation to get an equation for θ:

$$\left(\rho c^2 + \frac{4}{3}aT^4\right)\dot{\theta} = -\left[\nabla^2 p_1 + \nabla.\dot{\mathbf{q}} + \frac{2}{3}\lambda\nabla^2\theta\right] - \left(\rho c^2 + \frac{4}{3}aT^4\right)\nabla^2\theta. \tag{5.40}$$

(The coefficient of $\dot{\theta}$ is the relativistic term $(\mu + p)$ that comes into the relativistic versions of the momentum equation.)

5.2.1. *Dissipationless solution*

If we set $l_\gamma \to 0$ we easily recover the equation for adiabatic sound waves:

$$\ddot{\tau} - \frac{c^2}{3}\left(1 - \frac{1}{\psi}\right)\nabla^2\tau = 0 \tag{5.41}$$

which gives the familiar result for the adiabatic sound speed:

$$c_s^2 = \frac{c^2}{3}\left(1 - \frac{1}{\psi}\right) = \frac{c^2}{3}\left(1 + \frac{3\rho c^2}{4aT^4}\right) \tag{5.42}$$

5.2.2. *Dissipation - the Silk Mass*

If we keep the dissipation terms in, we get a somewhat lengthier expression

$$\ddot{\tau} - \frac{c^2}{3}\left(1 - \frac{1}{\psi}\right)\nabla^2\tau = \tag{5.43}$$

$$\frac{1}{9\psi^2} c l_\gamma \nabla^2 \tau - \frac{16}{81} \frac{\psi - 1}{\psi} c l_\gamma \nabla^2 \left[\dot{\tau} + \frac{4}{27} \frac{\psi - 1}{\psi^2} c l_\gamma \nabla^2 \tau \right] \tag{5.44}$$

The first term on the right is the damping due to thermal conduction and the second is the contribution from shear viscosity.

We can estimate the timescales for these processes acting on a perturbation of scale l:

$$t_{cond} \simeq \frac{9}{2\pi} \psi^2 \left(\frac{l}{l_\gamma} \right) \left(\frac{l}{c} \right) \simeq \frac{9}{2\pi} \left(1 + \frac{4\rho_r}{3\rho_m} \right)^2 t_{diff} \tag{5.45}$$

$$t_{visc} \simeq \frac{15}{4\pi} \left(\frac{\psi}{\psi - 1} \right) \left(\frac{l}{l_\gamma} \right) \left(\frac{l}{c} \right) \simeq \frac{15}{4\pi} \left(1 + \frac{3\rho_m}{4\rho_r} \right) t_{diff} \tag{5.46}$$

where the photon diffusion time is

$$t_{diff} \simeq \frac{l^2}{l_\gamma c}. \tag{5.47}$$

This is the time it takes for the photon to random walk a distance l when its mean free path is l_γ. What is interesting is that in the radiation dominated phases the viscosity is the prime damping agent, while in the matter dominated regime it is thermal conduction. In both regimes the damping time (the shorter of the two timescales) can be taken as the diffusion timescale t_{diff}.

We can deduce the mass associated with the characteristic damping scale by setting $t_{diff} = t_{exp}$, the cosmic expansion time. Then $l_{damp} \sim \sqrt{l_\gamma c t_{exp}}$ and this gives a typical damping mass

$$M_{damp} \simeq \sqrt{M_\gamma M_H} \tag{5.48}$$

where M_γ and M_H are respectively the masses enclosed within spheres whose radii are the photon mean free path and the horizon scale.

This gives a damping mass around $10^{12}\, M_\odot$ when evaluated at recombination and so sets a characteristic scale below which structure arising from adiabatic perturbations does not exist. The situation is however a little more complicated: the damping does not stop at the onset of recombination, but continues while the Universe recombines and is still partly neutral. The values of the transport coefficients are no longer valid in that regime.

5.3. *The optically thin regime*

The previous discussion is inapplicable when the scales in question become optically thin. There are several possibilities. The general way of dealing with the damping of perturbations since the days of Peebles and Yu (1970) has been to treat the matter and radiation as two fluids interacting through Thomson scattering. The equations can be written down, but the only way to make progress is to solve them on a computer. An alternative is to fix the diffusion coefficients. This can be done by working with the Boltzmann equation directly or by adopting the standard numerical gasdynamics fix of using a flux limiter.

The flux limiter is an interpolation formula between the thick and thin regimes (Levermore and Pomraning, 1981). Technically, the interpolation should be physically motivated, but it is common practise in numerical codes to take the simplest interpolant. In what follows we shall produce an interpolation formula and then provide a justification for it.

5.4. *Shear viscosity: flux limited approximation*

Let us focus on the drag force on a parcel of ionised plasma moving through the radiation field. Consider following modification to the viscous force on a parcel of plasma:

$$F_{visc} = \frac{1}{1+\dfrac{3l_\gamma^2}{5l^2}} \frac{4}{15} c l_\gamma \rho_r \nabla^2 v. \tag{5.49}$$

In the limit $l_\gamma \ll l$ this reduces to the viscous force (3.8), (5.30) on the parcel of plasma, while in the opposite limit it reduces to the Thomson drag force (3.8). Of course this requires that the Laplacian be replaced by v/l^2 in the optically thin limit. If it were physically motivated, (5.49) would be a candidate for the flux limited viscosity.

Some motivation that this is a reasonable interpolation formula comes from the paper of Jones and Wyse (1983) in which the collision term of the Boltzmann equation is modelled in a single scattering approach. An electron interacts with a photon that was scattered from a place where the physical properties of the plasma were slightly different. The model allows us to construct the radiation field seen by the electron irrespective of the optical depth of the medium.

In that paper the force exerted on an electron in the plasma parcel is calculated as being due to a number of effects. The electron is moving and sees a dipole component in the radiation distribution, the photon with which it interacts comes from a part of the gas that was itself moving and at a different temperature and so on. The contribution from the motion of the electron through the radiation is given by

$$F = -4B(\tau_0)\, \sigma_T \rho_r c^2 \, \frac{v}{c}, \qquad \tau_0 = \frac{l}{l_\gamma} \tag{5.50}$$

where

$$B(x) = \frac{1}{3} - x^3 \left[x^{-1} - \arctan x^{-1}\right] \tag{5.51}$$

The interpolation formula given by (5.51) differs by no more than a few percent from the interpolating function used in (5.49).

The other terms in the Jones-Wyse treatment of the Boltzmann equation give the thermal conductivity and other transport coefficients.

5.5. *Thermal conductivity: interpolation*

The simplest guess for a thermal flux limiter suggests that we modify (5.39) to read

$$\dot{\tau} - \frac{1}{3}\dot{\delta} = -\frac{1}{1+\dfrac{l_\gamma^2}{l^2}} \psi^{-1} c l_\gamma \nabla^2 \tau \tag{5.52}$$

In the absence of any notion of thermal conduction in the optically thin limit there is no better guess than this. It is at least typical of the numerical adjustment that is made in numerical gasdynamic codes for no other reason than it works.

The effect of the correction is that $\nabla^2 \tau$ in the heat conduction terms becomes τ/l_γ^2 in the optically thin regime, and therefore decreases as the photon mean free path gets longer. Heat conduction therefore rapidly becomes inefficient as the universe becomes transparent and the value of the conduction length at the onset of recombination is likely to be a good estimate. This contrasts with the viscous damping which continues into the recombination period.

5.6. So what happens to the pressure term?

Jones and Wyse (1983) provide in their solution the term which gives rise in the optically thick regime to the radiation pressure gradient. As the optical depth falls to unity, the pressure gradient becomes less effective in driving the electrons and we can write for the oscillation velocity of a sound wave in this regime:

$$\ddot{u} - \frac{c^2}{3}\frac{H(\tau_0)}{1 + \dfrac{3\rho_m}{4\rho_r}}u = 0 \tag{5.53}$$

ignoring the dissipative terms. Here τ_0 is the optical depth of the wavelength. This equation is precisely equation (5.41) apart from the optical depth correction to the sound speed. The function $H(\tau)$ is an analytic function which is unity at large optical depth and provides the appropriate behaviour as the optical depth goes to zero:

$$H(x) \to \begin{cases} 1 & x \to \infty \\ \frac{3}{2}\pi x^3 & \tau \to 0 \end{cases} \tag{5.54}$$

For the sake of curiousity, the function $H(x)$ is

$$H(x) = 6x^3[1 - x\arctan(1/x)]\arcsin\sqrt{\frac{1}{2}\left(1 - \frac{x}{\sqrt{1+x^2}}\right)} \tag{5.55}$$

(compare equation (5.51)). This is rather unedifying, but there are a number of suitable interpolation formulae that fit this.

6. What have we learned from this?

When we look at the details of the angular distribution of the cosmic microwave background temperature we are looking in detail at the recombination process itself, or at least at its immediate aftermath. If we are to interpret correctly what is going on we need a clear understanding of the physical processes taking place at that time.

We have seen that as the Universe approaches the recombination epoch, dissipative processes are beginning to play a role on scales not unlike those characterising galaxies and galaxy clusters. This is one of the major reasons we see expect to see little structure in the cosmic microwave background on the smallest (sub arcminute) scales.

There is also the Sakharov Oscillations effect: we expect to see peaks in the variance of the temperature fluctuations at well defined angular scales. These angular scales depend on the details of the cosmological model, and so it is by observing these peaks that we hope to infer details of the underlying cosmology. We are interested in their position and their relative separation. Of course, the higher-order peaks will be attenuated by the damping processes we have just been discussing.

The smaller scales, below the first peak of the temperature correlation function, are the very scales we observe in the galaxy distribution today. It is therefore a direct link between what structure we see locally and the initial conditions that generated that structure.

So perhaps it is clear why the correlation function of the microwave background anisotropy could well be regarded as "the Signature of the Universe".

Finally, the approach taken here differs from the usual Peebles-Yu two-fluid approach. That should not be taken to imply any doubts that the Peebles-Yu approach is essentially correct, on the contrary it would be hard to imagine where that could go wrong! The approach taken here is a gasdynamic approach - very much the way one might run a simulation on a computer if one starts in a hydrodynamic as opposed to plasma regime. Indeed, rather much as people running supernova explosion simulations have done with relative success. It is a method for handling finite optical depth effects within the framework of optically thick plasmas.

It has been a privilege to present a talk on this occasion. The work reviewed here goes back over a long period of time and my understanding of this issue has benefited from discussions with innumerable colleagues. However, at the start of my interest in this problem, two people, Kwing-Lam Chan and Rosie Wyse played a key role. Not only did I write papers with them, but they forced me to clarify what was going on. There is no experience in academia like that of working with ultra-smart students!

I thank Michael Rowan-Robinson for the hospitality of his department during the period when this article was written.

REFERENCES

ALBRECHT, A., COULSON, D., FERREIRA, P., & MAGUEIJO, J. 1996 *Phys. Rev. Lett.* **76**, 1413.

ARTEMOVA, J.V., BISNOVATY-KOGAN, G.S., BJORNSSON, G. & NOVIKOV, I.D. 1996 *Astrophys. J.* **456** 119.

BONDI, H. & GOLD, T. 1948 *Mon. Not. Roy. astr. Soc.* **108**, 252.

CHAN, K-L. & JONES, B.J.T. 1975 *Astrophys. J.* **200**, 454.

DICKE, R.H., PEEBLES, P.J.E., ROLL, P.G. AND WILKINSON, D.T. 1965 *Astrophys. J.* **142**, 414.

DOROSHKEVICH, A.G. & NOVIKOV, I.D. 1964 *Dokl. Acad. Nauk. SSSR.* **154**, 754.

HOYLE, F. 1948 *Mon. Not. Roy. astr. Soc.* **108** 372

HUBBLE, E. 1929 *Proc. Nat. Acad. Sci.* **15**, 168.

JONES, B.J.T. & WYSE, R.F.G. 1983 *Mon. Not. R. astr. Soc.* **205**, 983.

JØRGENSEN, H.E., KOTOK, E., NASELSKY, P. & NOVIKOV, I.D. 1993 *Mon. Not. Roy. astr. Soc.* **265** 261.

JØRGENSEN, H.E., KOTOK, E., NASELSKY, P. & NOVIKOV, I.D. 1995 *Astron. & Astrophys.* **294** 639.

LEVERMORE, C.D. & POMRANING, G.C. 1981 *Astrophys. J.* **248**, 321.

MASAKI, I. 1971 *Pub. Astr. Soc. Japan* **25**, 425

NASELSKY, P.D. & NOVIKOV, I.D.1993 *Astrophys. J.* **413** 14.

PEEBLES, P.J.E. 1980 *The Large Scale Structure of the Universe* Princeton University Press

PEEBLES, P.J.E. & DICKE, R.H. 1968 *Astrophys. J.* **154**, 891.

PEEBLES, P.J.E. & YU, J.T. 1970 *Astrophys. J.* **162**, 815.

PENZIAS, A.A. & WILSON, R.W. 1965 *Astrophys. J.* **142**, 419.

ROBERTSON, H.P. 1928 *Phil. Mag.* **5**, 835.

SACHS, R.K. & WOLFE, A.M. 1967 *Astrophys. J.* **147**, 73.

SILK, J. 1968 *Astrophys. J.* **151**, 459.

SLIPHER, V.M. 1915 *Popular Astron.* **23**, 21.

THOMAS, L.H. 1930 *Quart. J. math.* **1**, 329

ZEL'DOVICH, YA.B. & NOVIKOV, I.D. 1983 *The Structure and Evolution of the Universe* University of Chicago Press, Chicago

Theory of the Large-Scale Structure

By SERGEI F. SHANDARIN

Department of Physics and Astronomy, University of Kansas, Lawrence, KS, 66045, USA

Among many important contributions in relativistic astrophysics and cosmology Igor Novikov has obtained very significant results in theory of the structure of the universe. Here I review the current state of the theory of the large-scale structure. It focuses primarily on understanding the distribution of galaxies on scales of clusters and superclusters of galaxies (roughly $1 - 100$ Mpc). Although a complete theory must account for the entire evolution of the inhomogeneities from their origin to the formation of the observed structures I will concentrate only on its relatively small but very important and fascinating part: the nonlinear dynamics of cosmological gravitational clustering. The quantity and quality of observational data on the large scale galaxy distribution is improving very rapidly inspiring active theoretical work in this field. For more than two decades modeling and explaining the geometrical and statistical properties of the galaxy distribution in space has played the role of a crucial test for cosmological scenarios.

1. Introduction

1.1. *Historical Remarks*

Without aiming at completeness I would like to begin with mentioning a few works by Igor Novikov that in my opinion have played a very important role in the development of modern picture of the formation of the large-scale structure in the universe.

Igor Novikov was the first (Novikov 1964) who challenged the inference of classical Lifshitz's paper (Lifshitz 1946): *"we can apparently conclude that gravitational instability is not the source of condensation of matter into separate nebulae"*. He argued that the gravitational instability could in principle explain observed structures but the primordial seed fluctuations were necessary. At present the most popular and best developed class of theories of the structure formation is based on the gravitational instability paradigm.

Together with Doroshkevich and Zel'dovich (1967) Igor Novikov suggested that the heating and cooling of the pregalactic matter plays a crucial role in galaxy formation. Modern theories of galaxy formation – the explosion scenario (Ostriker & Cowie 1981) and the CDM (Cold Dark Matter) model (Blumenthal et al 1984) – employ this idea although in a significantly modified form.

In 1980 Igor Novikov with collaborators (Doroshkevich et al 1980) pioneered a few ideas that are actively used in the modern studies of the large-scale structure:

• A new numerical technique (the Particle – Mesh method) was adopted from plasma physics and proposed for simulating the cosmological gravitational clustering.

• A new method of generating the Gaussian initial conditions (with an arbitrary spectrum) for the N-body simulations was suggested. At present it is a standard component of almost every cosmological N-body code. Before the 1980 paper the initial conditions had been generated as a random Poisson process (e.g. Aarseth et al 1979, Gott et al 1979) implying a flat spectrum as well as a nonlinear amplitude on the scale of the mean separation of particles.

• For the first time it was demonstrated that the Zel'dovich approximation agreed reasonably well with the fully gravitational N-body simulations using identical initial conditions. In particular, it was shown that the first collapse in 2D is filament-like.

• The phenomenon of gravitational adhesion was demonstrated and studied for the first time. It was shown that the pancakes formed in a collisionless medium remain quite thin in the comoving coordinates.

• The paper also revealed the formation of the *connected network* structure in the process of cosmological gravitational clustering.

Although all the results were obtained in the two-dimensional simulation, the three-dimensional generalisation was straightforward and a similar three-dimensional simulation was done fairly quickly (Klypin & Shandarin 1983). In this talk I briefly review the modern developments of these ideas.

1.2. *Cosmological Gravitational Clustering Paradigm*

Understanding the formation of the large-scale structure in the Universe is one of the most important problems in modern cosmology (see e.g. Zel'dovich & Novikov 1983, Peebles 1993; Padmanabhan 1993; Coles & Luchin 1995). It is generally believed that the seed density fluctuations were generated at the inflationary stage when the universe underwent a period of accelerated expansion † The most developed hypothesis suggests that the primordial perturbations were small amplitude Gaussian random fields that arose as vacuum fluctuations during the inflation. As the universe evolved, the density perturbations grew by the amplitude as well as underwent various damping and dissipative processes depending on the scale of the perturbation. The signature of the primordial inhomogeneities [in the form of the angular variations of the temperature of the cosmic microwave background radiation ($\delta T/T$)] has been observed on very large scales in the COBE experiment (Lubin 1996). The COBE detectors had no sufficient angular resolution for detecting the temperature fluctuations on the scales of galaxies or even on the scales of superclusters. However, there is no doubt that the ongoing and/or future microwave background observations will eventually probe the primordial fluctuations on 100 Mpc scales which are already probed by the deep redshift surveys (Hu et al 1996). For an exhaustive review of the relationship between the density fluctuations and the observations of the angular variations of $\delta T/T$ (see Bond 1992).

After decoupling of baryonic matter and radiation at the redshift $z \approx 1400$ the density fluctuations grew in the linear regime until the first objects formed. As soon as the density fluctuation on some scale reaches nonlinearity ($\delta\rho/\rho \geq 1$), the physical processes become prohibitively complex for analytical treatment in a generic case. In addition, on scales of galaxies complex nonlinear physical processes like turbulence, cooling and heating of the gas, star formation and supernova explosions, are assumed to determine the shape and other properties of objects (e.g. Blumenthal et al 1984). However, on scales of superclusters these processes become considerably less important and it is often assumed that primarily gravity shapes the structures. A variety of observational evidence supports the hypothesis that most of the mass in the Universe is in the form of weakly interacting particles that feel only gravity (see e.g. Kolb & Turner 1990). Therefore, a simple physical model assuming that the Universe is filled with a cold dust-like matter interacting only via gravity is most often used for studying the large-scale structure formation. The baryonic component plays no significant dynamical role except on the scale of the luminous parts of galaxies or smaller scales.

On scales of superclusters the distribution of galaxies is probably quite similar to the mass distribution although the distributions are not identical. It is generally accepted that the larger the scale the more similar are the mass and galaxy distributions in space.

While the amplitude of the density fluctuations is small, their growth is adequately

† Various aspects of the epoch of inflation are discussed in this volume: Kofman (1996); Kolb (1996); Turner (1996).

described by the linear theory. However, as soon as the amplitude of the perturbations reaches $\sigma_\delta \equiv< [(\rho - \bar{\rho})/\bar{\rho}]^2 >^{1/2} \approx l_\delta/l_v$ (where ρ is the density field and $\bar{\rho}$ its average, l_δ and l_v are the scales of the initial density and velocity fields respectively) the Eulerian linear theory breaks down (see e.g. Shandarin 1994). In particular, in the case of the power law initial spectrum $P(k) \propto k^n$ with $n < -1$ the Eulerian linear theory breaks down at $\sigma_\delta \simeq l_\delta/l_v \ll 1$. On the other hand, the Lagrangian linear theory that takes into account the displacement of matter is always qualitatively correct until $\delta \simeq 1$.

Formal perturbative theories both Eulerian and Lagrangian usually address an issue of small nonlinear corrections in the essentially linear regime ($\sigma_\delta \ll 1$) (Buchert et al. 1994; Bouchet et al 1995; Catelan 1995). At $\sigma_\delta \approx 1$ all orders in the perturbative series become comparable and a moderate gain in accuracy in a generic case is not worth the mathematical complexity of the higher orders. In addition, the perturbative expansions are probably asymptotic series, and their convergence ranges are poorly known. Thus, the results at $\sigma_\delta \approx 1$ must be taken with great caution (Sahni & Shandarin 1996).

The routinely used method to deal with the dynamics at the strongly nonlinear stage is to run N-body simulations. The initial condition is usually generated as a realisation of a Gaussian random process with a given spectrum using the method originally suggested by Doroshkevich et al 1980 (in 2D) and Klypin & Shandarin 1983 (in 3D). In the N-body simulations of this type the gravitational force induced by the density distribution is calculated at each time step. The trajectory of every particle is integrated in a self-consistently varying gravitational field. Knowing the positions and velocities of particles at a given time one can easily compute the mass density smoothed on a specified scale. Then applying a chosen set of criteria (the simplest is the density threshold) one can identify "galaxies" and "clusters of galaxies" and compute their velocities. Comparing the statistical properties of the galaxy distribution in the N-body simulation with observations one can reject or corroborate the model. The major practical problems of the N-body technique are related to the limited dynamical range of the simulations: the particles used to approximate the essentially continuous medium of dark matter are usually huge and the volume of the simulation often is not sufficiently large.

From the theorist's point of view, the N-body simulation method *itself* does not bring satisfactory understanding of the nonlinear gravitational dynamics since it operates on a *microscopic* level (an accurate calculation of the forces between particles and integration of their trajectories). One way of acquiring better understanding of the structure formation consists in constructing approximations. The major goal of an approximation remains to find *macroscopic* principles that govern to a good approximation the dynamics of nonlinear gravitational clustering.

Long ago Zel'dovich (1970) suggested extrapolating the linear Lagrangian theory into the initial stage of the significantly nonlinear regime ($\sigma_\delta \approx 1$ but $\delta \gg 1$). The Zel'dovich approximation proved to be the most successful technique in gaining understanding of the large-scale structure formation (Zel'dovich & Novikov 1983; Shandarin & Zel'dovich 89; Padmanabhan 1993; Sahni & Coles 1995). In particular, it suggests a very simple principle: in a special reference frame (see bellow) the velocities remain approximately constant all the time before orbit crossings. Originally the Zel'dovich approximation was applied to the cosmological scenarios having no small scale power in the initial spectrum, e.g. the Hot Dark Matter model. However, cosmological observations almost certainly exclude such scenarios. Small scale initial fluctuations are required to explain the existence of quasars and galaxies at very high redshifts.

In this talk I concentrate on the nonperturbative approximations based on the Zel'dovich Approximation (ZA): the Truncated Zel'dovich Approximation (TZA) (Coles, Melott & Shandarin 1993; Melott, Pelman & Shandarin 1994), the Adhesion Approximation

(AA) (Gurbatov, Saichev, & Shandarin 1989; Kofman,Pogosyan & Shandarin 90; Kofman et al 1992), the COnserving Momentum Approximation (COMA) (Shandarin & Sathyaprakash 96) and the Peak Patch Picture (PPP) (Bond & Myers 1996; Bond, Kofman & Pogosyan 1996). Recently many other approximations have been suggested. However, it is worth noting that the standard Zel'dovich approximation outperforms all suggested *nonperturbative* approximations in the weakly nonlinear regime (for an exhaustive review see Sahni & Coles 1995).

2. Dynamical Equations

Describing the evolution of density inhomogeneities and motions in an expanding Universe is convenient in comoving coordinates $\mathbf{x}$ and in terms of peculiar velocities, $\mathbf{v}_p$

$$\mathbf{x} = \frac{1}{a(t)}\mathbf{r}, \quad \mathbf{v}_p = \frac{d\mathbf{r}}{dt} - H(t)\cdot\mathbf{r} = a(t)\frac{d\mathbf{x}}{dt}, \tag{2.1}$$

where $a = a(t)$ is a scalar factor describing the uniform expansion of the Universe and $H(t) = \dot{a}/a$ is the Hubble parameter [$H(\text{present time}) = H_0 = 100\,h\,\text{km}\,\text{s}^{-1}\text{Mpc}^{-1}$, $0.5 \le h \le 1$]. To simplify the discussion we will assume the pressure to be negligible $p = 0$ and the cosmological term $\Lambda = 0$. In a homogeneous Universe the comoving coordinates of a particle remain constant and the peculiar velocity is zero.

Density $\rho(\mathbf{x}, t)$, peculiar velocities $\mathbf{v}_p(\mathbf{x}, t)$, and gravitational potential perturbations $\phi(\mathbf{x}, t)$ are coupled by three equations (see e.g. Peebles 1980,)

$$\frac{\partial\rho}{\partial t} + 3H\rho + \frac{1}{a}\nabla\cdot(\rho\mathbf{v}_p) = 0, \tag{2.2}$$

$$\frac{\partial\mathbf{v}_p}{\partial t} + \frac{1}{a}(\mathbf{v}_p\cdot\nabla)\mathbf{v}_p + H\mathbf{v}_p = -\frac{1}{a}\nabla\phi, \tag{2.3}$$

$$\frac{1}{a^2}\nabla^2\phi = 4\pi G(\rho - \bar{\rho}), \tag{2.4}$$

where $\bar{\rho} = \bar{\rho}(t)$ is the mean density of the universe at time t.

For the further discussion it is convenient to rewrite the above equations (2.2 - 2.4) in a different form (Gurbatov, Saichev, & Shandarin 1989). Changing the time variable from t to $D \equiv D_g(t)$ ($D_g(t)$ is the growing mode of the linear theory) and rescaling the density, peculiar velocity and perturbation of the gravitational potential

$$\eta = \rho\cdot a^3, \tag{2.5}$$

$$\mathbf{v} = \frac{1}{a\dot{D}}\mathbf{v}_p = \frac{d\mathbf{x}}{dD}, \tag{2.6}$$

$$\varphi = \left(\frac{3}{2}\Omega_0\dot{a}^2 D\right)^{-1}\phi. \tag{2.7}$$

one easily obtains

$$\frac{\partial\eta}{\partial D} + \frac{\partial(\eta\, v_i)}{\partial x_i} = 0, \tag{2.8}$$

$$\frac{\partial v_i}{\partial D} + v_k\frac{\partial v_i}{\partial x_k} = -\frac{3}{2}\frac{\Omega_0}{D\cdot f^2}\left(\frac{\partial\varphi}{\partial x_i} + v_i\right), \tag{2.9}$$

$$\frac{\partial^2\varphi}{\partial x_i^2} = \frac{\delta}{D}, \tag{2.10}$$

where $f(t) = d\ln D/d\ln a$, $\Omega_0 = 8\pi G\bar{\rho}_0/3H_0^2$ is the dimensionless mean density, and summation over dummy indices is assumed. In the Einstein–de Sitter universe ($\Omega(t) = 1$) $f \equiv 1$ and $\varphi = (\frac{3}{2}\dot{a}^2 a)^{-1}\phi = const(t)\cdot\phi$ (Kofman & Shandarin 1988; Kofman 1988). The term proportional to the velocity in the Euler equation (2.9) describes the drag force due to the uniform expansion of the universe.

3. Zel'dovich Approximation

The right hand side of (2.9) vanishes in the linear theory since $v_i = -\partial\varphi/\partial x_i$ to linear order that greatly simplifies the Euler equation making velocity independent of other variables

$$\frac{dv_i}{dD} \equiv \frac{\partial v_i}{\partial D} + v_k\frac{\partial v_i}{\partial x_k} = 0. \tag{3.11}$$

Zel'dovich suggested to extrapolate this condition until the orbit crossing. The solution of (3.11) — the Zel'dovich approximation — looks like an inertial motion of noninteracting medium

$$\mathbf{x}(\mathbf{q}, D) = \mathbf{q} + D\cdot\mathbf{s}(\mathbf{q}), \tag{3.12}$$

where $\mathbf{x}$ and $\mathbf{q}$ are the Eulerian and Lagrangian coordinates, respectively, $s_i(\mathbf{q}) \equiv v_i(\mathbf{x}, 0) \equiv -\partial\Phi/\partial q_i$, and $\Phi(\mathbf{q})$ is the potential of the initial velocity field. The form of equation (3.11) and its solution (3.12) sometimes provokes an erroneous conclusion that the Zel'dovich approximation ignores gravity. Actually, the kinematic form of (3.12) results from the remarkable fact that the gravitational force generated by the density *inhomogeneities* exactly (to linear order) balances the drag force caused by the *uniform* expansion of the universe!

The Zel'dovich approximation (3.12) can also be viewed as a one-step mapping of a homogeneous (Lagrangian) space at the initial time ($D \to 0$) into the Eulerian space at a 'time' D. With the aid of the above mapping one can calculate the density at the final time D using the conservation of mass

$$\eta(\mathbf{x}, D)d^3x = \bar{\eta}d^3q. \tag{3.13}$$

In terms of the eigenvalues $\lambda_1(\mathbf{q})$, $\lambda_2(\mathbf{q}$ and $\lambda_3(\mathbf{q})$ of the initial deformation tensor field $d_{jk}(\mathbf{q}) = -\partial s_j/\partial q_k = \partial^2\Phi/\partial q_j\partial q_k$, the density can be expressed as

$$\eta(\mathbf{q}, D) = \frac{\bar{\eta}}{(1 - D\lambda_1)(1 - D\lambda_2)(1 - D\lambda_3)}, \tag{3.14}$$

thus predicting complex geometrical structures, the simplest of which are pancakes (Zel'dovich 1970; Arnold, Shandarin, & Zel'dovich 1982). The discussion of the shapes of the first collapsed objects can be found in Shandarin et al 1995 and Gurevich & Zybin (1995).

The velocity field remains constant in the Lagrangian space

$$\mathbf{v}(\mathbf{q}, D) = \frac{d\mathbf{x}}{dD} = \mathbf{s}(\mathbf{q}). \tag{3.15}$$

To find the density distribution in the Eulerian space one has to make use of eq. (3.12).

Summarising, the Zel'dovich approximation states that the motion of matter before the orbit crossing (in the single-stream flow regions) obeys a simple principle that can be expressed in two equivalent ways:

- the velocity equals the gravitational acceleration $v_i = -\partial\varphi/\partial q_i$,
- the velocity of each particle remains constant $v_i(\mathbf{q}, D) = s_i(\mathbf{q}) = -\partial\Phi/\partial q_i$.

Obviously there is no contradiction between the first and second formulations since a *non-inertial* reference frame has been used. The fact that the comoving coordinates $(\mathbf{x}, D)$ make a non-inertial reference frame becomes obvious if one considers a uniform universe: every particle in a Friedmann universe experience gravitational acceleration but in the comoving coordinates all particles are at rest.

3.1. *Cosmological Applications of ZA*

The Zel'dovich approximation was originally suggested for the adiabatic scenario of the structure formation in the baryon dominated universe (Zel'dovich & Novikov 1983). In 1980 it was adopted for the structure formation in the Hot Dark Matter scenario assuming that the dark matter was in the form of neutrinos of some sort with masses about $30eV$ (for review see Shandarin, Doroshkevich, & Zel'dovich 1983). In both models the spectrum in the linear regime has a sharp cutoff on the scale of superclusters ($\sim 10^{14} M_\odot$, here $M_\odot \approx 2 \times 10^{33} g$ is the mass of the Sun). As a result the first objects to form are the Zel'dovich pancakes of the corresponding sizes. Galaxies ($\sim 10^{11} M_\odot$) were supposed to form in the process of fragmentation of the pancakes. This picture is often referred to as a top-down scenario.

Unfortunately, the Zel'dovich approximation breaks down after the formation of the first caustics (pancakes) within the multistream flow regions. As a result the thickness of pancakes grows unbounded in the Zel'dovich approximation, which runs counter to the findings of N-body simulations showing that the thickness of pancakes remains considerably smaller than both the pancake diameters as well as the mean distance between pancakes (Doroshkevich et al 1980). This feature of the Zel'dovich approximation caused a serious problem in applying it to more realistic cosmological scenarios. For instance, in the Cold Dark Matter model the clumps with masses around $10^6\ M_\odot$ form first, then they cluster into galaxies ($\sim 10^{11} M_\odot$) in a hierarchical process, and finally clusters and superclusters of galaxies ($\sim 10^{14} M_\odot$) are assembled from galaxies. The process is known as the bottom-up scenario.

One way to deal with this problem consists in introducing smoothing of the initial conditions similar to that used in the standard peak theory (Press & Schechter 1974; Bardeen et al 1986).

4. Truncated Zel'dovich Approximation

The N-body simulations (Beacom et al 1991; Little et al 1991; Melott & Shandarin 93) as well as the adhesion approximation (Kofman et al. 1992) showed that the small-scale initial perturbations do not change much the overall large-scale distribution of mass although they make the appearance of the large structures clumpy. For every epoch the border between small and large scales can be roughly defined as the scale of nonlinearity $k_{n\ell}$ separating the scales in the linear (or quasi-linear) regime

$$\delta_{rms}(L) < 1 \quad \text{if} \quad L > k_{n\ell}^{-1} \tag{4.16}$$

from the scales in the strongly nonlinear regime

$$\delta_{rms}(L) > 1 \quad \text{if} \quad L < k_{n\ell}^{-1}. \tag{4.17}$$

The scale of nonlinearity $k_{n\ell}^{-1}$ can be found by solving the equation

$$\delta_{rms}^2(k_{n\ell}^{-1}) = 4\pi \cdot D^2 \int_0^{k_{n\ell}} P(k) k^2 dk = 1. \tag{4.18}$$

The truncated Zel'dovich approximation is simply the standard Zel'dovich approximation applied to the initial conditions smoothed with the scale of nonlinearity $k_{n\ell}(D)$; the time dependence of the scale of nonlinearity is determined by the initial spectrum and the cosmological model (Coles, Melott, & Shandarin 1993). Melott, Pelman & Shandarin (1994) showed that smoothing with a Gaussian filter slightly improves the approximation.

Making gravitationally bound clumps of mass assumes the transport of mass from one place to another. Therefore one can calculate and/or measure the rms displacement of mass in the process of building up the structure. Assuming that the displacements corresponding to the short wave perturbation $k \geq k_{n\ell}$ are averaged out in the clustering process one can give an analytic estimate of the rms displacement when the nonlinearity reaches a scale $k_{n\ell}$ (Shandarin 1993)

$$d_{rms}^2 = \frac{\int_0^{k_{n\ell}} P(k)dk}{\int_0^{k_{n\ell}} P(k)k^2dk}. \tag{4.19}$$

Since in the hierarchical clustering process the most of mass is in the clumps corresponding to the scale of nonlinearity ($M_{cl} \sim \bar{\rho} k_{n\ell}^{-3}$) d_{rms} also represents the average displacement of clumps. For instance, in the Cold Dark Matter model $d_{rms} \approx 9\ h^{-1} Mpc$.

In the case of the power law initial spectra

$$P(k) \propto k^n \ \ (k_1 \leq k \leq k_2), \tag{4.20}$$

one easily finds that

$$d_{rms} \sim k_{n\ell}^{-1}, \ \text{if} \ \ n > -1 \tag{4.21}$$

and

$$d_{rms} \sim \left(\frac{k_1}{k_{n\ell}}\right)^{\frac{n+1}{2}} k_{n\ell}^{-1}, \ \text{if} \ \ -3 < n < -1\ . \tag{4.22}$$

In the former case the characteristic displacement of mass roughly corresponds to the characteristic mass of nonlinear clumps $d_{rms} \sim k_{n\ell}^{-1}$, which intuitively is very clear. In the latter case d_{rms} can be easily much greater than $k_{n\ell}^{-1}$: $d_{rms} \gg k_{n\ell}^{-1}$, if $k_1 \ll k_{n\ell}$. If the characteristic displacement d_{rms} is greater than the typical distance between clumps it obviously means that the clumps themselves are displaced coherently by the distance d_{rms}. The perturbations having scales between the scale of nonlinearity $k_{n\ell}^{-1}$ and d_{rms} are in the quasi–linear regime: on the one hand the perturbations have a small amplitude $\delta(\ell) < 1$ ($k_{n\ell}^{-1} < \ell < d_{rms}$) and therefore must be described well by the linear theory, but on the other hand they are displaced from their initial positions by the distance $d_{rms} > k_{n\ell}^{-1}$ and the spatial structure must be different from the initial (Gaussian) field. This explains why the Eulerian linear theory can break down long before $\sigma_\delta \approx 1$.

It is interesting but not surprising that the scale at which the phases of the Fourier components can depart significantly from their original (linear) values is close to d_{rms} (Ryden & Gramann 1991).

4.1. *Cosmological Applications of TZA*

TZA proved to be a very effective instrument for the statistical studies of the distribution of galaxy clusters in space. The computation of the cluster correlations (Borgani et al 1994) and the moments of the cluster distribution (Plionis et al 1995) put important constraints on many cosmological models. Being computationally very fast the truncated Zel'dovich approximation allows to study much larger statistical ensemble of the models than the standard N-body technique.

The truncated Zel'dovich approximation has been used for analytic calculations of the nonlinear spectrum both in the Hot Dark Matter and Cold Dark Matter models (Schneider & Bartelmann 1995). As the authors stress this approach has an advantage of being free from the sampling problem as well as from the problem of the finite dynamic range.

Pioneered by Doroshkevich in 1970, the Zel'dovich approximation in the truncated form is now actively used for analytic calculations of the probability distribution functions in a weakly nonlinear regime (Padmanabhan & Subramanian 1993; Kofman et al 1994; Bernardeau & Kofman 1995).

A variant of TZA in the form of a differential equation in the Eulerian space — the Zel'dovich-Bernoulli equation — has been used in the POTENT procedure of reconstructing the density field from the peculiar velocity field (Nusser & Dekel 1992).

5. Adhesion Approximation

Another way of stabilising the thickness of pancakes is mock viscosity that mimics the effects of nonlinear gravitational dynamics in the multistream flow regions. In the Lagrangian linear theory and Zel'dovich approximation the right hand side of the Euler equation (2.9) vanishes. Constructing the adhesion approximation (see e.g. Gurbatov, Saichev & Shandarin 1989) added an artificial viscosity term into the Euler equation

$$\frac{\partial v_i}{\partial D} + v_k \frac{\partial v_i}{\partial x_k} = \nu \, \nabla^2 v_i, \qquad (5.23)$$

which became the three-dimensional Burgers equation for this particular choice of the viscosity. For potential flows equation (5.23) has an exact analytical solution for the velocity field

$$\mathbf{v}(\mathbf{x}, D) = \frac{\int d^3q \, \frac{(\mathbf{x}-\mathbf{q})}{D} \, \exp\left(\frac{1}{2\nu} S(\mathbf{q}, \mathbf{x}, D)\right)}{\int d^3q \, \exp\left(\frac{1}{2\nu} S(\mathbf{q}, \mathbf{x}, D)\right)} \qquad (5.24)$$

where

$$S(\mathbf{q}, \mathbf{x}, D) = \Phi(\mathbf{q}) - \frac{(\mathbf{x}-\mathbf{q})^2}{2D}, \qquad (5.25)$$

Equation (5.24) has an elegant geometrical interpretation which can be summarized as follows: whether or not a particle originally located at $\mathbf{q}$ is stuck within a pancake, filament or clump can be found by descending a paraboloid $P(\mathbf{q}; \mathbf{x}, D)$ with radius of curvature $2D$ and height P_0 onto the initial velocity potential Φ, in such a manner as to be tangential to the potential at $\mathbf{q}$. (The height of the paraboloid P_0 is a free parameter uniquely determined by the condition that $P(\mathbf{q}, \mathbf{x}, D)$ osculates $\Phi(\mathbf{q})$ at $\mathbf{q}$. If the paraboloid so constructed touches or intersects the potential at any other point $\mathbf{q}_1 \neq \mathbf{q}$, then we say that the particle in question has already entered a caustic, otherwise it has not. Thus, one can divide the Lagrangian space at any given time into *stuck* and *free* regions. Stuck regions correspond to particles already in caustics, whereas free regions are still expanding via the Zel'dovich relations (3.12) and correspond to voids. The apex of the paraboloid $\mathbf{x}$, which osculates the potential in two points (without intersecting elsewhere) describes the location of the pancakes, in three points - the filaments, and in four points - the compact clumps of matter in the Eulerian space. The geometrical technique gives the skeleton of the large-scale structure – a geometrical construction showing the positions of clumps, filaments and pancakes.

Another method of solving Burgers' equation developed by Weinberg & Gunn (1990b) allows to calculate the particle distributions similar to the N-body simulations.

The adhesion approximation like the Zel'dovich approximation assumes that before entering the multistream flow regions particles move along straight lines with a constant speed. After entering the multistream flow region the trajectory of a particle become a random walk which can be approximately described by Burgers' equation of the nonlinear diffusion.

The adhesion approximation was thoroughly compared with N-body simulations and proved to be quite accurate in general description of the nonlinear stage (Kofman, Pogosyan & Shandarin 1990; Weinberg & Gunn 1990b; Nusser & Dekel 1990; Kofman et al. 1992; Melott, Shandarin, & Weinberg 1994)

5.1. *Cosmological Applications of AA*

The adhesion approximation has played an important role in establishing the universality of the cellular and network structures in cosmological gravitational clustering (Gurbatov, Saichev, & Shandarin 1989; Nusser & Dekel 1990; Kofman et al. 1992). The statistical analysis of nonlinear density distributions in an ensemble of N-body simulations of the power law models (the initial spectrum is assumed to be a simple power law $P_{in} \propto k^n$) with $n = 1, 0, -1, -2$ has confirmed the inferences from AA (Yess & Shandarin 1996). In the early 1990s the AA simulations (Weinberg & Gunn 1990a,b) were the largest simulations of the large-scale galaxy distributions. In particular, they revealed spectacular 'spherical shells' in the galaxy distribution in redshift space. They were later observed in the deep redshift surveys: first probably detected by Broadhurst et al. (1990) and then observed by Landy et al. (1996).

Recently, the adhesion approximation has been used as a dynamical model for the analytic calculations of the mass distribution function (Vergassola et al 1994; Cavaliere et al. 1996). Sahni, Sathyaprakash & Shandarin (1994) have used AA for the study of the statistics of voids in the Cold Dark Matter model. In particular, they predict that voids can be populated by substructure such as mini-pancakes and mini-filaments.

The adhesion approximation stresses the importance on the initial gravitational potential as a primary quantity determining some properties of the large-scale structure (Kofman & Shandarin 1988; Kofman 1988). Sahni, Sathyaprakash & Shandarin (1994) showed that the size of voids strongly correlates with the value of the primordial gravitational potential at void centers.

The major limitation of applying AA to the realistic cosmological models seems to do with the lack of simple efficient numerical codes. Both the geometrical code (Kofman, Pogosyan & Shandarin 1990) and the particle code (Weinberg & Gunn 1990b) are relatively complex. In the next section I will discuss a new very simple and extremely efficient numerical technique to simulate a slightly modified adhesion model.

6. COnserving Momentum Approximation (COMA)

As in the Zel'dovich and adhesion approximations this approximation assumes that the gravitational force exerted on a particle before orbit crossing is approximated by the velocity ($\partial\varphi/\partial x_i \approx -v_i(\mathbf{x}, D)$) which sets the right hand side of the Euler equation (2.9) to zero. On small scales the approximation assumes the collision of particles resulting in local exchange of the generalised momentum defined as

$$\mathbf{p} \equiv \eta \mathbf{v} = \frac{a^2}{\dot{D}} \rho \mathbf{v}_p, \tag{6.26}$$

which mimics the effects of gravity in dense regions (Shandarin & Sathyaprakash 1996). In contrast with the adhesion approximation COMA assumes physical viscosity described by the equation similar to the Navier-Stokes equation (here we ignore the 2nd viscosity)

$$\frac{\partial v_i}{\partial D} + v_k \frac{\partial v_i}{\partial x_k} = \frac{1}{\eta} \frac{\partial}{\partial x_k} [\mu (\frac{\partial v_i}{\partial x_k} + \frac{\partial v_k}{\partial x_i} - \frac{2}{3} \delta_{ik} \frac{\partial v_l}{\partial x_l})]. \tag{6.27}$$

where μ is the dynamical viscosity.

The qualitative change of the type of the particle trajectory after entering a multistream region (from a straight line to a random walk) is mainly caused by the gravitational interaction with the other particles within the same region of multistream flow. Therefore this interaction must conserve the momentum on the scale of the multistream flow region. Thus the major difference of the COMA with the adhesion approximation in the form of Burgers' equation consists in that it conserves the momentum locally while the standard adhesion approximation does not.

The numerical code realising the model assumes that the collisions of the particles are perfectly inelastic† and operates with the density and velocity distributions on a cubic grid in the Eulerian space: $\eta = \eta(\mathbf{x}, D)$ and $\mathbf{v} = \mathbf{v}(\mathbf{x}, D)$. At each time step it calculates the flows of mass and momentum from a mesh cell to its neighbors and computes the new density and velocity fields on the grid using a cloud-in-cell algorithm thereby explicitly conserving the mass and the generalised momentum.

The major features of the COMA code are:

- it is *universal* in terms of the initial and boundary conditions as well as the shape of the computational box,
- *extremely simple* both conceptually and practically,
- *very economical* in terms of memory – just four arrays, each as big as the size of the box, are sufficient to implement the algorithm,
- *very fast*,
- local and therefore *100% parallelisable.*

In the system of coordinates employed, particles evolve simply by conserving the mass and momentum; as a consequence one circumvents the problem of having to compute the gravitational force after each time step. Generalisation of the algorithm to deal with particles, rather than densities, is straightforward but would require additional memory.

The first comparison of the COMA with N-body simulations (Shandarin & Sathyaprakash 1996) shows that the vast majority of density peaks in the N-body simulations have their counterparts in the approximate model often shifted by a couple of mesh sites and of slightly lower amplitudes. Comparisons of the power spectra show that the COMA does not have enough power on small scales which is obviously related to a relative smoothness of the density fields in the approximate model. Similarly, the density distribution functions show that the high densities ($\rho > 10 - 30$ depending on the stage) are less abundant in the approximate model.

7. Peak Patch Picture

A model labeled as the Peak Patch Picture is an attempt to unify the Zel'dovich approximation with the standard peak theory (Bond & Myers 1996). It splits the initial displacement field (proportional to the initial velocity field) into a background field smoothed over a large scale and a fluctuating small-scale field. The background field represents the initial condition for the Zel'dovich approximation similar but not identical to the truncated Zel'dovich approximation. The major difference is that TZA assumes that the scale of smoothing is universal throughout the space and close to the scale of nonlinearity $k_{n\ell}^{-1}$ (Coles, Melott, & Shandarin 1993; Melott, Pelman & Shandarin 1994); by contrast, PPP uses quite an elaborate scheme employing the locally adaptive filter that selects the patches – progenitors of the mass clumps. The small-scale dynamics (within the patches) is described by the homogeneous ellipsoidal collapse model that

† A generalisation to a partially inelastic collisions is obvious and straightforward.

has been shown to be superior of the Zel'dovich approximation. Contrary to the previously discussed approximations, PPP addresses the issue of the internal dynamics of the clumps which is certainly a step forward.

The model has been tested against the cosmological N-body simulations and showed an adequately close agreement.

7.1. *Cosmological Applications of PPP*

Similarly to the truncated Zel'dovich approximation the peak-patch method has been used for the studies of clusters of galaxies. However, the better treatment of the small-scale dynamics has allowed to study the X-ray luminosity of the clusters. The model has been used for making deep field X-ray and Sunyaev-Zel'dovich maps. Combined with X-ray temperature data the model puts constraints on the amplitude and the slope of the primordial spectrum of perturbations.

Another interesting application of PPP is the study of superclusters and the large-scale structure as a whole (Bond, Kofman, & Pogosyan 1996). The model introduces a new probabilistic interpretation of filaments as bridges of the Zel'dovich-mapped correlation functions constrained by shear at two high density peaks – progenitors of the clusters of galaxies.

8. Discussion

At present there is no exact analytic solution to the problem of the gravitational clustering in the universe even in the frame of the simplest mathematical model discussed above. The equations are nonlinear and realistic initial conditions are random (probably Gaussian) fields having in addition a rather complex spectrum. However, the nonlinear approximations outlined in the talk give a very good qualitative (in some respects also quantitative) description of the nonlinear stage.

The truncated Zel'dovich approximation is the simplest nonlinear model which reproduce the overall density distribution on large scales. The advantage of the approximation is its simplicity and the local character. However, it does not reproduce the clumpy distribution within large voids which is qualitatively incorrect for initial spectra having small scale power. However, it can be successfully used for studies of the distribution of the clusters of galaxies, which does not require a resolution better than $5 - 10\ h^{-1}Mpc$ (Borgani et al 1994; Plionis et al 1995).

Both the adhesion approximation and COMA produce much clumpier distributions than the truncated Zel'dovich approximation. This is reflected in better correspondence of the power spectrum and density distribution function to the N-body simulations. However, in the particle realisation (Weinberg & Gunn 1990b) of the adhesion model the density distribution is still smoother than in the N-body simulations (see also Melott, Shandarin, & Weinberg 1994).

The adhesion approximation based on Burgers' equation has an advantage of admitting an analytic solution (5.24) for generic initial conditions. However, its practical implementation is not very easy. Also the viscosity term does not describe physical viscosity and the model does not conserve linear momentum.

COMA probably is physically slightly better approximation because it conserves momentum on the scale of the multistream flow regions as well as globally. It also better suits practical cosmological applications since the the code is simple and effective but it does not have an analytical solution.

A great change in the character of the motion of a particle after entering the multistream flow region leaves very little chance that the perturbative theory will be able to describe this transition. Thus, some sort of statistical approach is probably unavoidable. The adhesion approximation and COMA are attempts to do that.

The peak-patch picture method combines the Zel'dovich approximation and the peak theory. Its advantage is a better description of the gravitational dynamics within the clumps but at the cost of considerably greater complexity.

All above approximation can be also viewed as the smoothing procedures in the *Lagrangian* space followed by the mapping into the Eulerian space. The truncated Zel'dovich approximation represents smoothing with a spatially uniform filter that roughly equals the scale of nonlinearity (4.18). Other approximations use adaptive filters of different types. After filtering the dynamical mapping into the Eulerian space is made by means of the Zel'dovich equation (3.12).

The analytic and semi-analytic models of nonlinear gravitational clustering based on the Zel'dovich approximation have recently reached the level when they can explain qualitatively (in some cases quantitatively) major features of the large-scale galaxy distribution. In particular, these models predict that highly anisotropic shapes of superclusters (filaments and sheets) as well as their connectedness are natural generic consequences of cosmological gravitational clustering resulting from primordial density fluctuations of Gaussian type. The difference between different scenarios is mainly quantitative. Deriving the observational predictions from the models is a challenge for theorists. Ironically enough, history shows that they did not make many interesting observational predictions from simple Zel'dovich's theory even though the theory itself (properly applied) allowed those predictions. Measuring the statistical parameters of the large-scale structure in existing and especially upcoming large redshift surveys (e.g. Sloan Digital Sky Survey Gunn & Weinberg 1995) and comparing them with the theoretical predictions will be an exciting field of cosmological research in coming years. Igor Novikov's pioneer works in this field have been greatly responsible for this inspiration.

I thank Lev Kofman, for many discussions of the issues related to this talk. I am grateful for financial support from NASA Grant NAGW-3832, NSF Grants AST-9021414, the University of Kansas GRF 96 grant, and in particular to the organizers of the meeting.

REFERENCES

AARSETH, S.J., GOTT, J.R., & TURNER, E. 1979 *Astrophys. J.* **288**, 664.

ARNOLD, V.I., SHANDARIN, S.F., & ZEL'DOVICH, YA.B. 1982 *Geophys. Astrophys. Fluid Dyn.* **20**, 111.

BARDEEN, J.M., BOND, J.R., KAISER, N., & SZALAY, A.S. 1986 *Astrophys. J.* **304**, 15.

BEACOM, J., DOMINIK, K., MELOTT, A.L., PERKINS, S., & SHANDARIN, S.F. 1991 *Astrophys. J.* **372**, 351.

BERNARDEAU, F., & KOFMAN, L. 1995 *Astrophys. J.* **443**, 479.

BLUMENTHAL, G.R., FABER, S.M., PRIMACK, J.R., & REES, M.J. 1984 *Nature* **311**, 517.

BOND, J.R. 1992 *in Proc. of the IUCAA Dedication Ceremonies* Pune, India, Dec. 28-30, 1992, ed. T.Padmanabhan, Wiley.

BOND, R., & MYERS, S. 1996 *Astrophys. J. Suppl.* **103**, 1.

BOND, R., KOFMAN, L., & POGOSYAN, D. 1996 *Nature* **380**, 603.

BORGANI, S., COLES, P., & MOSCARDINI, L. 1994 *MNRAS* **271**, 223.

BOUCHET, F.R., COLOMBI, S., HIVON, E., & JUSZKIEWICZ, R. 1995 *Astron. Astrophys.* **296**, 575.

Broadhurst, T.J., Ellis, R., Koo, D., & Szalay, A. 1990 *Nature* **343**, 726.

Buchert, T., Melott, A.L., & Weiss, A.G. 1994 *Astron. Astrophys.* **288**, 349.

Catelan, P. 1995 *MNRAS* **276**, 115.

Cavaliere, A., Menci, N., & Tozzi, P. 1996 *Astrophys. J.* in press.

Coles, P., & Luchin, F. 1995 *Cosmology: The Origin and Evolution of Cosmic Structures* John Wiley & Sons, Chichester.

Coles, P., Melott, A., & Shandarin, S.F. 1993 *MNRAS* **260**, 765.

Doroshkevich, A.G. 1970 *Astrophysica* **6**, 320.

Doroshkevich, A.G., Zel'dovich, Ya.B., & Novikov, I.D. 1967 *Sov. Astron.* **11**, 233.

Doroshkevich, A.G., Kotok, E., Novikov, I.D., Polyudov, A., Shandarin, S.F., & Sigov, Yu. 1980 *MNRAS* **192**, 321.

Gott, J.R., Turner, E., & Aarseth, S. 1979 *Astrophys. J.* **234**, 13.

Gunn, J., & Weinberg, D. 1995 *in Wide-Field Spectroscopy and the Distant Universe*, eds. S.J.Maddox ans A. Aragón – Salamanca, World Scientific.

Gurbatov, S.N., Saichev, A.I., & Shandarin S.F. 1989 *MNRAS* **236**, 385.

Gurevich, A.V., & Zybin, K.P. 1995 *Sov. Phys. Usp.* **165**, 723.

Hu, W., Sugiyama, N., & Silk, J. 1996 *Nature* to be published.

Klypin, A., & Shandarin, S.F. 1983 *MNRAS* **204**, 891.

Kofman, L. 1988 *in Morphological Cosmology* Proc. of the XIth Cracow Cosmological School, eds. P.Flin and H.W. Duerbeck.

Kofman, L.A. 1996 *this volume.*

Kofman, L., & Shandarin, S.F. 1988 *Nature* **334**, 129.

Kofman, L., Pogosyan, D., & Shandarin, S.F. 1990 *MNRAS* **242**, 200.

Kofman, L., Pogosyan, D., Shandarin, S.F., & Melott, A.L. 1992 *Astrophys. J.* **393**, 450.

Kofman, L., Bertschinger, E., Gelb, J.M., Nusser, A., & Dekel, A. 1994 *Astrophys. J.* **420**, 44.

Kolb, R. 1996 *this volume.*

Kolb, R., & Turner, M. 1990 *The Early Universe* Addison-Wesley Publ. Comp.

Landy, S.D., Shectman, S.A., Lin, H., Kirshner, R.P., Oemler, A.A., & Tucker, D. 1996 *Astrophys. J. Lett.* **456**, L1.

Lifshitz, E.M. 1946 *ZhETF* **16**, 587.

Little, B., Weinberg, D.H., & Park, C. 1991 *MNRAS* **253**, 295.

Lubin, Ph. 1996 *this volume.*

Melott, A.L., & Shandarin, S.F. 1993 *Astrophys. J.* **410**, 469.

Melott, A.L., Pelman, T., & Shandarin, S.F. 1994 *MNRAS* **269**, 626.

Melott, A.L., Shandarin, S.F., & Weinberg, D.H. 1994 *Astrophys. J.* **428**, 28.

Novikov, I.D. 1964 *ZhETF* **46**, 686.

Nusser, A., & Dekel, A. 1990 *Astrophys. J.* **362**, 14.

Nusser, A., & Dekel, A. 1992 *ApJ* **391**, 443.

Ostriker, J. & Cowie, L. 1981 *Astrophys. J.* **243**, L127.

Padmanabhan, T. 1993 *Structure Formation in the Universe* Cambridge University Press, Cambridge.

Padmanabhan, T., & Subramanian, K. 1993 *Astrophys. J.* **410**, 482.

Peebles, P.J.E. 1980 *The Large-Scale Structure of the Universe*, Princeton University Press, Princeton.

Peebles, P.J.E. 1993 *Principles of Physical Cosmology*, Princeton University Press, Princeton.

Plionis, M., Borgani, S., Moscardini, L., Coles, P. 1995 *Astrophys. J.* **441**, L57.

Press, W., & Schechter, P. 1974 *Astrophys. J.* **187**, 425.

Ryden, B., & Gramann, M. 1991 *Astrophys. J. Lett.* **383**, L33.

Sahni, V., & Coles, P. 1995 *Physics Reports* **262**, 1.

Sahni, V., & Shandarin, S.F. 1996 *MNRAS* to be published.

Sahni, V., Sathyaprakash, B., & Shandarin, S.F. 1994 *Astrophys. J.* **431**, 20.

Schneider, P., & Bartelmann, M. 1995 *MNRAS* **273**, 475.

Shandarin, S.F. 1993 *in Cosmic Velocity Fields* eds. F.R.Bouchet and M. Lachièze-Rey, Edition Frontieres, p.383.

Shandarin, S.F. 1994 *Physica D* **77**, 342.

Shandarin, S.F. & Sathyaprakash, B. 1996 *Astrophys.J. Lett.* to be published.

Shandarin, S.F., & Zel'dovich, Ya.B. 1989 *Rev. Mod. Phys.* **61**, 185.

Shandarin, S.F., Doroshkevich, A.G., & Zel'dovich, Ya.B. 1983 *Sov. Phys. Usp.* **26**, 46.

Shandarin, S.F., Melott, A.L., McDavitt, K., Pauls, J.L., & Tinker, J. 1995 *Phys. Rev. Lett.* **75**, 7.

Turner, M. 1996 *this volume.*

Vergassola, M., Dubrulle, B., Frisch, U., & Noullez, A. 1994 *Astron. Astrophys.* **289**, 325.

Weinberg, D.H., & Gunn, J. 1990a *Astrophys.J. Lett.* **352**, L25.

Weinberg, D.H., & Gunn, J. 1990b *MNRAS* **247**, 260.

Yess, C., & Shandarin, S.F. 1996 *Astrophys.J.* **465**, July 1 issue.

Zel'dovich, Ya. B. 1970 *Astron. Astrophys.* **5**, 84.

Zel'dovich, Ya.B. & Novikov, I.D. 1983 *The Structure and Evolution of the Universe* University of Chicago Press, Chicago, London.

The Origin of Matter in the Universe: Reheating after Inflation

By LEV A. KOFMAN

Institute for Astronomy, University of Hawaii, 2680 Woodlawn Dr., Honolulu, HI 96822, USA

In the inflationary scenario all the matter constituting the universe was created in the process of reheating after inflation. Recent development of the theory of reheating is briefly reviewed. The list of topics discussed includes elementary (perturbative) theory of reheating; quantum field theory in a time-varying background; parametric resonance and explosive particle creation; non-thermal phase transitions from reheating; baryogenesis from reheating; residual oscillations of the scalar field, and other cosmological applications.

1. Introduction

As a graduate student, I studied particle creation from strong gravitational and electric fields. I was impressed by a paper which Ya. Zel'dovich (1972) devoted to J. Wheeler's 60th birthday volume. In particular, Zel'dovich quoted there the "Novikov paradox": let an electron-positron pair be created in a strong electric field. The total momentum of the pair must be equal to zero, but the total energy is not zero. Query: how can this result be Lorentz-invariant? Answer: the result indeed is Lorentz-invariant, if the space-time interval between the events of the electron and positron creation is space-like. Hence, on the classical level, the two events are causally independent. The "Novikov paradox" is a fair example of how Igor treats complex physical problems such as quantum creation of particles. Therefore, taking occasion to contribute to Igor Novikov's 60th birthday volume, I would like to consider another application — indeed, one of the most spectacular ones — of the quantum theory of particle creation that emerges together with the inflationary scenario. In this scenario almost all matter constituting the Universe at the subsequent radiation-dominated stage was created through reheating after inflation.

The term "reheating" is an anachronism inherited from the first inflationary models in which the Universe was hot before inflation and was (re-)heated again after inflation. In modern versions of inflationary cosmology the pre-inflationary hot stage is no longer necessary (Linde 1990). It is assumed that the Universe initially expands quasi-exponentially in a vacuum-like state with a vanishing entropy and particle number density. At the stage of inflation, all energy was concentrated in a classical slowly moving inflaton field ϕ. Soon after the end of inflation, when an observable universe was the size of a dime, the inflaton field began to oscillate near the minimum of its effective potential $V(\phi)$. An almost homogeneous inflaton field $\phi(t)$ coherently oscillated with a very large amplitude of the order of the Planck mass $\phi \sim M_p$. The interaction of the inflaton field with other elementary particles led to creation of many ultra-relativistic particles from the classical inflaton oscillations. Gradually, the inflaton field decayed and transferred all its energy non-adiabatically to the created particles. They interacted with each other and came to a state of thermal equilibrium at some temperature T_r, which was called the reheating temperature. An enormous total entropy of the observable universe (or the total number of particles inside the horizon), $S \sim 10^{88}$, was thus produced from the reheating after inflation.

The reheating is an intermediate stage between the inflation and the radiation dominated Friedmann expansion. Therefore the reheating is associated with a number of issues of the Big Bang scenario: the large entropy problem, the baryogenesis problem, the problem of relic monopoles, the problem of topological defects, the primordial black holes problem, etc.

The elementary theory of reheating, based on perturbation theory, was developed right after the first models of inflation were suggested. Various aspects of the theory of reheating were further elaborated by many authors. The elementary theory found its place in the textbooks on inflation (Linde 1990; Kolb & Turner 1990). Still, the general scenario of reheating in inflationary cosmology was absent, and a number of important problems remained unresolved. In particular, reheating in chaotic inflation theory, which covers many popular models of inflation, remained virtually unexplored. Recently, new effects in particle creation, arising beyond the perturbation theory, were found, together with further constraints on the elementary theory itself. These effects significantly alter the reheating scenarios. In this chapter we will briefly review the theory of the reheating after inflation. We will discuss different approaches to the theory both from a quantitative but simplified point of view, and within rigorous qualitative formalism. This presentation is based on an ongoing collaboration with Andrei Linde and Alexei Starobinsky.

2. Evolution of the Inflaton Field

Consider a simple chaotic inflation scenario. Among all the fields $(\phi, \chi, \psi, A_i, h_{ik}, ...)$ which are present at the relevant energy scales, the main contribution to gravity comes from the condensate of the scalar field ϕ, (or perhaps from other effectively similar condensates). In the fundamental Lagrangian $\mathcal{L}(\phi, \chi, \psi, A_i, h_{ik}, ...)$ for large energy density, we can retain for the cosmological applications only gravity and the dominant scalar field

$$\mathcal{L}(R, \phi) = -\frac{M_p^2}{16\pi} R + \frac{1}{2}\phi_i \phi^i - V(\phi). \tag{2.1}$$

The evolution of the FRW universe is described by the Einstein equation

$$H^2 = \frac{8\pi}{3M_p^2}\left(\frac{1}{2}\dot{\phi}^2 + V(\phi)\right), \tag{2.2}$$

where $H = \dot{a}/a$. The Klein-Gordon equation for $\phi(t)$ is

$$\ddot{\phi} + 3H\dot{\phi} + V_{,\phi} = 0 . \tag{2.3}$$

For sufficiently large initial values of $\phi > M_p$, the "drag" term $3H\dot{\phi}$ in (2.3) dominates $\ddot{\phi}$; the potential term in (2.2) dominates over the kinetic term. This is the inflationary stage when the universe expands (quasi-)exponentially, $a(t) = a_0 \exp\left[\int dt H(t)\right]$. However, with a decrease of the field ϕ the "drag" term becomes less and less important, and inflation terminates at $\phi \lesssim M_p/2$. After a short stage of the fast rolling down, the inflaton field rapidly oscillates around the minimum of $V(\phi)$ with the initial amplitude $\phi_0 \sim 0.1 M_p$. Although this value is below the magnitude needed for inflation, it is still very large by the standards of familiar physics.

The character of the classical oscillations of the homogeneous scalar field depends on the shape of its potential $V(\phi)$ around the minimum. We will consider two models: the quadratic potential $V(\phi) = \frac{1}{2} m_\phi \phi^2$ and the potential $V(\phi) = \frac{1}{4}\lambda\phi^4$.

For the quadratic potential the solutions of eqs. (2.2) and (2.3) at the stage of oscillations are

$$\phi(t) \approx \phi_0(t) \cdot \sin(m_\phi t) , \quad \phi_0(t) = \frac{M_p}{\sqrt{3\pi}} \cdot \frac{1}{m_\phi t} . \tag{2.4}$$

The scalar factor averaged over several oscillations is $a(t) \approx a_0 t^{2/3}$. Oscillations of ϕ in this theory are sinusoidal, with the the amplitude $\phi_0(t) \sim 0.1 M_p a^{-3/2}$ decreasing as the universe expands. The energy of the field ϕ decreases in the same way as the density of nonrelativistic particles of mass m_ϕ, $\epsilon_\phi = \frac{1}{2}\dot{\phi}^2 + \frac{1}{2}m_\phi^2\phi^2 \sim a^{-3}$. Hence the coherent oscillations of the homogeneous scalar field correspond to the the matter-dominated effective equation of state with a vanishing pressure.

For the theory with the potential $V(\phi) = \frac{1}{4}\lambda\phi^4$ it is more convenient to express the solutions of eqs. (2.2) and (2.3) via the conformal time $\eta = \int \frac{dt}{a(t)}$:

$$\phi(\eta) \approx \phi_0(\eta) \cdot cn\left(\frac{\omega}{c}\eta, \frac{1}{\sqrt{2}}\right) , \quad \phi_0(\eta) = \sqrt{\frac{3}{2\pi}}M_p \cdot \frac{c}{\omega\eta} . \tag{2.5}$$

The oscillations in this theory are not sinusoidal, but given by elliptic function $cn(\frac{\omega}{c}\eta, \frac{1}{\sqrt{2}})$, with the period of oscillations $2\pi/\omega$, where $c \approx 0.85$ and the effective frequency of oscillations is $\omega = c\sqrt{\lambda}a\phi_0 \sim 0.1\sqrt{\lambda}M_p$. With a good accuracy one can write $\phi(\eta) \approx \phi_0(\eta)\sin(c\sqrt{\lambda}a\phi_0\eta)$. The scalar factor is $a(t) \propto \eta(t) \propto \sqrt{t}$. The amplitude of oscillations $\phi_0(t) \sim 0.1M_pa^{-1}$ decreasing as the universe expands. The energy of the field ϕ decreases in the same way as the density of relativistic particles: $\epsilon_\phi = \frac{1}{2}\dot{\phi}^2 + \frac{1}{4}\lambda\phi^4 \sim a^{-4}$. The effective equation of state corresponds to the radiation dominated plasma with pressure $p_\phi \approx \frac{1}{3}\epsilon_\phi$.

3. Elementary Theory of Reheating

The amplitude of the oscillations gradually decreases not only because of the expansion of the universe, but also because of the energy transfer to particles created by the oscillating field. To describe the decay of the inflaton oscillations, we shall remember again the rest of the fundamental Lagrangian $\mathcal{L}(\phi, \chi, \psi, A_i, h_{ik}, ...)$ which includes the other fields and their interaction with the inflaton field. The inflaton field ϕ after inflation may decay into bosons χ and fermions ψ due to the interaction terms $-\frac{1}{2}g^2\phi^2\chi^2$ and $-h\bar{\psi}\psi\phi$, or into its own Bose quanta $\delta\phi$ due to the self-interaction $\lambda\phi^2\delta\phi^2$, or due to the gravitational interaction (like in the Starobinsky inflationary model). Here λ, g and h are small coupling constants. In the case of spontaneous symmetry breaking, the term $-\frac{1}{2}g^2\phi^2\chi^2$ gives rise to the term $-g^2\sigma\phi\chi^2$. We will assume for simplicity that the bare masses of the fields χ and ψ are very small, so that one can write $m_\chi(\phi) \approx g\phi$, $m_\psi(\phi) \approx |h\phi|$.

An elementary theory of reheating based on the perturbation theory was developed by Dolgov & Linde (1982) and by Abbot, Fahri & Wise (1982) for the new inflationary scenario. At the same time the theory of reheating was constructed in the Starobinsky (1982) model. Let us briefly recall the elementary theory of reheating. We consider for simplicity the classical field $\phi(t)$ with the mass m_ϕ oscillating near the minimum of the quadratic potential. A homogeneous scalar field oscillating with frequency $\omega = m_\phi$ can be interpreted as a collection of a number of ϕ-particles with zero momenta. The coherent wave of particles at rest with energy ϵ_ϕ corresponds to particle density $n_\phi = \epsilon_\phi/m_\phi$. From another point of view, n_ϕ oscillators of the same frequency m_ϕ, oscillating coherently with the same phase, can be viewed as a single homogeneous wave $\phi(t)$. Based on that interpretation, the effects related to the particle production can be incorporated into the equation of motion (2.3) for the inflaton field by means of the polarisation operator (Linde 1990):

$$\ddot{\phi} + 3H\dot{\phi} + \left(m_\phi^2 + \Pi(\omega)\right)\phi = 0 . \tag{3.6}$$

Here $\Pi(\omega)$ is the flat space polarisation operator for the field ϕ with the four-momentum $k_i = (\omega, 0, 0, 0)$, $\omega = m_\phi$.

The real part of $\Pi(\omega)$ gives only a small correction to m_ϕ^2, but when $\omega \geq \min(2m_\chi, 2m_\psi)$, the polarisation operator $\Pi(\omega)$ acquires an imaginary part $\mathrm{Im}\,\Pi(\omega)$. We will assume that $m_\phi^2 \gg H^2, m_\phi^2 \gg \mathrm{Im}\,\Pi$. The first condition is automatically satisfied after the end of inflation; the second usually is also true. The solution of (3.6) that generalises the solution (2.4) and describes damped oscillations of the inflaton field is

$$\phi \approx \frac{M_p}{\sqrt{3\pi} m_\phi t} \exp\left(-\frac{1}{2}\Gamma t\right) \sin\left(m_\phi t\right) , \tag{3.7}$$

where Γ is the total decay rate of ϕ-particles. Here we used the relation $\mathrm{Im}\,\Pi = m_\phi \Gamma$, which follows from unitarity. Thus, eq. (3.7) implies that the amplitude of oscillations of the field ϕ decreases as $\phi_0(t) \approx 0.1 M_p a^{-3/2} \exp\left(-\frac{1}{2}\Gamma t\right)$ due to particle production which occurs during the decay of the inflaton field, as well as due to the expansion of the universe.

For a phenomenological description of this effect one can just add an extra friction term $\Gamma\dot\phi$ to the classical equation of motion of the field ϕ, instead of adding the polarisation operator (Kolb & Turner 1990):

$$\ddot\phi + 3H\dot\phi + \Gamma\dot\phi + m_\phi^2 \phi = 0 . \tag{3.8}$$

At the stage of oscillations the solution of this equation is similar to (3.7).

One should note, that this equation, just as equation (3.7), is valid only under the conditions $m_\phi \gg H$, $m_\phi \gg \Gamma$, and only at the stage of rapid oscillations of the field ϕ near the minimum of $V(\phi)$. This equation cannot be used to investigate the stage of slow and fast rolling of the field ϕ during inflation.

Suppose that Γ^{-1} is much less than the typical expansion time of the Universe H^{-1}, the energy density of the inflaton field decreases exponentially within the time Γ^{-1}, $\epsilon_\phi \approx \frac{1}{2} m_\phi^2 \phi_0^2 \exp(-\Gamma t)$. This is exactly the result one would expect on the basis of the interpretation of the oscillating field ϕ as a coherent wave consisting of decaying ϕ-particles.

The rate of decrease of the energy of oscillations coincides with the decay rate of ϕ-particles which is well-known from the perturbation theory. The rates of the processes $\phi \to \chi\chi$ and $\phi \to \psi\psi$ (for $m_\phi \gg 2m_\chi, 2m_\psi$) plus gravitational decay are given by

$$\Gamma_{\phi\to\chi\chi} = \frac{g^4\sigma^2}{8\pi m_\phi} , \quad \Gamma_{\phi\to\psi\psi} = \frac{h^2 m_\phi}{8\pi} , \quad \Gamma_g = \frac{m_\phi^3}{8\pi M_p^2} . \tag{3.9}$$

Reheating completes at the moment t_r when the rate of expansion of the universe given by the Hubble constant $H = \sqrt{\frac{8\pi\rho}{3M_p^2}} \sim t_r^{-1}$ becomes equal to the total decay rate Γ. The energy density of the universe at the moment $t_r = \Gamma^{-1}$ is

$$\epsilon(t_r) \simeq \frac{3\Gamma^2 M_p^2}{32\pi} . \tag{3.10}$$

If the particles produced by the decaying inflaton field interact with each other strongly enough, then the thermodynamic equilibrium sets in quickly after the decay of the inflaton field, and the matter acquires a temperature T_r. The energy density of the Universe dominated by the ultrarelativistic particles in a state of thermal equilibrium is

$$\epsilon(T_r) \simeq \frac{\pi^2 g_*}{30} T_r^4 , \tag{3.11}$$

where a numerical factor $g_*(T_r) \sim 10^2 - 10^3$ depends on the number of ultra-relativistic degrees of freedom. Comparing (3.10) and (3.11) we get the following result for the reheating temperature:

$$T_r \simeq 0.1\sqrt{\Gamma M_p}\ . \tag{3.12}$$

Note that T_r does not depend on the initial value of the field ϕ, it is completely determined by the parameters of the underlying elementary particle theory via the decay rate Γ.

In order to get numerical estimates for the duration of reheating t_r, decay rate of the inflaton field and the reheating temperature T_r, one should know the mass of the inflaton field and the coupling constants. We shall take into account several constraints on these in the framework of perturbation theory (Kofman, Linde & Starobinsky 1996b). The coupling constants of interaction of the inflaton field with matter cannot be too large, otherwise the radiative corrections would alter the shape of the inflaton potential. The necessary condition for the decay of inflaton oscillations is $\omega > 2m_\chi, 2m_\psi$. Parameters of the inflaton potential are restricted by the constraints on the amplitude of cosmological fluctuations. All together it allows us to update a constraint on the elementary theory: the largest possible total decay rate in the perturbation theory is $\Gamma < 10^{-20}M_p$. For instance, for the quadratic inflaton potential, it takes at least 10^{14} oscillations to transfer the energy of inflaton oscillations into the created particles. This is a very strong condition, which makes reheating very slow. From (3.12) one can obtain the general bound on the reheating temperature in the model of slow reheating:

$$T_r < 10^9\,\text{GeV}\ . \tag{3.13}$$

This is a very small temperature, at which the standard mechanism of baryogenesis in the GUTs cannot work; in this case we definitely need a theory of a low-temperature baryogenesis. At such a temperature no cosmologically interesting heavy strings, monopoles and textures can be produced, because the generation of the GUT topological defects requires high temperature phase transitions at $T_{GUT} \sim 10^{16}$ GeV . For the massive inflaton field in the slow reheating scenario the post-inflationary matter-dominated phase lasts sufficiently long to enhance the gravitational instability of density fluctuations. This leads to the formation of primordial black holes for some specific spectra of the initial density fluctuations.

4. Quantum Field Theory in a Time-dependent Background

Surprisingly, it was found recently that typically the transition between inflation and the hot Big Bang universe may be very different from what the elementary theory predicts (Kofman, Linde & Starobinsky 1994; Shtanov, Traschen & Brandenberger 1995; Boyanovsky *et al* 1995a; Yoshimura 1995). The elementary theory of reheating might be still applicable if the inflaton field can decay into fermions only, with a small coupling constant $h^2 \ll m_\phi/M_p$. This theory also can provide a qualitatively correct description of particle decay at the last stages of reheating. However, the perturbation theory is inapplicable to the description of the first stages of reheating, and that makes the whole process quite different. In what follows we will primarily consider the theory of the first stages of reheating. We will begin with the theory of a massive scalar field ϕ decaying into particles χ, and then consider the $\frac{\lambda}{4}\phi^4$ theory.

For simplicity, we consider here the interaction $-\frac{1}{2}g^2F(\phi)\chi^2$ between the *classical* inflaton field ϕ and the *quantum* scalar field χ with the Lagrangian:

$$\mathcal{L}(\chi) = \frac{1}{2}\left(\chi_{,k}\chi^{,k} - m_\chi^2\chi^2 + \xi R\chi^2 - g^2F(\phi)\chi^2\right)\ . \tag{4.14}$$

Here ξ is a coupling constant of interaction with the space-time curvature R. We also will distinguish two cases, the four-legs interaction $g^2F(\phi)\chi^2 = g^2\phi^2\chi^2$, and the three-legs interaction $g^2F(\phi)\chi^2 = 2g^2\sigma\phi\chi^2$ which arises when symmetry is broken. The Heisenberg representation of the quantum scalar field $\hat{\chi}$ is

$$\hat{\chi}(t,\vec{x}) = \frac{1}{(2\pi)^{3/2}} \int d^3k \left(\hat{a}_{\vec{k}} \chi_{\vec{k}}(t)\, e^{-i\vec{k}\vec{x}} + \hat{a}^+_{\vec{k}} \chi^*_{\vec{k}}(t)\, e^{i\vec{k}\vec{x}} \right) , \tag{4.15}$$

where $\hat{a}_{\vec{k}}$ and $\hat{a}^+_{\vec{k}}$ are the annihilation and creation Bose-operators. For the flat Friedmann background with a scalar factor $a(t)$ the temporal part of the eigenfunction with momentum $\vec{k}$ obeys the equation

$$\ddot{\chi}_k + 3\frac{\dot{a}}{a}\dot{\chi}_k + \left(\frac{k^2}{a^2} + m_\chi^2 - \xi R + g^2F(\phi) \right) \chi_k = 0 \, . \tag{4.16}$$

Let us use the conformal time η and introduce the new function $f_k(\eta) = a(\eta)\chi_k(\eta)$. Then instead of (4.16) we have

$$f_k'' + \Omega_k^2 f_k = 0 \, , \tag{4.17}$$

where prime stands for $\frac{d}{d\eta}$, and $\Omega_k^2 = (m_\chi a)^2 + k^2 + g^2a^2F(\phi) + a^2(1/6-\xi)R$. This equation describes the oscillators with a variable frequency $\Omega_k^2(t)$ due to the time-dependence of the background field $\phi(\eta)$ and $a(\eta)$. At $\eta \to -\infty$ we choose the positive-frequency solution $f_k(\eta) \simeq \frac{e^{-ik\eta}}{\sqrt{2k}}$. We expect the quantum effects of χ-particle creation and vacuum polarisation to take place as the inflaton field is varying during the slow and fast-rolling down and oscillations after inflation. The problem is to find a general solution of the equation (4.17) and regularise the formal expressions for the values of the energy-density $\langle \epsilon_\chi \rangle$ and the vacuum expectation $\langle \chi^2 \rangle$.

We follow a physically transparent treatment of the equation (4.17) with an arbitrary time dependence of the classical background field, which was originally developed by Zeldovich and Starobinsky (1972) for the problem of particle creation in varying strong gravitational fields. In terms of classical waves of the χ-field, quantum effects occur due to the departure from the initial positive-frequency solution. Therefore one may represent solutions of eq.(4.17) as a product of its solutions in the adiabatic approximation, $\exp\left(\pm i \int d\eta\, \Omega_k\right)$, multiplied by some functions $\alpha(\eta)$ and $\beta(\eta)$:

$$f_k(\eta) = \frac{\alpha_k(\eta)}{\sqrt{2\Omega_k}}\, e^{-i\int \Omega_k d\eta} + \frac{\beta_k(\eta)}{\sqrt{2\Omega_k}}\, e^{+i\int \Omega_k d\eta} \, . \tag{4.18}$$

An additional condition on the functions α and β can be imposed by taking the derivative of the expression (4.18), while considering α and β time-independent. Then the expression (4.18) is a solution of equation (4.17) if the functions α_k, β_k satisfy the equations

$$\alpha_k' = \frac{\Omega_k'}{2\Omega_k} e^{+2i\int \Omega_k d\eta}\, \beta_k \, , \qquad \beta_k' = \frac{\Omega_k'}{2\Omega_k} e^{-2i\int \Omega_k d\eta}\, \alpha_k \, . \tag{4.19}$$

Normalisation gives $|\alpha_k|^2 - |\beta_k|^2 = 1$, and the initial conditions at $\eta \to -\infty$ are $\alpha_k = 1$, $\beta_k = 0$. The coefficients $\alpha_k(\eta)$ and $\beta_k(\eta)$ in our case coincide with the coefficients of the Bogoliubov transformation of the creation and annihilation operators which diagonalises the Hamiltonian of the $\hat{\chi}$-field at each moment of time η.

The regularised vacuum expectation values for the χ^2, energy and particle number densities in terms of $\alpha_k(\eta)$ and $\beta_k(\eta)$ are given correspondingly by

$$\langle \chi^2 \rangle = \frac{1}{2\pi^2 a^2} \int_0^\infty dk\, k^2 \frac{1}{\Omega_k} \left(|\beta_k|^2 + \mathrm{Re}\,(\alpha_k\, \beta_k^*\, e^{-2i\int \Omega_k d\eta}) \right)_{reg} , \tag{4.20}$$

$$\langle \epsilon_\chi \rangle = \frac{1}{2\pi^2 a^4} \left(\int_0^\infty dk\, k^2 \Omega_k\, |\beta_k|^2 \right)_{reg} , \quad \langle n_\chi \rangle = \frac{1}{2\pi^2 a^4} \int_0^\infty dk\, k^2 |\beta_k|^2 \, . \tag{4.21}$$

The formal expressions for the vacuum expectation values should be renormalised. The WKB expansion of the solution of eqs. (4.19) provides a natural scheme of regularisation (Zel'dovich & Starobinsky 1972). Thus, the final result is expressed via the coefficients of the Bogoliubov transformation.

The equation of motion of the inflaton field with the feedback of the quantum effects is

$$\ddot{\phi} + 3H\dot{\phi} + V_{,\phi} + \frac{1}{2}g^2\langle\chi^2\rangle F_{,\phi} = 0 \ , \tag{4.22}$$

where the vacuum average is given by (4.20). The quantum effects contribute to the effective mass of the inflaton field, e.g., for the massive scalar field $m^2_{eff} = m^2_\phi + \frac{1}{2}g^2\langle\chi^2\rangle F_{,\phi}$. The interactions with fermions $-h\bar{\psi}\psi\phi$ and the self-interaction $-\lambda\phi^2\delta\phi^2$, where $\delta\phi = \hat{\phi} - \phi$, result in extra terms $h < \bar{\psi}\psi >$ and $3\lambda\phi^2\langle\delta\phi^2\rangle$ in (4.22).

With the self-consistent equations (4.20) and (4.22) based on the rigorous quantum field theory in the time varying background, one can investigate the role of the quantum effects in different stages of the evolution of the inflaton field. There are three main quantum effects in an external classical background: vacuum polarisation, particle creation and generation of long-wavelength semi-classical fluctuations. Different effects dominate at different stages of the inflationary scenario. During inflation the long-wavelength fluctuations are generated for the minimally coupled light fields with $\xi = 0$, but this effect is irrelevant to the reheating. The vacuum polarisation and particle production could occur at the next stage of the fast rolling down of the inflaton field to the minimum of its potential $V(\phi)$ at the end of inflation. However, these effects are negligible and do not alter the dynamics of the inflaton field. At the stage of oscillations of the inflaton field the only important effect is the creation of particles due to the non-adiabatic change of $\Omega_k(\eta)$. The particle production leads to the generation of $\langle\chi^2\rangle$, $\langle\epsilon_\chi\rangle$ and the damping of the inflaton oscillations.

However, the term $\frac{1}{2}g^2\langle\chi^2\rangle F_{,\phi}$ in the equation (4.22) which describes the back reaction of the quantum effects on the evolution of the inflaton field is not reduced to the phenomenological "friction" term $\Gamma\dot{\phi}$, first suggested by Albrecht *et al* (1982). This oversimplified ansatz is widely used to investigate all the stages of the evolution of the inflaton field. Many authors used this method in order to understand whether reheating may slow down the rolling of the field ϕ and, consequently, to support inflation. Remember that one of the main reasons why inflation is possible is the presence of the friction term $3H\dot{\phi}$ in the equation of motion of the scalar field in an expanding Universe. The presence of a similar term due to reheating could support inflation even in the models where inflation would otherwise be impossible. As we have seen, the damping of the scalar field during slow and fast rolling down cannot be described in such a simple way. In our study (Kofman, Linde & Starobinsky 1996b) we did not find any possibility to support inflation by reheating. As an exception, the friction term can correctly describe damping of oscillations in the case where the perturbation theory for interaction is applicable.

It is instructive to return in the framework of the rigorous approach to the perturbation theory. Assuming small $|\beta_k| \ll 1$, from eqs. (4.19) one can obtain an iterative solution:

$$\beta_k \simeq \frac{1}{2}\int_{-\infty}^{\eta} d\eta' \, \frac{\Omega'_k}{\Omega_k} \, \exp\left(-2i\int_{-\infty}^{\eta'} d\eta'' \Omega_k(\eta'')\right) \ . \tag{4.23}$$

Using expression for $\Omega_k(\eta)$, we obtain that $\beta_k(\eta)$ is proportional to the integral $g^2\int_{-\infty}^{\eta} d\eta' \, e^{-2ik\eta'} \, (a^2F(\phi))'$, where $\phi(t)$ is given by the oscillating solutions (2.4) or (2.5), and $F(\phi)$ depends on the type of the interaction. This integral is evaluated by the method of stationary phase (e.g., Starobinsky 1982). To compare the results with these derived in the previous section, let us focus on the case of massive scalar field decaying via the three-legs interaction $g^2\sigma\phi\chi^2$. In this case the dominant contribution is given by the integration near η_k, where $a(\eta_k) = \frac{2k}{m_\phi}$. This corresponds to the creation of a pair of χ-particles with momentum $k = \frac{1}{2}a(\eta_k)m_\phi$ from an inflaton with the four-momentum $(m_\phi, 0, 0, 0)$ at the moment η_k.

Substituting this estimation for β_k into (4.21), it is easy to calculate the decay rate of the inflaton field. For the three-leg interactions $g\sigma\phi\chi$ or $h\phi\bar{\psi}\psi$ it is indeed reduced to the formulas (3.9). For the four-leg interaction $\phi\phi \to \chi\chi$, we have creation of a pair of χ-particles with momentum $k = a(\eta_k)m_\phi$ from a pair of massive inflatons with the four-momentum $(m_\phi, 0, 0, 0)$ at the moment η_k. The decay rate of the massive inflaton field in this case rapidly decreases with the expansion of the Universe as $\frac{1}{a^4}\frac{d}{dt}(a^4\epsilon_\chi) \propto a^{-6}$. Therefore the complete decay of the massive inflaton field in the theory with no spontaneous symmetry breaking or with no interactions with fermions is impossible! This observation gives rise to an unexpected possibility of having the residual scalar field oscillations as a dark matter candidate. Indeed, if the inflaton oscillations correspond to a matter dominated equation of state, the inflaton field by itself, or other scalar fields with similar properties can be cold dark matter candidates, even if they strongly interact with each other. However, this possibility requires a fine tuning. More immediate application of our result is that it allows one to rule out the inflationary models which predict too small or too large value of the present radiation density parameter Ω_r.

The fact that the number of particles $|\beta_k(\eta)|^2$ with the momentum k is predominantly generated at the certain moment η_k gives us a proper physical interpretation how the particles are created from the coherent oscillations of the homogeneous inflaton field: a pair of particles is created at the instant of the resonance $k = \omega a(\eta)/2$ between k and the external frequency ω (which is equal m_ϕ for the massive inflaton). As the universe expands, that particular mode is redshifted from the resonance, and creation of particles with the momentum k is terminated leaving $n_k = |\beta_k(\eta)|^2 \ll 1$. At an arbitrary moment η there is a corresponding momentum k resonating with the external frequency.

5. Parametric Resonance: the Stage of Preheating

The new interpretation, introduced in the last section, that particles are created at the instant of resonance with the inflaton oscillations gives us a hint for the more general case where n_k is not assumed to be small.

Let us assume there is no symmetry breaking before reheating. We will thus onsider oscillations (2.4) of the massive scalar field ϕ that decay into light particles χ due to the interaction $-\frac{1}{2}g^2\phi^2\chi^2$ and oscillations (2.5) in the theory $\lambda\phi^4$ decaying due to the self-interaction $-\frac{3}{2}\lambda\phi^2\delta\phi^2$. In the first case the equation (4.16) for quantum fluctuations $\chi_k(t)$ can be rewritten in the form

$$\frac{d^2(a^{3/2}\chi_k)}{dt^2} + \left(\frac{k^2}{a^2(t)} + g^2\phi_0^2\,\sin^2(m_\phi t) + \Delta\right)(a^{3/2}\chi_k) = 0\ , \tag{5.24}$$

where ϕ_0 stands for the amplitude of oscillations of the field ϕ, defined in the formula (2.4), $\Delta = m_\chi^2 + (9/2 - 3\xi)H^2$. As we shall see, the main contribution to χ-particle production is given by excitations of the field χ with $k/a > m_\phi$, which is much greater than H at the stage of oscillations. Therefore, in the first approximation we may neglect the expansion of the Universe, taking $a(t)$ as a constant and omitting the term Δ in (5.24).

In the case of theory with the $\frac{\lambda}{4}\phi^4$ potential the role of quantum field plays the quantum fluctuations $\delta\phi$. In the equations (4.16) for the quantum fluctuations we have to substitute $\chi_k \to \delta\phi_k$, $g^2 \to 3\lambda$. Since in this theory the amplitude $\phi_0(t) \propto a^{-1}(t)$, it is convenient to use the conformal time η and equations for fluctuations in the form (4.17). For the oscillations (2.5) the exact equation for quantum fluctuations $\delta\phi$ can be reduced to the Lame equation. For simplicity here we will use an approximate equation

$$\frac{d^2(a\delta\phi_k)}{d\eta^2} + \left[k^2 + 3\lambda a^2\phi_0^2 \sin^2(c\sqrt{\lambda}a\phi_0\eta)\right](a\delta\phi_k) = 0 \ , \tag{5.25}$$

where $c \approx 0.85$ [see (2.5)]. The crucial observation is that the equations (5.24) and (5.25), which describe the fluctuations in the models, can be reduced to the well-known Mathieu equation:

$$\frac{d^2 y_k}{dz^2} + (A(k) - 2q\cos 2z)\, y_k = 0 \ , \tag{5.26}$$

with the periodic effective frequency $\Omega_k^2 = A(k) - 2q\cos 2z$. For the first model (5.24), $y_k = a^{3/2}\chi_k$, $A(k) = k^2/(\omega^2 a^2) + 2q$, $q = g^2\Phi^2/(4\omega^2)$, $z = \omega t$ and the frequency of the inflaton oscillations is $\omega = m_\phi$. For the second model (5.25), $y_k = a\delta\phi_k$, $A \approx k^2/\omega^2+2.08$, $q \approx 1.04$, $z = \omega\eta$ and the frequency of inflaton oscillations is $\omega = c\sqrt{\lambda}a\phi_0$.

An important property of solutions of the equation (5.26) is the existence of an exponential instability $y_k \propto \exp\left[\mu_k^{(n)} z\right]$ within the set of resonance bands of frequencies $\Delta k^{(n)}$ labeled by an integer index n. This instability corresponds to exponential growth of occupation numbers of quantum fluctuations $n_{\vec{k}}(t) \propto \exp(2\mu_k^{(n)} m_\phi t)$ that may be interpreted as particle production. The simplest way to analyse this effect is to study the stability/instability chart of the Mathieu equation (MacLachlan 1961).

We will distinguish different regimes of resonance which correspond to the different regions of the parameters of the instability chart:

Perturbation theory: $q \ll 1$; $A = l^2$, $l = 1, 2, \ldots$; $2\pi\mu_k \ll H/\omega \ll 1$; $n_k \ll 1$. If no expansion of the universe, fluctuations y_k are slowly generated at the discrete modes k. The expansion quickly redshifts them from the resonance bands. The net effect is reduced to the creation of $n_k \ll 1$ particles with continuous spectrum, which we considered in the previous section.

From this point of view the perturbative regime for the Starobinsky model was considered by Starobinsky (1982) and for the chaotic inflation by Kofman, Linde & Starobinsky (1996b).

Narrow resonance: $q \ll 1$; $A \simeq l^2$, $l = 1, 2, \ldots$; $2\pi\mu_k \le H/\omega \ll 1$; $n_k \le O(1)$. The universe expansion redshifts the momentum k/a from the narrow resonance band not too fast, so a significant number of particles is generated at the resonance mode. This is an intermediate regime between the perturbation theory and the regime of the broad parametric resonance, where expansion of the universe is a subdominant effect.

The narrow resonance was mentioned by Dolgov & Kirilova (1990) for the new inflationary scenario. The importance of this regime for that model was first recognised by Traschen & Brandenberger (1990), but for several reasons their final results were not quite correct. A detailed theory of particle creation in the narrow resonance regime in an expanding universe was developed for the chaotic scenario by Kofman, Linde & Starobinsky (1994; 1996b), see also Shtanov, Traschen & Brandenberger (1995) and Kaiser (1995). In order to investigate the explosive particle production from reheating, many authors considered the model of the narrow resonance without taking into account expansion of the universe because it made investigation simpler (Yoshimura 1995; Boyanovsky *et al.* 1995a,b). However for some range of parameters many vital features of the reheating may disappear in this approximation. In particular, the effects in the model studied by Son (1996) disappear in an expanding universe.

Broad resonance: $q \simeq O(1)$; $A \ge 2q$; $2\pi\mu_k \sim O(1)$; $n_k \gg 1$. Creation of particles in the regime of a broad resonance is very different from the previous cases. In particular, it proceeds during a tiny part of each oscillation of the field ϕ when $1-\cos z \sim q^{-1}$ and the induced effective mass of the fluctuations y_k is less than m_{eff}. As a result, the number of the Bose particles grows exponentially fast. This regime occurs only if parameter q is not small. A typical energy E of a particle produced at this stage is determined by equation $A - 2q \simeq 4\sqrt{q}$. The line $A = 2q$ divides the region of the broad resonance from the tachyonic regime where $A \leq 2q$; $\mu_k \geq \pi^{-1}$. The line $A = 2q$ also corresponds to the momentum $k = 0$. Near the line $A = 2q$ typically $\mu_k \sim 0.175$ in the instability bands, with the maximal value about 0.28.

This regime is the most difficult to study. The investigation can be advanced with the following method: consider eq. (5.26) as the equation for the propagation of the wave $y_k(z)$ backward in time. The evolution of $y_k(z)$ can be estimated for each scattering instance, then the net effect is the product of the successive scattering matrixes (Kofman, Linde & Starobinsky 1996b). The results for particle production based on this method were reported by Kofman, Linde & Starobinsky (1994). This stage was also studied by Boyanovsky *et al.* (1995a) and Fujisaki *et al.* (1995).

We shall focus on the most interesting regime of the broad resonance. For our first model (5.24) the necessary condition for it is $m_\phi \ll gM_p$. A typical energy E of a particle produced at this stage is $E \sim \sqrt{gm_\phi M_p}$. As we will see, the broad resonance (preheating) ends up within the short time $t_{ph} \sim m_\phi^{-1} \ln(m_\phi/g^5 M_p)$. As a results, at the end of this stage the occupation number of created particles $n_k \sim exp(2\mu_k m_\phi t)$ with the energy $k = E$ is extremely large:

$$n_E \sim \frac{1}{g^2} \gg 1 \,. \tag{5.27}$$

For another model (5.25), the parameter $q = 1.04$ corresponds to the broad resonance regime. Looking at the instability chart, we see that the resonance occurs in the second band. The typical energy E of a created particles is $E \sim \sqrt{\lambda} a\phi_0$. In the second band maximal value of the coefficient μ_k is approximately 0.07. Note that the rigorous Lame equation for the fluctuations in the $\lambda\phi^4$ theory gives a twice smaller value of $\mu_k \approx 0.036$. As long as the back-reaction of created particles is small, expansion of the Universe does not shift fluctuations away from the resonance band, and the number of particles $n_k \sim \exp(\sqrt{\lambda}\phi_0 t/10)$. The broad resonance in this model ends up within the time interval $t_{ph} \sim M_\phi^{-1}\lambda^{-1/2}|\ln\lambda|$. At the end of this stage the occupation number is

$$n_E \sim \frac{1}{\lambda} \gg 1 \,. \tag{5.28}$$

In addition to this models, the field ϕ in the theory $\lambda\phi^4$ may decay into χ-particles due to the interaction $-\frac{1}{2}g^2\phi^2\chi^2$. This is the leading process for $g^2 \gg \lambda$. The parametric resonance is broad. The values of the parameter μ_k along the line $A = 2q$ do not change monotonically, but typically for $q \sim g^2/\lambda \gg 1$ they are 3 to 4 times greater than the parameter μ_k for the decay of the field ϕ into its own quanta. Therefore, the resonance in this theory is very efficient.

The stage of broad parametric resonance, which leads to the explosive particle creation, lasts typically for a few tens of oscillation periods (for several periods in the last model). However, this does not mean that the process of reheating has been completed. Instead of created particles in the thermal equilibrium, by the end of this stage one has particles far away from equilibrium but with extremely large mean occupation numbers (5.27), (5.28). It is therefore convenient to divide the whole process of reheating into three different stages. In the first stage, which cannot be described by the elementary theory of reheating, the classical coherently oscillating inflaton field ϕ decays into bosons due to parametric resonance. In many models the resonance is broad, and the process occurs extremely rapidly (explosively). Because of the Pauli exclusion principle, there is no explosive creation of fermions. To distinguish this stage from the subsequent stages when particles decay or interact, we call it *pre-heating*. The second stage is the evolution of the system after preheating. During this stage the particles produced from the preheating, can decay into other particles and self-interact. The last stage is the stage of thermalisation, which can be described by standard methods.

6. Back Reaction of Created Particles

Creation of particles leads to several effects which can change the dynamics of the system. In particular, it terminates the broad resonance particle production. First, the energy from the homogeneous field $\phi(t)$ is transferred to the created particles. The amplitude of the classical oscillations ϕ_0 is therefore decreasing faster than it would decrease due to the expansion of the universe only. Second, as it follows from (4.22), fluctuations contribute to the effective frequency of the inflaton oscillations: $\omega_{eff}^2 = m_\phi^2 + g^2\langle\chi^2\rangle$ or $\omega_{eff}^2 = c^2\lambda\phi^2 + 3\lambda\langle\,\delta\phi^2\rangle$. Third, fluctuations also can contribute to the effective mass of fluctuations themselves. As a result, the q parameters in the Mathieu equation are changing in the two models correspondingly as $q = g^2\phi_0^2/4\omega_{eff}^2$ and $q = 3\lambda a^2\phi_0^2/4\omega_{eff}^2$. The $A(k)$ parameters are changing as $A(k) = k^2/(\omega_{eff}^2 a^2) + 2q$, and $A(k) = (k^2 + 3\lambda a^2\langle\,\delta\phi^2\rangle)/\omega_{eff}^2 + 2q$. This moves the resonance momenta k from their initial positions on the instability chart. The parameter q is decreasing, which dynamically shifts the parameters towards the narrower resonance.

As a result, the creation of particles in the broad resonance regime is terminated. A simple criterion for the end of the preheating is the condition that the energy density of produced fluctuations should be of the same order as the energy density of the scalar field. For the first model $\langle\chi^2\rangle \sim \phi_0^2$ while for the second model $\langle\,\delta\phi^2\rangle \sim \phi_0^2$. The timing of the preheating t_{ph}, given above, was estimated from this criterion.

After the copious particle production around the resonance mode E has slowed down, the inflaton field still continues to transfer its energy into the energy of fluctuations. Son (1996) noticed that self-interaction of the $\delta\phi$ particles in the $\lambda\phi^4$ theory might be important since they have very large occupation number density $n_E \sim 1/\lambda$ at the resonance mode. Indeed, the usual estimation for the relaxation time τ for the dilute gas of the $\delta\phi$ particles due to the self-interaction has limited relevance when the occupation number density of particles is large. The scattering of created particles, however, does not eliminate the preheating regime. The particles created at the resonance occupy only very narrow shells in the momentum space, corresponding to successive resonance bands. After preheating, most of the particles are located in the first shell of the radius E and the width $\sim 0.1E$ (in addition, their occupation number n_k sharply falls towards the edges of the shell). Two particles from this shell can be scattered inside the shell. The rate of redistribution of particles within the shell is indeed much larger than τ^{-1}, but it does not terminate the resonance effect since particles remain inside the resonance shell. Two particles from the resonance shell can be scattered out of it. However, this process is slow compared with the preheating time t_{pr}. There is an interesting possibility that some of the particles from the first resonance shell are scattered into other resonance shells. However, the whole process of rescattering and further creation of particles in the narrow resonance after preheating is rather complicated. Khlebnikov & Tkachev (1996) performed first numerical simulation of the nonlinear evolution of the inflaton field in $\lambda\phi^4$ theory with rescattering. Based on the result of Polarski & Starobinsky (1996) that the particles created with the large occupation number can be viewed as the superposition of the classical waves with random phases, they modelled preheating and scattering of ϕ particles as the evolution of classical random gaussian field with the spectrum which has a sharp spike with the amplitude (5.28) at the mode E superimposed on the homogeneous background. Numerical simulation confirms the features of preheating derived analytically. They also show the further growth of the fluctuations $\langle \delta\phi^2 \rangle$ and reveal their spectrum after the preheating in this toy model. We note that in the more general case, when inflaton field can decay into other bosons (e.g., due to $g^2\phi^2\chi^2$-interaction) the rescattering is less significant. We also do not find evidence of the strong Bose condensations at $k = 0$.

7. Non-Thermal Phase Transitions from After Inflation

As we have already seen, the reheating scenario depends on the type and strength of interactions of the fields involved. In this section we show that the reheating also strongly depends on the structure of the elementary particle theory. So far we considered theories with no symmetry breaking. In the case with symmetry breaking, where the amplitude of oscillations ϕ_0 is in the beginning much greater than σ, the theory of decay of the inflaton field is the same as in the case considered above. The most important part of pre-heating occurs at this stage. When the amplitude of the oscillations becomes smaller than $m_\phi/\sqrt{\lambda}$ and the field begins oscillating near the minimum of the effective potential at $\phi = \sigma$, particle production due to the narrow parametric resonance typically becomes weak. The main reason for this is related to the backreaction of particles created at the preceding stage of pre-heating on the rate of expansion of the universe and on the shape of the effective potential. However, importance of spontaneous symmetry breaking for the theory of reheating should not be underestimated, since it gives rise to the interaction term $g^2\sigma\phi\chi^2$ or $\lambda\sigma\phi\delta\phi^2$ which is linear in ϕ. Such terms are necessary for a complete decay of the inflaton field in accordance with the perturbation theory.

However, in the theories where preheating is possible one may expect many unusual phenomena. One of the most interesting effects is the possibility of specific non-thermal post-inflationary phase transitions which occur after preheating (Kofman, Linde & Starobinsky 1996a, Tkachev 1996). These phase transitions in certain cases can be much more pronounced that the standard high temperature cosmological phase transitions (Kirzhnits & Linde 1972, Linde 1990). They may lead to copious production of topological defects and even to a secondary stage of inflation after reheating.

We first recall the theory of phase transitions in theories with spontaneous symmetry breaking. For simplicity let us consider the theory of scalar fields ϕ the effective potential

$$V(\phi, \chi) = \frac{\lambda}{2}(\phi^2 - \sigma^2)^2 . \tag{7.29}$$

$V(\phi)$ has a minimum at $\phi = \sigma$, and a maximum at $\phi = 0$ with the curvature $V_{\phi\phi} = -m^2 = -\lambda\sigma^2$. This effective potential acquires corrections due to quantum (or thermal) fluctuations of the scalar fields $\Delta V = \frac{3}{2}\lambda\langle(\delta\phi^2)\rangle\phi^2$. In the large temperature limit $\langle(\delta\phi)^2\rangle = \frac{T^2}{12}$. The effective mass squared of the field ϕ

$$m^2_{\phi,eff} = -m^2 + 3\lambda\phi^2 + 3\lambda\langle(\delta\phi)^2\rangle \tag{7.30}$$

becomes positive and symmetry is restored (i.e. $\phi = 0$ becomes the stable equilibrium point) for $T > T_c$, where $T_c^2 = 4m^2/\lambda \gg m^2$. At this temperature the energy density of the gas of ultrarelativistic particles is given by $\rho = g_*(T_c)\pi^2 T_c^4/30 = 8\pi^2 m^4 N(T_c)/(15\,\lambda^2)$.

The theory of cosmological phase transitions is an important part of the theory of the evolution of the universe, and during the last twenty years it was investigated in grat detail. However, it was typically assumed that the phase transitions occured in the state of thermal equilibrium. We are now going to show that similar phase transitions may occur even much more efficiently prior to thermalisation due to the anomalously large expectation values $\langle(\delta\phi)^2\rangle$ produced during the first stage of reheating after inflation.

We will first consider the model (7.29) with the amplitude of spontaneous symmetry breaking $\sigma \ll M_p$. As we have seen in the previous sections, during preheating inflaton oscillations transfer most of its energy $\sim \lambda M_p^4$ to its long-wave fluctuations $\langle(\delta\phi)^2\rangle \sim M_p^2$ in the regime of broad parametric resonance.

The crucial observation is the following. If the initial energy density $\sim \lambda M_p^4$ were instantaneously thermalised, the reheating temperature $T_r \sim \lambda^{1/4} M_p$ would be much greater than the typical particle energy after preheating $E_\phi \sim \sqrt{\lambda} M_p$, and the magnitude of fluctuations $\langle(\delta\phi)^2\rangle \sim T_r^2/12 \sim \sqrt{\lambda} M_p^2$ would be much smaller than the magnitude of non-thermalised fluctuations $\langle(\delta\phi)^2\rangle \sim M_p^2$. Thus after the first stage of reheating, the non-thermalised fluctuations of the scalar field ϕ are much greater than the thermalised ones. Thermal fluctuations would lead to symmetry restoration in our model only for $\sigma \lesssim T_r \sim \lambda^{1/4} M_p$. Meanwhile, according to (7.30), the non-thermalised fluctuations $\langle(\delta\phi)^2\rangle \sim M_p^2$ may lead to symmetry restoration even if the symmetry breaking parameter σ is as large as M_p. Thus, the non-thermal symmetry restoration occurs even in those theories where the symmetry restoration due to high temperature effects would be impossible. Later on, $\langle(\delta\phi)^2\rangle \propto a^{-2}$, $E_\phi \propto a^{-1}$ because of the expansion of the universe (as far as $E_\phi \gg m$). This leads to the phase transition with symmetry breaking at the moment $t = t_c \sim \sqrt{\lambda} M_p m^{-2}$ when $m_{\phi,eff} = 0$, $\langle(\delta\phi)^2\rangle = \sigma^2/3$ and $E_\chi \sim m$. Note that the homogeneous component $\phi(t)$ at this moment is significantly less than $\sqrt{\langle(\delta\phi)^2\rangle}$ due to its further decay after preheating.

The mechanism of symmetry restoration described above is very general. In particular, it explains a surprising behaviour of oscillations of the scalar field found numerically in the $O(N)$-symmetric model of Boyanovsky *et al.* (1995a). It is important that during the interval between preheating and the establishment of thermal equilibrium the universe could experience a series of phase transitions which we did not anticipate before. For example, cosmic strings and textures, which could be an additional source for the formation of the large scale structure of the universe, should have $\sigma \sim 10^{16}$ GeV. To produce them by thermal phase transitions in our model one should have the temperature after reheating greater than 10^{16} GeV, which is extremely hard to obtain (Kofman & Linde 1987). Meanwhile, as we now see, fluctuations produced at preheating may be quite sufficient to restore the symmetry. Then the topological defects are produced in the standard way when the symmetry breaks down again. In other words, production of superheavy topological defects can be easily compatible with inflation.

On the other hand, the topological defect production can be quite dangerous. For example, the model (7.29) of a one-component real scalar field ϕ has a discrete symmetry $\phi \to -\phi$. As a result, after the phase transition induced by fluctuations $\langle(\delta\phi)^2\rangle$ the universe may become filled with domain walls separating phases $\phi = +\sigma$ and $\phi = -\sigma$. This is expected to lead to a cosmological disaster. Investigation shows that in the theory (7.29) with $\sigma \ll 10^{16}$ Gev fluctuations destroy the coherent distribution of the background oscillations of ϕ and divide the universe in an equal number of domains with $\phi = \pm\sigma$, which leads to the domain wall problem. This mean that in consistent inflationary models of the type of (7.29) one should have either no symmetry breaking or $\sigma \geq 10^{16}$ Gev.

Similar effects of the nonthermal phase transitions occur in the models where the symmetry breaking occurs for fields other than the inflaton field ϕ (Kofman, Linde & Starobinsky 1996a). The simplest model has an effective potential (e.g., Kofman & Linde 1987; Linde 1994):

$$V(\phi,\chi) = \frac{\lambda}{4}\phi^4 + \frac{\alpha}{4}\left(\chi^2 - \frac{M^2}{\alpha}\right)^2 + \frac{1}{2}g^2\phi^2\chi^2 \, . \tag{7.31}$$

We will assume here that $\lambda \ll \alpha < g^2$, so that at large ϕ the curvature of the potential in the χ-direction is much greater than in the ϕ-direction. In this case at large ϕ the field χ rapidly rolls toward $\chi = 0$. An interesting feature of such models is the symmetry restoration for the field χ for $\phi > \phi_c = M/g$, and symmetry breaking when the inflaton field ϕ becomes smaller than ϕ_c. As emphasized by Kofman & Linde (1987), such phase transitions may lead to formation of topological defects without any need for high-temperature effects.

We would now like to point out some other specific features of such models. If the phase transition discussed above occurs during inflation (i.e. if $\phi_c > M_p$ in our model), then no new phase transitions occur in this model after reheating. However, for $\phi_c \ll M_p$ the situation is much more complicated. First of all, in this case the field ϕ oscillates with the initial amplitude $\sim M_p$ (if $M^4 < \alpha\lambda M_p^4$). This means that each time when the absolute value of the field becomes smaller than ϕ_c, the phase transition with symmetry breaking occurs and topological defects are produced. Then the absolute value of the oscillating field ϕ again becomes greater than ϕ_c, and symmetry is restored again. However, this regime does not continue for too long a time. Within several oscillations, quantum fluctuations of the field χ will be generated with the dispersion $\langle(\delta\chi)^2\rangle \sim g^{-1}\sqrt{\lambda}M_p^2$. For $M^2 < g^{-1}\sqrt{\lambda}\alpha M_p^2$, these fluctuations will keep the symmetry restored. The symmetry breaking will be finally completed when $\langle(\delta\chi)^2\rangle$ becomes small enough. Riotto and Tkachev (1996) noted that in the model (7.31) in the case of the strong self-interaction of χ particles, $\lambda \ll g^2 \ll \alpha$, the presence of a huge occupation number of the created χ particles $n_E \sim 1/g^2$ would require treatment beyond the one-loop approximation. It can be shown, however, that in this case the preheating stops earlier when $n_E \sim 1/\alpha$ (Kofman, Linde & Starobinsky 1996b).

One may imagine even more complicated scenario when oscillations of the scalar field ϕ create large fluctuations of the field χ, which in their turn interact with the scalar fields Φ breaking symmetry in GUTs. We would then have phase transitions in GUTs induced by the fluctuations of the field χ. Note that in the models considered in this section the field χ does not oscillate near $\chi = 0$ prior to the phase transition, since such oscillations are damped out during the long stage of inflation prior to the phase transition. Thus oscillations of the field χ in the theory (7.31) definitely do not suppress the topological defect production. This means that no longer can the absence of primordial monopoles be considered as an automatic consequence of inflation. To avoid the monopole production one should use the theories where quantum fluctuations produced during preheating are small or decoupled from the GUT sector. This condition imposes additional constraints on realistic inflationary models.

8. Conclusions

We have found that the decay of the inflaton field typically begins with the explosive production of particles during the stage of preheating in the regime of broad parametric resonance. During the second stage, the inflaton field decays further in the regime of narrow resonance, and particles produced at this stage decay into other particles and interact among themselves. The third stage is the thermalisation. The two last stages require much more time that the stage of preheating. Typically, coupling constants of interaction of the inflaton field with matter are extremely small, whereas coupling constants involved in the decay of other particles can be much greater. As a result, the reheating temperature T_r given by (3.12) will not necessarily be defined by the perturbative decay rate of the inflaton oscillations (3.9). T_r can be much higher than the typical temperature $T_r \lesssim 10^9$ GeV, which would be obtained if we neglected the stage of parametric resonance.

A specific feature of the preheating is that bosons produced at this stage are far from thermal equilibrium and typically have enormously large occupation numbers. This may have very interesting applications to cosmology. There is a new class of phase transitions that may occur at the intermediate stage between the end of inflation and the establishment of thermal equilibrium. These phase transitions take place even in the theories where the scale of spontaneous symmetry breaking is comparable to M_p and where the reheating temperature is very small. Therefore, phase transitions of the new type may have dramatic consequences for inflationary models and the theory of physical processes in the very early universe.

The preheating may help with the baryogenesis problem (Kofman, Linde & Starobinsky 1996a; Yoshimura 1996). In the models of GUT baryogenesis, it was assumed that GUT symmetry was restored by high-temperature effects, since otherwise the density of X, Y, and superheavy Higgs bosons would be very small. This condition is hardly compatible with inflation. It was also required that the products of decay of these particles should stay out of thermal equilibrium, which is a very restrictive condition. In our case, the superheavy particles responsible for baryogenesis can be abundantly produced by parametric resonance, and they as well as the products of their decay will be out of thermal equilibrium until the end of reheating.

Another consequence of the resonance effects is a fast change of the equation of state from a vacuum-like one to the equation of state of relativistic matter. This leads to suppression of primordial black holes formation that could be produced after inflation.

The preheating stage may appear not only in the models where the reheating occurs due to the decay of the homogeneous scalar field. Kolb & Riotto (1996) suggested that preheating may also occur in first-order inflation. It would be very interesting to investigate this problem, as well as many other exciting possibilities which the new theory of reheating may offer us.

REFERENCES

ABBOTT, L.F., FARHI, E. & WISE, M. 1982 Particle Production in the New Inflationary Cosmology. *Phys. Lett.* **117B**, 29–34.

ALBRECHT, P.J., STEINHARDT, M.S. TURNER AND F. WILCZEK 1982 Reheating an Inflationary Universe. *Phys. Rev. Lett.* **48**, 1437–1440.

BOYANOVSKY, D., DE VEGA, H.J., HOLMAN, R., LEE, D.S. & SINGH, A. 1995a Dissipation via Particle Production in Scalar Field Theories. *Phys. Rev.* **D51**, 4419–4444.

BOYANOVSKY, D., D'ATTANASIO, M., DE VEGA, H.J., HOLMAN, R., LEE, D.-S. & SINGH, A. 1995b Reheating and Thermalisation: Linear vs. Non-linear Relaxation. *Phys. Rev.* **D52**, 6805–6827.

DOLGOV, A.D. & LINDE, A.D. 1982 Baryon Asymmetry in Inflationary Universe. *Phys. Lett.* **116B**, 329–334.

DOLGOV, A.D. & KIRILOVA, D.P. 1990 Production of Particles by a Variable Scalar Field. *Sov. Nucl. Phys.* **51**, 172–177.

FUJISAKI, H., KUMEKAWA, K., YAMAGUCHI, M. & YOSHIMURA, M. 1995 Particle Production & Dissipative Cosmic Field. *hep-ph/9508378.*

KAISER, D. 1995 Post-Inflation Reheating in an Expanding Universe. *Phys. Rev.* **53**, 1776–1783.

KHLEBNIKOV, S. & TKACHEV, I. 1996 Classical Decay of Inflaton. *hep-ph/9603378.*

KIRZHNITS, D.A. & LINDE, A.D. 1972 Macroscopic Consequences of the Weinberg-Salam Model. Phys. Lett. **42B**, 471–474.

KOFMAN, L. & LINDE, A. 1987 Generation of the Density Fluctuations in the Inflationary Cosmology. *Nucl. Phys.* **B282**, 555–585.

KOFMAN, L., LINDE, A. & STAROBINSKY, A. 1994 Reheating after Inflation *Phys. Rev. Lett.* **73**, 3195–3198.

KOFMAN, L., LINDE, A. & STAROBINSKY, A. 1996a Nonthermal Phase Transitions after Inflation. *Phys. Rev. Lett.* **76**, 1011–1014.

KOFMAN, L., LINDE, A. & STAROBINSKY, A. 1996b Theory of Reheating after Inflation. *In preparation.*

KOLB, E. & RIOTTO, A. 1996 Preheating and Symmetry Restoration in Collisions of Vacuum Bubbles. *astro-ph/9602095.*

KOLB, E. & TURNER, M. 1990 *The Early Universe.* Addison-Wesley, Redwood.

LINDE, A.D. 1990 *Particle Physics and Inflationary Cosmology* . Harwood, Chur, Switzerland.

LINDE, A.D. 1994 Hybrid Inflation. *Phys. Rev.* **D49**, 748–754.

MAC LACHLAN, N.W. 1961 *Theory and Application of Mathieu functions.* Dover.

POLARSKI, D. & STAROBINSKY, A. 1996 Semiclassicality and Decoherence of Cosmological Perturbations. *Class. Quant. Grav.* **13**, 377–392.

SHTANOV, YU., TRASCHEN, J. & BRANDENBERGER, R. 1995 Universe Reheating after Inflation *Phys. Rev.* **D51**, 5438–5455.

SON, D. 1996 Reheating and Thermalisation in a Simple Scalar Model. *hep-ph/9604340.*

STAROBINSKY, A.A. 1982 Nonsingular Model of the Universe with the Quantum-Gravitational De Sitter Stage and its Observational Consequences. In *Quantum Theory of Gravity II, Moscow 1981* (ed. Markov, M.A. *et al.*) pp.58–72.

TKACHEV, I. 1995 Phase Transition and Preheating. *hep-th/9510146.*

TRASCHEN, J. & BRANDENBERGER, R. 1990 Particle Production during Out-of-Equilibrium Phase Transition. *Phys. Rev.* **D42**, 2491–2504.

YOSHIMURA, M. 1995 Catastrophic Particle Production under Periodic Perturbations. *Progr. Theor. Phys.* **94**, 873–898.

YOSHIMURA, M. 1996 Baryogenesis and Thermal History after Inflation. *hep-ph/9605246.*

ZEL'DOVICH, YA.B. 1972 The Creation of Particles and Antiparticles in Electric and Gravitational Fields. In *Magic without Magic: J.A.Wheeler* (ed. J. Klauder), pp. 277–288. Freeman.

ZEL'DOVICH, YA. B. & STAROBINSKY, A.A. 1972 Particle Production and Vacuum Polarisation in an Anisotropic Gravitational Field. *Sov.Phys–JETP* **34**, 1159–1166.

New Guises for Old Dark-Matter Suspects

By EDWARD W. KOLB

NASA/Fermilab Theoretical Astrophysics Group
Fermi National Accelerator Laboratory, Box 500, Batavia, Illinois, 60510, USA

The three most popular suspects for particle dark matter are massive neutrinos, axions, and supersymmetric relics. There are now well developed early-universe plots capable of placing each of the suspects at the scene today to be the dark matter of the universe. Because Igor Novikov has contributed in a fundamental way to the idea of a neutrino-dominated universe and knows that story well, I thought he would like to hear something about the other two suspects. But rather than retelling the standard story, I will point out that there may be new twists to the dark matter story, and that old, familiar suspects may hide in unfamiliar guises.

1. Introduction

One of the simplest questions one might ask about the universe is, "What is it made of?" It is somewhat of an embarrassment for modern cosmologists to be forced to answer the question with the shrug of the shoulders. Although today we see more of the universe than ever before, everything we do see reinforces the fact that there is more to the universe than meets the eye. On scales as small as our galaxy to scales as large as the Hubble radius, most of the mass of the universe seems to be invisible to us. The nature of the ubiquitous dark matter is perhaps the most fundamental question in cosmology today.

Of particular interest to cosmologists interested in the early universe is the possibility that some fraction of the dark matter is non-baryonic, in the form of some species of elementary particle. In most talks on particle dark matter, the first step in the investigation is to round up the usual suspects for identification. At the top of the most-wanted list is the neutrino. Of all the suspects for dark matter, the neutrino is the only one *known* to exist. A mass for the neutrino even as small as a few eV would mean that the neutrino is dynamically important in understanding the formation and evolution of structure in the present universe. Furthermore, without a known principle to require the neutrino to be massless, the neutrino cannot use the alibi that it is naturally massless.†

There is still reasonable doubt about neutrinos as dark matter because it is very hard to construct a scenario where light neutrinos, by themselves, are responsible for structure formation. The problem is that by the time neutrinos would come to dominate the mass of the universe and structure would begin to form, they would have traveled a rather large distance since they first decoupled in the early universe. Because of collisionless phase mixing (or Landau damping), initial perturbations on scales smaller than the free-streaming length, $20\,\mathrm{Mpc} \times (30\,\mathrm{eV}/m_\nu)$ for neutrinos of mass m_ν, would be suppressed. Such a dark matter species with a relatively large velocity just before matter domination, like a light neutrino species, is known as hot-dark matter, or HDM for short.

In order to develop large-scale structure from small initial seeds, most cosmologists favor a type of dark matter known as cold dark matter, or CDM. The matter is called cold, because at the time when the universe first became matter dominated the dark matter was extremely nonrelativistic, and the collisionless damping length was smaller than any scale of astrophysical interest. The list of dark-matter suspects is very long, exceeding in number even the number of characters in a classic Russian novel. But two

† There is no symmetry principle that demands the neutrino must be massless in the way that gauge invariance implies that the photon must be massless.

on the CDM list might be characterized as "prime" suspects: axions and supersymmetric relics. These are the two I will discuss in this talk.

Before starting, it is useful to dispense with the preliminary definitions. It is convenient to express the present mass density of a particle species in terms of its contribution to something known as the critical density. If the density of a matter-dominated universe is greater than the critical density, the universe will eventually recollapse, while if the density is less than the critical density the universe will expand forever. In terms of Hubble's constant H_0 and Newton's constant G, the critical density is $\rho_C \equiv 3H_0^2/8\pi G = 1.88h^2 \times 10^{-29}\text{g cm}^{-3}$. The annoying factor of h accounts for the imprecision in our knowledge of Hubble's constant: $H_0 = 100h$ km s^{-1} Mpc^{-1}. The energy density in some species "i" is given as a ratio to the critical density, and denoted as $\Omega_i \equiv \rho_i/\rho_C$.

2. Supersymmetric relics

2.1. *Supersymmetry and Supersymmetry breaking*

There are two fundamental reasons for believing that nature is supersymmetric. The first reason is that supersymmetry can rescue the standard electroweak model from the embarrassment of finely tuned coupling constants. The standard electroweak model employs fundamental scalars, usually refereed to as "Higgs" scalars, to break the gauge symmetry spontaneously. But scalar particles have very bad ultraviolet behavior, which has the effect of dragging the electroweak Higgs mass up to the mass scale of any encompassing theory, such as a grand-unified theory. Thus, unless coupling constants are very finely tuned or some other dynamics enters the picture, light scalar masses (of order the electroweak scale) would not be possible. Supersymmetry (SUSY) is an example of "some other dynamics." Because of the relative factor of -1 between fermionic and bosonic loops, the addition of fermionic loops can mitigate the bad ultraviolet behavior of scalar loops. The way to realize this possibility is if for every boson there is a corresponding fermion appearing in the calculation of the quantum corrections to the Higgs mass.

This correspondence between fermions and bosons implies that both fermions and bosons appear in multiplets, and they are transformed into each other by supersymmetry transformations. Thus, SUSY is intimately related to Poincare symmetry. In fact, the commutator of SUSY transformations generates the momentum operator. SUSY is the only known way to unify spacetime and the internal symmetries of the S-matrix. Thus, SUSY seems to be a fundamental part of any attempt to unify gravity with the fundamental forces. This aesthetic reason is the second motivation for SUSY.

The particular realization of SUSY I will consider is the supersymmetric extension of the standard model. Although this model, the minimal supersymmetric standard model (MSSM), has many parameters in addition to the plethora of parameters of the non-symmetric version of the standard model, it is sufficiently restrictive to have some predictive power.

Of course in nature SUSY is broken—there is no massless fermionic photon for instance. I will return to the question of SUSY breaking in a moment. But the first relevant issue for SUSY dark matter is the existence of something known as R-parity, a discrete multiplicative symmetry. The R-parity of a particle is given in terms of its spin S, baryon number B and lepton number L, by $R = (-1)^{3(B-L)+2S}$. Known particles all have even R-parity, while their SUSY partners are all R-odd particles. If R-parity is conserved, then the lightest R-odd color singlet (LROCS) must be stable—and hence a candidate for dark matter. There are inconveniences with any theory if R-parity is broken. Rapid proton decay, for instance. If one works hard enough these difficulties can be overcome, but the assumption of exact R-parity is very attractive and naturally leads to dark matter candidates. So I will assume exact R parity.

Now let's return to the issue of SUSY breaking. The details of SUSY breaking will determine the identity of the LSP, as well as its mass and interaction strength. Unfortunately, essentially nothing is known about SUSY breaking. The only reasonable constraint one might imagine is that any SUSY-breaking Lagrangian terms must have mass dimension less than four.† Since we have no other choice, let's consider all possible dimension-two and dimension-three SUSY-breaking terms consistent with gauge and Lorentz symmetry:

$$\begin{aligned}\Delta\mathcal{L}(\text{SUSY BREAKING}) &= -m_1^2|H_1|^2 - m_2^2|H_2|^2 - m_{12}^2(H_1H_2 + H_1^*H_2^*)\\ &-\widetilde{Q}^\dagger_{Li}(M^2_{\widetilde{Q}})_{ij}\widetilde{Q}_{Lj} - \widetilde{u}^\dagger_{Ri}(M^2_{\widetilde{u}})_{ij}\widetilde{u}_{Rj} - \widetilde{d}^\dagger_{Ri}(M^2_{\widetilde{d}})_{ij}\widetilde{d}_{Rj} - \widetilde{L}^\dagger_{Li}(M^2_{\widetilde{L}})_{ij}\widetilde{L}_{Lj} - \widetilde{e}^\dagger_{Ri}(M^2_{\widetilde{e}})_{ij}\widetilde{e}_{Rj}\\ &-H_2\widetilde{Q}_{Li}(h_uA_u)_{ij}\widetilde{u}_{Rj} - H_1\widetilde{Q}_{Li}(h_dA_d)_{ij}\widetilde{d}_{Rj} - H_1\widetilde{L}_{Li}(h_eA_e)_{ij}\widetilde{e}_{Rj}\\ &-\frac{1}{2}M_1\overline{\widetilde{B}}\,\widetilde{B} - \frac{1}{2}M_2\overline{\widetilde{W}^a}\widetilde{W}^a - \frac{1}{2}M_3\overline{\widetilde{G}^a}\widetilde{G}^a.\end{aligned} \tag{2.1}$$

The tilde superscript denotes the SUSY partner of familiar particles: $\widetilde{e}$ is the selectron, $\widetilde{d}$ is the down squark, $\widetilde{G}^a$ are the gluino fields, $\widetilde{W}^a$ and $\widetilde{B}$ are the supersymmetric partners of the familiar $SU(2)$ and $U(1)$ gauge fields, and $\widetilde{Q}$ and $\widetilde{L}$ are $SU(2)$ doublets containing the SUSY partners of left-handed quarks and leptons. The fields H_1 and H_2 are the two Higgs necessary in SUSY. The parameters $M^2_{..}$ and $A_{..}$ are 3×3 symmetric matrices. The matrix $A_{..}$ has mass dimension one. Note that the operators appearing in the first two lines of (2.1) are operators of mass dimension two, while the last two lines contain operators of mass dimension three.

The usual procedure is to choose a set of parameters including the constants appearing in (2.1), requiring that the resulting low-energy theory leads to the usual standard model. The choice of these mass parameters, along with the Higgsino mass parameter μ, results in a mass matrix for the neutralinos: the bino $\widetilde{B}$, the zino $\widetilde{W}^3$, and the Higgsinos $\widetilde{H}_1^0$ and $\widetilde{H}_2^0$. In terms of the mass of the Z, the weak mixing angle θ_W, and $\tan\beta$ (the ratio of the vacuum expectation values of the two Higgs fields responsible for electroweak symmetry breaking), the neutralino mass matrix in the basis ($\widetilde{B}$, $\widetilde{W}^3$, $\widetilde{H}_1^0$, $\widetilde{H}_2^0$) is given as

$$\begin{pmatrix} M_1 & 0 & -m_Z\cos\beta\sin\theta_W & m_Z\sin\beta\sin\theta_W\\ 0 & M_2 & m_Z\cos\beta\cos\theta_W & -m_Z\sin\beta\cos\theta_W\\ -m_Z\cos\beta\sin\theta_W & m_Z\cos\beta\cos\theta_W & 0 & -\mu\\ m_Z\sin\beta\sin\theta_W & -m_Z\sin\beta\cos\theta_W & -\mu & 0\end{pmatrix}. \tag{2.2}$$

The SUSY-breaking masses M_1 and M_2 are commonly assumed to be of order m_Z or larger, and if the SUSY model is embedded in a grand unified theory, then $3M_1/M_2 = 5\alpha_1/\alpha_2$. If we assume the relation between M_1 and M_2, then there are three parameters in the neutralino mass matrix: μ, $\tan\beta$, and $M_{1/2}$ (the zino-bino mass parameter).

Now the game is set: for a given set of parameters, diagonalize the mass matrix, find the mass of the lightest supersymmetric particle and its field content (of course in general it is a linear combination of the four neutralino fields), determine its annihilation cross section, and put the above information into the freeze-out machinery to determine the relic abundance.

Different groups who study the problem come up with slightly different composite

† Dimension-four SUSY-breaking terms suffer from the bad ultraviolet behavior we are trying to fix.

sketches for the dark matter suspect. Some believe the particle content is mostly Higgsino, while some find mostly bino. However just about all groups find a relatively large mass for the suspect, between 30 and several hundred GeV.

In this part of the talk† I would like to propose a different picture for the wanted poster: a particle of low mass (500 MeV to 1.6 GeV) and "photino-like," with an $SU(2) \times U(1)$ content almost identical to the photon.

The motivation for this light photino comes from our lack of knowledge about SUSY-breaking. Referring to the SUSY-breaking terms in (2.1) we see that there are dimension-two and dimension-three terms. There are theoretical reasons to believe that dimension-three terms might be much smaller than the dimension-two terms. It appears difficult to break SUSY dynamically in a way that produces dimension-three terms while avoiding problems associated with the addition of gauge-singlet superfields. In models where SUSY is broken dynamically or spontaneously in the hidden sector and there are no gauge singlets, all dimension-three SUSY-breaking operators in the effective low-energy theory are suppressed compared to SUSY-breaking scalar masses by a factor of $\langle\Phi\rangle/m_{Pl}$, where $\langle\Phi\rangle$ is the vacuum expectation value of some hidden-sector field and m_{Pl} is the Planck mass. Thus, dimension-three terms may not contribute to the low energy effective Lagrangian. This would imply that at the tree level the gluino is massless, and the neutralino mass matrix is given by (2.1) with vanishing (00) and (11) entries. However, non-zero contributions to the gluino mass and the neutralino mass matrix come from two sources: radiative corrections such as the top-stop loops for the gluino and neutralinos, and "electroweak" loops involving higgsinos and/or winos and binos for the neutralinos (but not for the gluino).

The generation of radiative gaugino masses in the absence of dimension-three SUSY breaking was studied by Farrar and Masiero. They found (taking $\mu \gtrsim 40$ GeV) that as the typical SUSY-breaking scalar mass, M_0, varies between 100 and 400 GeV, the gluino mass ranges from about 700 to about 100 MeV, while the photino mass ranges from around 400 to 900 MeV. This estimate for the photino mass should be considered as merely indicative of its possible value, since an approximation for the electroweak loop is strictly valid only when μ and M_0 are much larger than m_W. The other neutralinos are much heavier, and the production rates of the photino and the next-lightest neutralino in Z^0 decay are consistent with all known bounds.

The conclusion is that light gluinos and photinos are quite consistent with present experiments, and result in a number of striking predictions. One prediction is that the photino $\widetilde{\gamma}$ should be the relic R-odd particle, even though it may be more massive than the gluino. This is because below the confinement transition the gluino is bound with a gluon into a color-singlet hadron, the R^0, whose mass (which is in the 1 to 2 GeV range when the gluino is very light) is greater than that of the photino. In this scenario, LSP is an ambiguous term: the gluino is lighter than the photino, although the photino is lighter than the R^0. As discussed above, a more relevant term would be LROCS—*lightest R-odd color singlet.*

However, models with light gauginos were widely thought to be disallowed because it had been believed that in such models the relic density of the lightest neutralino would exceed cosmological bounds unless R-parity would be violated allowing the relic to decay. In the next subsection I will rehash that argument, and then point out how the restriction can be evaded if the R^0 mass is close to the $\widetilde{\gamma}$ mass (here "close" means within a factor of two).

† Reference to all the material presented here can be found in Farrar and Kolb (1996) and Chung, Farrar, and Kolb (1996)

2.2. *Self-annihilation and freeze out*

The reaction rates that determine freeze out will depend upon the $\widetilde{\gamma}$ and R^0 masses, the cross sections involving the $\widetilde{\gamma}$ and R^0, and possibly the decay width of the R^0 as well. In turn the cross sections and decay width also depend on the masses of the $\widetilde{\gamma}$, $\widetilde{g}$ and R^0, as well as the masses of the squarks and sleptons. We will denote the squark/slepton masses by a common mass scale (expected to be of order 100 GeV). Even if the masses were known and the short-distance physics specified, calculation of the width and some of the cross sections would be no easy task, because one is dealing with light hadrons. Fortunately, our conclusions are reasonably insensitive to individual masses, lifetimes, and cross sections, but depend crucially upon the R^0-to-$\widetilde{\gamma}$ mass ratio, denoted by r. When we do need an explicit value of the photino mass m, the R^0 mass M, or the squark/slepton mass $M_{\widetilde{S}}$, we will parameterize them by the dimensionless parameters μ_8, r, and μ_S:

$$m = 800\mu_8\,\mathrm{MeV}; \qquad M = rm; \qquad M_{\widetilde{S}} = 100\mu_S\,\mathrm{GeV}. \tag{2.3}$$

The standard procedure for the calculation of the present number density of a thermal relic of the early universe is to assume that the particle species was once in thermal equilibrium until at some point the rates for self-annihilation and pair-creation processes became much smaller than the expansion rate, and the particle species effectively froze out of equilibrium. After freeze out, its number density decreased only because of the dilution due to the expansion of the universe.

Since after freeze out the number of particles in a *comoving* volume is constant, it is convenient to express the number density of the particle species in terms of the entropy density, which in a comoving volume element is also constant for most of the history of the universe. This number-density-to-entropy ratio is usually denoted by Y. If a species of mass m is in equilibrium and nonrelativistic, Y is given simply in terms of the mass-to-temperature ratio $x \equiv m/T$ as $Y_{EQ}(x) = 0.145(g/g_*)x^{3/2}\exp(-x)$, where g is the number of spin degrees of freedom, and g_* is the total number of relativistic degrees of freedom in the universe at temperature $T = m/x$. Well after freeze out $Y(x)$ is constant—we will denote this asymptotic value of Y as Y_∞.

If self annihilation determines the final abundance of a species, Y_∞ can be found by integrating the Boltzmann equation (dot denotes d/dt) $\dot{n}+3Hn = -\langle|v|\sigma_A\rangle\left(n^2 - n_{EQ}^2\right)$, where n is the number density, n_{EQ} is the equilibrium number density, H is the expansion rate of the universe, and $\langle|v|\sigma_A\rangle$ is the thermal average of the annihilation rate.

There are no general closed-form solutions to the Boltzmann equation, but there are reliable, well tested approximations for the late-time solution, i.e., Y_∞. Then with knowledge of Y_∞, the contribution to Ωh^2 from the species can easily be found. Let us specialize to the survival of photinos assuming self-annihilation determines freeze out.

Calculation of the relic abundance involves first calculating the freeze-out value of x, known as x_f, where the abundance starts to depart from the equilibrium abundance. Using standard approximate solutions to the Boltzmann equation gives $x_f = \ln(0.0481 m_{Pl} m\sigma_0) - 1.5\ln[\ln(0.0481 m_{Pl} m\sigma_0)]$, where we have used $g=2$, $g_*=10$, and parameterized the nonrelativistic annihilation cross section as $\langle|v|\sigma_A\rangle = \sigma_0 x^{-1}$. Using the diagram shown in Figure 1, $\sigma_0 = 2\times 10^{-11}\mu_8^2\mu_S^{-4}\mathrm{mb}$, which leads to $x_f \simeq 12.3 + \ln(\mu_8^3/\mu_S^4)$. The value of x_f determines Y_∞:

$$Y_\infty = \frac{2.4x_f^2}{m_{Pl} m\sigma_0} \simeq 7.4\times 10^{-7}\mu_8^{-3}\mu_S^4. \tag{2.4}$$

Once Y_∞ is known, the present photino energy density can be easily calculated: $\rho_{\widetilde{\gamma}} = mn_{\widetilde{\gamma}} = 0.8\,\mu_8\mathrm{GeV}\cdot Y_\infty\,2970\,\mathrm{cm}^{-3}$. When this result is divided by the critical density,

$\rho_C = 1.054h^2 \times 10^{-5}$ GeV cm^{-3}, the fraction of the critical density contributed by the photino is found to be $\Omega_{\widetilde{\gamma}}h^2 = 2.25 \times 10^8 \mu_8 Y_\infty$. For Y_∞ in (2.4), $\Omega_{\widetilde{\gamma}}h^2 = 167\mu_8^{-2}\mu_S^4$.

The age of the universe restricts $\Omega_{\widetilde{\gamma}}h^2$ to be less than one, so for $\mu_S = 1$, the photino must be more massive than about 10 GeV or so if its relic abundance is determined by self-annihilation.

2.3. R^0-*catalyzed freeze out*

Farrar and I pointed out that for models in which both the photino and the gluino are light, freeze out is not determined by photino self annihilation, but by $\widetilde{\gamma} \longleftrightarrow R^0$ interconversion. The basic point is that since the R^0 has strong interactions, it will stay in equilibrium longer than the photino, even though it is more massive. As long as $\widetilde{\gamma} \longleftrightarrow R^0$ interconversion occurs at a rate larger than H, then through its interactions with the R^0, the photino will be able to maintain its equilibrium abundance even after self annihilation has frozen out.

Griest and Seckel discussed the possibility that the relic abundance of the lightest species is determined by its interactions with another species. They concluded that the mass splitting between the relic and the heavier particle must be less than 10% for the effect to be appreciable. We find that $\widetilde{\gamma} \longleftrightarrow R^0$ interconversion determines the $\widetilde{\gamma}$ relic abundance even though the R^0 may be twice as massive as the $\widetilde{\gamma}$. The difference arises because Griest and Seckel assumed that all cross sections were roughly the same order of magnitude. But in our case the R^0 annihilation is about 10^{12} times larger than other relevant cross sections.

I will now consider in turn the reactions we found to be important in our scenario. The diagrams for the individual constituent processes can be found in Figure 1. However, as we shall see, it is not a simple task to go from the constituent diagrams to the cross sections and decay width.

$\underline{\widetilde{\gamma}\widetilde{\gamma} \rightarrow X}$: For photino self-annihilation at low energies the final state X is a lepton-antilepton pair, or a quark-antiquark pair which appears as light mesons. The process involves the t-channel exchange of a virtual squark or slepton between the photinos, producing the final-state fermion-antifermion pair. (See the upper third of Figure 1.) In the low-energy limit where the mass $M_{\widetilde{S}}$ of the squark/slepton is much greater than $\sqrt{s}$, the photino-photino-fermion-antifermion operator appears in the low-energy theory with a coefficient proportional to $e_i^2/M_{\widetilde{S}}^2$, with e_i the charge of the final-state fermion. Also, as there are two identical fermions in the initial state, the annihilation proceeds as a p-wave, which introduces a factor of v^2 in the low-energy cross section. The resultant low-energy photino self-annihilation cross section is:

$$\langle |v| \sigma_{\widetilde{\gamma}\widetilde{\gamma}} \rangle = 8\pi\alpha_{EM}^2 \sum_i q_i^4 \frac{m^2}{M_{\widetilde{S}}^4} \frac{v^2}{3} \simeq 2.0 \times 10^{-11}\ x^{-1}\ \left[\mu_8^2 \mu_S^{-4}\right] \text{ mb}, \tag{2.5}$$

where we have used for the relative velocity $v^2 = 6/x$ with $x \equiv m/T$, and q_i is the magnitude of the charge of a final-state fermion in units of the electron charge. For the light photinos we consider, summing over e, μ, and three colors of u, d, and s quarks leads to $\sum_i q_i^4 = 8/3$.

$\underline{R^0 R^0 \rightarrow X}$: In R^0 self-annihilation, at the constituent level the relevant reactions are $\widetilde{g}+\widetilde{g} \rightarrow g+g$ and $\widetilde{g}+\widetilde{g} \rightarrow q+\bar{q}$ (see the middle third of Figure 1), which are unsuppressed by any powers of $M_{\widetilde{S}}$, so the cross section should be typical of strong interactions. In the nonrelativistic limit we expect the R^0R^0 cross section to be comparable to the $\bar{p}p$ cross section, but with an extra factor of v^2, accounting for the fact that there are identical fermions in the initial state so annihilation proceeds through a p-wave. There is some en-

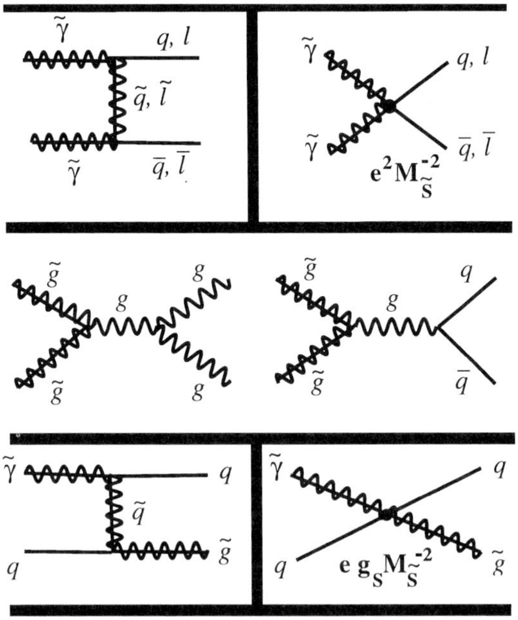

FIGURE 1. Feynman diagrams for the constituent processes determining the relic photino abundance. The top left diagram is for $\tilde{\gamma}$ self annihilation, and on the top right is the effective low-energy operator for that process. The two diagrams in the middle are the diagrams for R^0 self annihilation. Finally in the lower left-hand corner is the diagram for the $\tilde{\gamma} \longleftrightarrow R^0$ interconversion processes (all interconversion processes can be obtained from crossings of this diagram), and on the lower right-hand side is the effective low-energy operator.

ergy dependence to the $\bar{p}p$ cross section, but it should be sufficient to consider $\langle |v|\sigma_{R^0R^0}\rangle$ to be a constant, approximately given by $\langle |v|\sigma_{R^0R^0}\rangle \simeq 100v^2\,\text{mb} = 600\,x^{-1}r^{-1}\,\text{mb}$, where we have used for the relative velocity $v^2 = 6T/M = 6/(rx)$, with $x \equiv m/T$.

We should note that the thermal average of the cross section might be even larger if there are resonances near threshold. In any case, this cross section should be much larger than any cross section involving the photino, and for the relatively small values of r we employ it will ensure that the R^0 remains in equilibrium longer than the $\tilde{\gamma}$, greatly simplifying our considerations.

$\underline{\tilde{\gamma}R^0 \to X}$: This is an example of a phenomenon known as co-annihilation whereby the particle of interest (in our case the photino) disappears by annihilating with another particle (here, the R^0). Of course co-annihilation also leads to a net decrease in R-odd particles. We can estimate the cross section for $\tilde{\gamma}R^0 \to X$ in terms of the $\tilde{\gamma}$ self annihilation cross section by comparing the lower third of Figure 1 to the upper third:

$$\langle |v|\sigma_{\tilde{\gamma}R^0}\rangle \simeq \frac{\alpha_S}{\alpha_{EM}}\frac{4}{3}\frac{2}{8/3}\frac{M}{m}\frac{3}{v^2}\,\langle |v|\sigma_{\tilde{\gamma}\tilde{\gamma}}\rangle, \tag{2.6}$$

where the ratio of α's arises because the short-distance operator for co-annihilation is proportional to $e_i^2 g_S^2$ rather than e_i^4, the second factor is the color factor coming from the gluino coupling, and the third factor comes from the ratio of $\sum_i q_i^2/\sum_i q_i^4$ for the participating fermions. We have replaced m^2 appearing in (2.5) by mM, although the actual dependence on m and M may be more complicated. Finally, the annihilation is s-wave so there is no $v^2/3$ suppression as in photino self-annihilation.

Although the short-distance physics is perturbative, the initial gluino appears in a light hadron, and there are complications in the momentum fraction of the R^0 carried

Table I: Cross sections and the decay width used in the calculation of the relic photino abundance. The dimensionless parameters μ_8 and μ_S were defined in (2.3), and $\mathcal{F}(r)$ was defined below (2.9). The coefficients A, B, and C reflect uncertainties involving the calculation of hadronic matrix elements.

Process		Cross section or width
R^0 self annihilation:	$\langle \lvert v \rvert \sigma_{R^0 R^0} \rangle$	$600\ x^{-1}\ r^{-1}$ mb
$\widetilde{\gamma}$ self annihilation:	$\langle \lvert v \rvert \sigma_{\widetilde{\gamma}\widetilde{\gamma}} \rangle$	$2.0 \times 10^{-11}\ x^{-1}\ \left[\mu_8^2 \mu_S^{-4}\right]$ mb
co-annihilation:	$\langle \lvert v \rvert \sigma_{\widetilde{\gamma} R^0} \rangle$	$1.5 \times 10^{-10}\ r\ \left[\mu_8^2 \mu_S^{-4} A\right]$ mb
R^0 decay:	$\Gamma_{R^0 \to \widetilde{\gamma}\pi\pi}$	$2.0 \times 10^{-14}\ \mathcal{F}(r)\ [\mu_8^5 \mu_S^{-4} B]$ GeV
$\widetilde{\gamma}$ - R^0 conversion:	$\langle \lvert v \rvert \sigma_{R^0 \pi} \rangle$	$1.5 \times 10^{-10}\ r\ \left[\mu_8^2 \mu_S^{-4} C\right]$ mb

by the gluino and other non-perturbative effects. For our purposes it will be sufficient to account for the uncertainty by including in the cross section an unknown coefficient A, leading to a final expression

$$\langle |v| \sigma_{\widetilde{\gamma} R^0} \rangle \simeq 1.5 \times 10^{-10}\ r\ \left[\mu_8^2 \mu_S^{-4} A\right]\ \text{mb}. \tag{2.7}$$

$R^0 \to \widetilde{\gamma}\pi^+\pi^-$: In what we call interconversion processes, there is an R-odd particle in the initial as well as in the final state. Although the reactions do not of themselves change the number of R-odd particles, they keep the photinos in equilibrium with the R^0s, which in turn are kept in equilibrium through their self annihilations. An example is R^0 decay. It can occur via, e.g., the gluino inside the initial R^0 turning into an antiquark and a virtual squark, followed by squark decay into a photino and a quark. In the low-energy limit the quark–antiquark–gluino–photino vertex can be described by the same type of four-Fermi interaction as in co-annihilation (see Figure 1). One expects on dimensional grounds a decay width $\Gamma_0 \propto \alpha_{EM} \alpha_S M^5 / M_{\widetilde{S}}^4$. The lifetime of a free gluino to decay to a photino and massless quark-(anti)quark pair was computed by Haber and Kane. However this does not provide a very useful estimate when the gluino mass is less than the photino mass.

In an attempt to account for the effects of gluino-gluon interactions in the R^0, which is necessary for even a crude estimate of the R^0 lifetime, Farrar developed a picture based on the approach of Altarelli et al.: The R^0 is viewed as a state with a massless gluon carrying momentum fraction x, and a gluino carrying momentum fraction $(1-x)$,† having therefore an effective mass $M\sqrt{1-x}$. The gluon structure function $F(x)$ gives the probability in an interval x to $x + dx$ of finding a gluon, and the corresponding effective mass for the gluino. One then obtains the R^0 decay width (neglecting the mass of final state hadrons):

$$\Gamma(M, r) = \Gamma_0(M, 0) \int_0^{1-r^{-2}} dx\ (1-x)^{5/2} F(x)\ f(1/r\sqrt{1-x}), \tag{2.8}$$

where in this expression the factor $\Gamma_0(M, 0)$ is the rate for a gluino of mass M to decay to a massless photino, and $f(y) = [(1-y^2)(1+2y-7y^2+20y^3-7y^4+2y^5+y^6) + 24y^3(1-y+y^2)\log y]$ contains the phase space suppression which is important when

† Of course there should be no confusion with the fact that in the discussion of the R^0 lifetime we use x to denote the gluon momentum fraction whereas throughout the rest of the paper x denotes m/T.

the photino becomes massive in comparison to the gluino. Modeling $K^{\pm}$ decay in a similar manner underestimates the lifetime by a factor of 2 to 4. This is in surprisingly good agreement; however caution should be exercised when extending the model to R^0 decay, because kaon decay is much less sensitive to the phase-space suppression from the final state masses than the present case, since the range of interest will turn out to be $r \sim 1.2 - 2.2$. For r in this range, taking $F(x) \sim 6x(1-x)$ leads to an approximate behavior

$$\Gamma_{R^0 \to \widetilde{\gamma}\pi^+\pi^-} = 2.0 \times 10^{-14}\ \mathcal{F}(r)\ \mathrm{GeV}\ [\mu_8^5 \mu_S^{-4} B]\ , \tag{2.9}$$

where $\mathcal{F}(r) = r^5(1-r^{-1})^6$, and the factor B reflects the overall uncertainty. We believe a reasonable range for B is $1/30 \lesssim B \lesssim 3$. Lattice QCD calculation of the relevant hadronic matrix elements would allow a more reliable determination of B.

$\underline{R^0\pi \to \widetilde{\gamma}\pi}$: The short-distance subprocess in this reaction is $q + \widetilde{g} \to q + \widetilde{\gamma}$, again described by the same low-energy effective operator as in co-annihilation and R^0 decay. At the hadronic level the matrix element for $R^0\pi \to \widetilde{\gamma}X$ is the same as for $R^0\widetilde{\gamma} \to \pi X$ for any X, evaluated in different physical regions. Thus the difference between the various cross sections is just due to the difference in fluxes and final state phase space integrations, and variations of the matrix element with kinematic variables. Given the crude nature of the analysis here, and the great uncertainty in the overall magnitude of the cross sections, incorporating the constraints of crossing symmetry are not justified at present. We will therefore use the same form as for (2.7), letting C parameterize the hadronic uncertainty in this case: $\langle |v|\sigma_{R^0\pi}\rangle \simeq 1.5 \times 10^{-10}\, r\ \left[\mu_8^2 \mu_S^{-4} C\right]$ mb.

This completes the discussion of the lifetimes, cross sections, and their uncertainties. The results are summarized in Table I.

Once the cross sections and decay rate is known, one can develop the Boltzmann equations for the system and numerically solve them to find the relic abundance. This is being studied by Chung, Farrar, and Kolb. But for the purpose of illustrating the main issues, it will suffice to compare equilibrium reaction rates to the expansion rate near freeze out.

To obtain an estimate of when the rates will drop below the expansion rate, we will assume all particles are in LTE (local thermodynamic equilibrium). In LTE a particle of mass m in the nonrelativistic limit has a number density

$$n = \frac{g}{(2\pi)^{3/2}}(mT)^{3/2}\exp(-m/T) = \frac{g}{(2\pi)^{3/2}}(T/m)^{3/2}m^3\exp(-m/T). \tag{2.10}$$

Here g counts the number of spin degrees of freedom, and will be 2 for the R^0 and the $\widetilde{\gamma}$. Of course all rates are to be compared with the expansion rate. In the radiation-dominated universe with $g_* \sim 10$ degrees of freedom

$$H = 1.66 g_*^{1/2} T^2/m_{Pl} = 2.8 \times 10^{-19} x^{-2}\left[\mu_8^2\right]\ \mathrm{GeV}. \tag{2.11}$$

There are two striking features apparent when comparing the magnitudes of the equilibrium reaction rates in Table II. The first feature is that the numerical factor in $R^0R^0 \to X$ is enormous in comparison to the other numerical factors. This simply reflects the fact that R^0 annihilation proceeds through a strong process, while the other processes are all suppressed by a factor of $M_{\widetilde{S}}^{-4}$.

The other important feature is the exponential factors of the rates. They will largely determine when the photino will decouple, so it is worthwhile to examine them in detail.

The exponential factor in $\widetilde{\gamma}\widetilde{\gamma} \to X$ is simply $\exp(-m/T)$, which arises from the equilibrium abundance of the $\widetilde{\gamma}$. It is simple to understand: the probability of one $\widetilde{\gamma}$ to find another $\widetilde{\gamma}$ with which to annihilate is proportional to the photino density, which contains a factor of $\exp(-m/T) = \exp(-x)$ in the nonrelativistic limit.

Table II: The ratio of the equilibrium reaction rates to the expansion rate for the indicated reactions. Shown in $[\cdots]$ is the scaling of the rates with unknown parameters characterizing the cross sections and decay width.

Process	Γ_{EQ}/H	Scaling
$\widetilde{\gamma}\widetilde{\gamma} \rightarrow X$	$1.2 \times 10^7 \; x^{-1/2} \; \exp(-x)$	$[\mu_8^3 \mu_S^{-4}]$
$R^0 R^0 \rightarrow X$	$3.5 \times 10^{20} x^{-1/2} \, r^{1/2} \; \exp(-rx)$	$[\mu_8]$
$\widetilde{\gamma} R^0 \rightarrow X$	$8.9 \times 10^7 \; x^{1/2} \, r^{5/2} \; \exp(-rx)$	$[\mu_8^3 \mu_S^{-4} A]$
$\widetilde{\gamma}\pi^+\pi^- \rightarrow R^0$	$7.1 \times 10^4 \; x^2 \, r^{3/2} \, \mathcal{F}(r) \; \exp[-(r-1)x]$	$[\mu_8^3 \mu_S^{-4} B]$
$\widetilde{\gamma}\pi \rightarrow R^0 \pi$	$9.6 \times 10^6 \; x^{1/2} \, r^{5/2} \; \exp[-(r-1)x] \; \exp(-0.175 \mu_8^{-1} x)$	$[\mu_8^{3/2} \mu_S^{-4} C]$

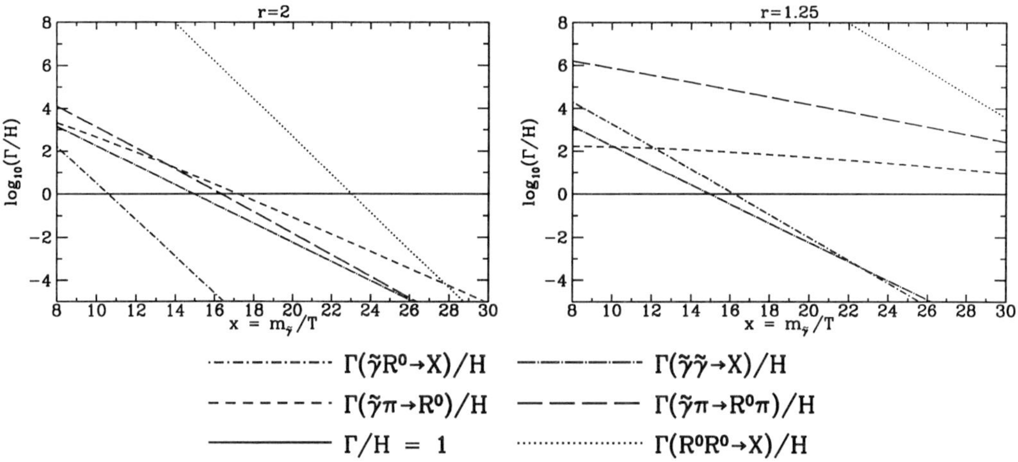

FIGURE 2. Equilibrium reaction rates divided by H for $r = 1.25$ and $r = 2.0$, assuming $\mu_8 = \mu_S = 1$, and that the factors $A = B = C = 1$. The rates can be easily scaled for other choices of the parameters.

The similar exponential factor in $R^0 R^0 \rightarrow X$ is also easy to understand. An R^0 must find another R^0 to annihilate, and that probability is proportional to $\exp(-M/T) = \exp(-rx)$.

The process $\widetilde{\gamma} R^0 \rightarrow X$ is also exothermic, so the only exponential suppression is the probability of a $\widetilde{\gamma}$ locating the R^0 for co-annihilation, proportional to the equilibrium number density of R^0, which is proportional to $\exp(-M/T) = \exp(-rx)$

In $\widetilde{\gamma}\pi^+\pi^- \rightarrow R^0$ the exponential factor is $\exp[-(M-m)/T] = \exp[-(r-1)x]$, which is just the "$Q$" value of the decay process.

For the process $\widetilde{\gamma}\pi \rightarrow R^0\pi$, it is necessary for the collision to have sufficient center-of-mass energy to account for the $\widetilde{\gamma}$–R^0 mass difference, which accounts for a factor of $\exp[-(M-m)/T]$. The number density of target pions is $\exp(-m_\pi/T)$, so this factor is also present. Combining the two factors leads to the overall factor appearing in Table II: $\exp[-(M-m+m_\pi)/T] = \exp[-(r-1)x]\exp(-0.175\mu_8^{-1}x)$.

Graphs of the reaction rates as a function of r is shown in Figure 2. There are several things to notice from the graphs: 1) The R^0 self-annihilation rate is always larger than the other rates. This means that the assumption that the R^0 is in equilibrium during $\widetilde{\gamma}$

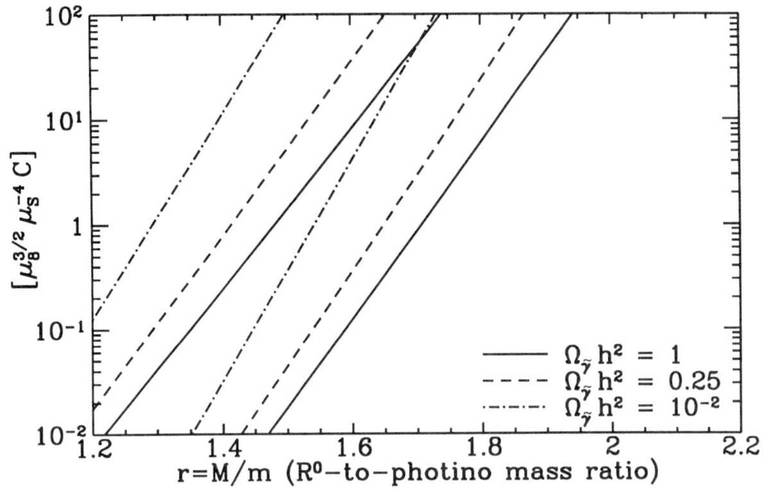

FIGURE 3. Assuming $\tilde{\gamma}$ freeze out is determined by $\tilde{\gamma} \leftrightarrow R^0$ interconversion, this figures shows as a function of r the values of $[\mu_8^{3/2}\mu_S^{-4}C]$ required to give the indicated values of $\Omega_{\tilde{\gamma}}h^2$. The uncertainty band is generated by allowing μ_8 to vary independently over the range $0.5 \leq \mu_8 \leq 2$.

freeze out is a good approximation for the values of r considered here. 2) Even for r as large as $r = 2$, the interconversion rates seem to be (slightly) more important than $\tilde{\gamma}$ self annihilation in keeping photinos in equilibrium. 3) The process $\tilde{\gamma}\pi \longleftrightarrow R^0\pi$ seems to be the most important process for $r < 2$. Detailed numerical integration of the Boltzmann equation by Chung, Farrar, and Kolb confirms this. 4) The freeze-out temperature is *very* sensitive to the value of r. This traces to the exponential sensitivity of the reaction rates upon r. We will make use of this last feature to find a cosmologically acceptable range of r.

Assuming that $\tilde{\gamma}\pi \longleftrightarrow R^0\pi$ does determine the $\tilde{\gamma}$ relic abundance, Figure 3 shows the sensitivity of $\Omega_{\tilde{\gamma}}h^2$ to r. From this figure we can draw some very interesting conclusions. If we assume that $[\mu_8^{3/2}\mu_S^{-4}C] < 10^2$, then r must be less than 1.9. If we demand that the relic photinos are dynamically important (say $\Omega_{\tilde{\gamma}}h^2 \geq 10^{-2}$) then $r > 1.2$ if $[\mu_8^{3/2}\mu_S^{-4}C] > 0.1$. Finally, if we choose our best guess $[\mu_8^{3/2}\mu_S^{-4}C] \sim 1$ and $\Omega_{\tilde{\gamma}}h^2 \sim 0.25$, then the allowed range of r is $1.4 \lesssim r \lesssim 1.6$.

2.4. *Testing the scenario*

Direct detection of light photinos is not easy. The interaction cross section decreases with photinos mass, and more importantly, the kick they would give to a massive target nucleus also decreases with decreasing photino mass.

The case for light photinos hinges upon laboratory experiments. The scenario depends upon the existence of the R^0, the gluino—gluon bound state,† with a mass roughly 1.5 times the photino mass. If the R^0 can be discovered (and after all, the discovery of a 1.5 GeV hadron does not sound impossible), from its decay one can learn the $\tilde{\gamma}$ mass, and hence r, as well as the parameters of the short-distance matrix element. While there is no shortage of candidates for relic dark matter particle species, this proposal extends the idea that photinos may be the dark matter to a previously excluded mass range by incorporating new reactions that determine the photino relic abundance. If this scenario

† In fact, one suggested title for this talk is "The R^0: sglueball or screwball?"

is correct, direct and indirect detection of dark matter might be even more difficult than anticipated. However the scenario requires the existence of low-mass hadrons, which can be produced and detected at accelerators of moderate energy. Thus particle physics experiments will either disprove this scenario, or make light photinos the leading candidate for dark matter.

So the next step is obvious: find the R^0:

3. Relic axions

3.1. *The axion*

The axion story begins with PQ symmetry breaking, which occurs when a complex scalar field $\vec{\phi}$ with non-zero PQ charge develops a vacuum expectation value. This PQ symmetry breaking can be modeled by considering a potential of the standard form $V(\vec{\phi}) = \lambda\left(|\vec{\phi}|^2 - f_a^2/2\right)^2$. The axion is the Nambu–Goldstone degree of freedom resulting from spontaneous breaking of the global symmetry. However, since the PQ symmetry is anomalous, it is broken explicitly by QCD instanton effects, leading to a mass for the axion. In general the instanton effects respect a residual Z_N symmetry, and the axion develops a potential due to instanton effects of the form $V(a) = m_a^2(f_a/N)^2[1 - \cos(Na/f_a)]$, where $m_a = \Lambda_{QCD}/f_a$. The axion field is often represented in terms of an angular variable $\theta \equiv Na/f_a$, and if θ is taken as the dynamical variable, its potential (for $N = 1$) is

$$V(\theta) = m_a^2(T)f_a^2(1 - \cos\theta). \tag{3.12}$$

The above description is only valid at zero temperature. Because QCD instantons are large, with a size set by $\Lambda_{\rm QCD}^{-1}$, their effects are strongly suppressed at high temperatures. Thus, after PQ symmetry breaking at $T \sim f_a$, but before QCD effects are important around T_{QCD}, the axion is effectively massless. For $T \gg \Lambda_{\rm QCD}$, the temperature dependence of the axion mass scales as $m_a^2(T) \propto (T/T_*)^{-n}$, where $n = 7.4 \pm 0.2$

3.2. *The standard scenario*

When the field $\theta(x)$ is created during the Peccei-Quinn symmetry breaking phase transition at $T \sim f_a$, it should be uncorrelated on scales larger than the Hubble radius at that time.

The usual assumption in axion cosmology is that PQ symmetry is broken before or during inflation, and the reheat temperature after inflation is too low to restore the PQ symmetry. This would mean that inflation will "smooth" the axion field on scales larger than the Hubble radius today. While this may indeed be the case, I will not make that assumption here. Since Tkachev and I first did the work reported here, Kofman, Linde, and Starobinsky, as well as Tkachev, have made the interesting observation that symmetry restoration and breaking in the *preheating* phase of inflation may have interesting effects. Just one of the interesting effects may be to restore PQ symmetry after inflation, even if the reheat temperature is too low to do so. This means that PQ symmetry will be restored and broken after inflation, producing a spatially dependent axion field. While the dust has yet to settle on the issue, it is clear that it is naïve to think that inflation automatically results in a smooth axion field. So it is worthwhile to explore the possibility that the axion field is misaligned on scales larger than the Hubble radius at the time of the QCD transition when the axion mass switches on. In this part of the talk† I would like to propose another different picture for the wanted poster: axions that are clumped today within our galaxy, rather than uniformly distributed.

Assuming an initially chaotic axion field after the PQ transition, as the temperature decreases and the Hubble radius grows the field becomes smooth on scales up to the Hubble radius. This continues until $T = T_1 \sim 1$ GeV when the axion mass "switches on," i.e., when $m_a(T_1) \approx 3H(T_1)$, and the axion mass begins to become important in the equations of motion. Coherent axion oscillations then transform fluctuations in the initial amplitude of the axion field into fluctuations in the axion density.

Since the initial amplitude of the coherent axion oscillations on the scale of the Hubble radius is uncorrelated, one expects that typical positive density fluctuations on this scale will satisfy $\rho_a \approx 2\bar{\rho}_a$, where $\bar{\rho}_a$ is mean cosmological density of axions. These overdense regions were called axion miniclusters by Hogan and Rees. At the temperature of equal matter and radiation energy density, $T_{\rm EQ} = 5.5\,\Omega_a h^2\,{\rm eV}$, non-linear fluctuations will separate out as miniclusters with $\rho_{\rm MC} \approx 10^{-14}\,{\rm g\,cm^{-3}}$. The minicluster mass will be of the order of the axion mass within the Hubble radius at temperature T_1, $M_{\rm MC} \sim 10^{-9}\,M_\odot$. The radius of the cluster is $R_{\rm MC} \sim 10^{13}$cm, and the gravitational binding energy will result in an escape velocity of $v_e/c \sim 10^{-8}$.

It is easy to understand how a minicluster forms. First consider the evolution of the background axion density. In the limit that the initial misalignment angle is constant on all scales, the axion energy density scales with temperature as $\bar{\rho}_a(T) = 3T_{\rm EQ}s/4$ for $T \ll \Lambda_{\rm QCD}$, where $s \propto T^3$ is the entropy density and $T_{\rm EQ}$ is the temperature of equal matter and radiation energy densities.

Now suppose that there is a region with axion over-density, $\rho_a = (1+\Phi)\bar{\rho}_a$. Then the matter density in that region will dominate the radiation density at a temperature $T_\Phi = (1+\Phi)T_{\rm EQ}$. If Φ is larger than unity, then that region will separate out from the cosmological expansion, gravitationally collapse, relax, and form a minicluster with the approximate density it had at T_Φ. A detailed study of this leads to a final minicluster density of

$$\rho_{mc} \simeq 140\Phi^3(1+\Phi)\bar{\rho}_a(T_{\rm eq}) \approx 3\times 10^{-14}\Phi^3(1+\Phi)\left(\Omega_a h^2\right)^4 {\rm g\ cm^{-3}}. \tag{3.13}$$

Even a relatively small increase in Φ is important because the final density depends upon Φ^4 for $\Phi \gtrsim 1$.

3.3. *Non-linear axion dynamics and production of miniclusters*

Igor Tkachev integrated the field equations near the QCD transition with various choices for initial conditions where the axion field is inhomogeneous on scales larger than the Hubble radius.

In our numerical investigations of the dynamics of the axion field around the QCD epoch we found that as the oscillations commence, important, previously neglected, non-linear effects can result in the formation of transient soliton-like objects we called axitons. The non-linear effects result in regions with Φ much larger than unity, possibly as large as several hundred, leading to enormous minicluster densities. We also found that the minicluster mass scale is set by the total mass in axions within the Hubble radius at a temperature around $T = T_1 \approx 1$ GeV when axion mass is equal to H. Since this temperature is somewhat higher than $\Lambda_{\rm QCD}$, the effect is to lower the minicluster mass from the estimate of Hogan and Rees to about $10^{-12}M_\odot$.

† Reference to all the material presented here can be found in Kolb and Tkachev (1993, 1994,1996)

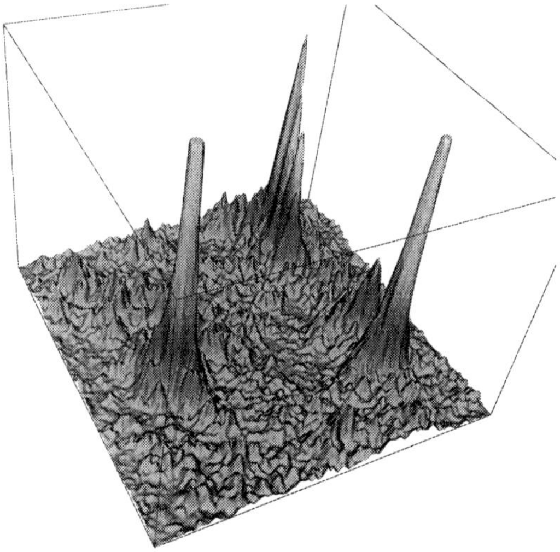

FIGURE 4. This is a two-dimensional slice through a three dimensional box of Tkachev's simulation showing the distribution of axion energy densities for $T \gg T_{\rm eq}$. The height of the plot corresponds to $\Phi = \delta\rho_a/\bar{\rho}_a = 20$, and the width to a length of $4H^{-1}(T_1)$ (this corresponds to a comoving length of about 0.25 pc). The miniclusters are clearly seen.

3.4. *Detecting axion miniclusters*

If a large fraction of axions end up in miniclusters (and in our numerical simulations that is what we found), then the local axion density may be very clumpy. This would mean that direct searches for axions would not be very efficient unless we just happened to be passing through a minicluster.

However it may be possible to detect miniclusters through femtolensing or picolensing. The basic idea is that if gamma-ray bursts are at cosmological distances, then miniclusters could act as gravitational lenses. For source and lens of cosmological distances, the Einstein ring radius is $R_E \sim 5 \times 10^{10}(M/10^{-12}M_\odot)^{1/2}$cm. This means that if two gamma-ray burst detectors are separated by distances larger than R_E, typically one detector will be within the Einstein ring and the other will not. Measurement of different fluxes in two widely separated detectors would extend the detectable range of lens masses to $M < 10^{-7}\,M_\odot$. If the angular separation of the images of a point-like source is comparable to the wavelength of the light, interference between the lensed images will lead to a distinctive fringe pattern both in coordinate space and in energy spectrum. The interference pattern in the energy spectrum will depend upon the location of the detector. Since well separated detectors will see different patterns, there is a distinctive signature of the effect.

4. Conclusions

Although there are solid, well motivated scenarios that lead to one of several suspects forming cold dark matter, one must explore vigorously the possibility of a non-standard scenario for standard dark-matter suspects (such as SUSY and axions). It would be a crime if we fail to apprehend dark matter just because a prime suspect appears in an unfamiliar guise.

I would like to acknowledge collaboration with Glennys Farrar and Daniel Chung on the light-photino work reported here, and collaborations with Igor Tkachev on the axion-minicluster work. This work was supported by the Department of Energy and by NASA under Grant NAG5–2788.

REFERENCES

FARRAR, G. R. & KOLB, E. W. 1996 Light photinos as dark matter. *Phys. Rev. D* **53**, 2990–3001.

CHUNG, D. J. H., FARRAR, G. R. & KOLB, E. W. 1996 Light photinos as dark matter II. *In preparation.*

KOLB, E. W. & TKACHEV, I. I. 1993 Axion miniclusters and bose stars. *Phys. Rev. Lett.* **71**, 3051–3054.

KOLB, E. W. & TKACHEV, I. I. 1994 Non-linear axion dynamics and formation of cosmological pseudo-solitons. *Ap. Phys. Rev. D* **49**, 5040–5049.

KOLB, E. W. & TKACHEV, I. I. 1996 Femtolensing and picolensing by axion miniclusters. *Ap. J. Lett.* **400**, L25–L29.

On the Unification of Gravitational and Inertial Forces

By D. LYNDEN-BELL

Institute of Astronomy, Cambridge CB3 0HA and

Physics Dept., The Queen's University, Belfast†, BT 7

Einstein's principle of equivalence showed that gravitational forces could be regarded as due to the use of locally accelerated axes instead of freely falling inertial ones. Nevertheless, Mach's idea that inertia, like gravity, arises from some kind of interaction between masses is not yet generally accepted. Here is is shown that such a unification is not achieved in open universes but is a property of closed universes with hyperspherical topology. Thus for Mach's principle to be true we need $\widetilde{\Omega} \equiv \rho/\rho_c > 1$.

---

1. An amusing problem with a bearing on Mach's Principle

Since this is Igor Novikov's 60th birthday celebration we should start with some light-hearted amusement and for physicists; this means a delicate but simple problem that has implications of deeper truth. Consider a massive spherical shell M of radius 'a' with a relativistically significant internal gravitational potential so that GM/a is of order c^2. Somewhere within this shell place a small mass m attached to which is a thread of cotton which stretches radially through a small hole in the shell and extends far away to where Igor pulls it with a weak tension T (see Fig. 1). Is the acceleration of the end of the cotton at Igor greater than, equal to, or less than T/m? It is assumed that the acceleration is slow so the motion of the thread is slow and changes in thread length are negligible, likewise the total mass of the thread is small compared with m.

Three possible though seriously incomplete replies to this question which illustrate the possibilities are:

1. Energy in a deep potential well is worth less than energy at infinity by the redshift factor familiar from photons. Hence the inertia of the mass m inside the sphere should be smaller by that factor and the acceleration will therefore be greater than T/m.

2. Space within the sphere is flat; space at infinity is also flat. No gravitational field acts on any of the masses involved; the acceleration will therefore equal T/m.

3. Inertia is due to other masses in accordance with Mach's ideas. Thus the proximity of the heavy sphere will increase the inertia of m; the acceleration will therefore be less than T/m.

At this point in the conference a vote was taken among participants. There was one vote for acceleration $> T/m$, 17 votes for $= T/m$ and 27 votes for $< T/m$.

We may solve the problem as follows. First we must find whether a thread at equilibrium has a constant tension in a strong gravity field. This is most readily done by using the principle of virtual work. Consider a small shift, δx, in the thread along its own length. The work done by Igor pulling it is $T\delta x$. If T' is the tension of the thread at m, then the work done there is $T'\delta x$ but this is in the local energy units. Those are devalued by the redshift factor $(1-2GMc^{-2}/a)^{1/2}$ due to the potential well, so matching the work done when they are all put into the international system of energy units defined at infinity (Euro-joules), we find

† Current address

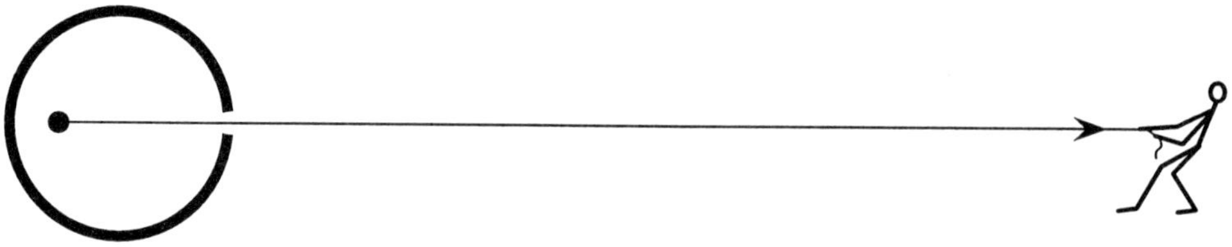

FIGURE 1.

$$T = T'(1 - 2GMc^{-2}/a)^{1/2} \tag{1.1}$$

Notice we have taken an equal shift of the thread at each end. This preserves its proper length. Thus the tension felt at m will be larger. Inside the shell space is flat, so we may use Newton's law

$$m\ d^2x/dt'^2 = T' \tag{1.2}$$

where t' is the local time.

The metric everywhere is, writing $\hat{\mathbf{r}} = (\sin\theta\cos\phi, \sin\theta\sin\phi, \cos\theta)$

$$ds^2 = \begin{cases} (1 - 2GMc^{-2}/r)dt^2 - dr^2/(1 - 2GMc^{-2}/r) - r^2 d\hat{\mathbf{r}}^2, & r \geq a \\ dt'^2 - dr^2 - r^2 d\hat{\mathbf{r}}^2 & r \leq a \end{cases}$$

matching the metrics on $r = a$ where the dr terms vanish we find

$$dt'^2 = (1 - 2GMc^{-2}/a)dt^2 \tag{1.3}$$

Hence substituting 1.3 and 1.1 into 1.2 we find

$$d^2x/dt^2 = (1 - 2GMc^{-2}/a)^{1/2} T/m \tag{1.4}$$

Since the string only accelerates slowly, all points may be taken to accelerate together in global t time so Igor observes a slow acceleration $< T/m$ and deduces that the effective mass observed from outside has been increased by the proximity of the massive shell surrounding it. Whereas this may accord with Mach's ideas we note that an experiment done within the sphere would give no such effect. As we are within the Universe the effect is hardly what Mach and Boltzmann were hoping for.

2. History

Even in Newton's day his use of Absolute Space provoked considerable controversy which may be characterised as Newton v. Leibniz. To Newton his Absolute Space was a fundamental *reality*. To Leibniz (1712) it was a useful invention for simplifying calculations of the relative motions of bodies.

No-one believes Cartesian coordinates are other than useful inventions. Is Absolute

Space in that category? Newton knew that his concept of Absolute Space involved presuppositions so he prepared the ground in his Scholium on space and time (1687): "It is indeed a matter of great difficulty to discover and effectually to distinguish, the true motions of particular bodies from the apparent; because the parts of that immovable space in which those motions are performed do by no means come under the observation of our senses. Yet the thing is not altogether desperate, ...". He then imagines all the stars and the Earth removed so that there are no external markers of directions in space and considers two globes at relative rest joined by a thread. He points out that if the two globes are in reality rotating about one another in absolute space, then the centrifugal force will cause a tension in the thread. By changing this rotation of the spheres to minimise the tension in the thread, it will be possible to reduce it to zero and the spheres will then be non-rotating in absolute space. He deduces that rotation is absolute and independent of the presence or absence of other bodies to mark it by. Later he drives home this point with his famous bucket experiment. By considering the 4 cases in which: i) both bucket and water are at rest; ii) the bucket rotates but the water does not; iii) both bucket and water rotate together and iv) the water rotates although the bucket is at rest. He establishes that the concavity of the upper surface of the water depends on its rotation alone and is independent of the rotation or non-rotation of the bucket.

Foucault's pendulum was unknown in Newton's day but it provides a yet more sensitive test of rotation. If the pendulum is isolated from the air and is set swinging at the North pole then it swings straight in space leaving the Earth rotating underneath it. Thus relative to an Earth-bound observer the plane of the pendulum seems to rotate. The pendulum does not swing straight relative to the Earth but does it swing straight relative to i) the Sun, ii) the Galaxy or to iii) the distant galaxies? To the accuracy of the observations i) and ii) are now known to be false and it is the distant galaxies that appear to have fixed directions in inertial axes. Using the Solar System as a giant gyroscope, astronomers have shown that no Coriolis forces are needed in solar system dynamics in the frame defined by the extragalactic nebulae. The accuracy of their determination is to better than three rotations since the Big Bang. While this is a remarkably high accuracy Hawking's (1969) indirect method, based on the isotropy of the Cosmic Background Radiation gives a far more precise result. The lack of a significant quadrupole distortion of the radiation implies that the universe has rotated by less than 10^{-12} rotations† since the last scattering at $z \sim 1000$! From this it is natural to deduce that

THE UNIVERSE DOES NOT ROTATE

Within their realm of validity Newton's laws are wonderfully successful in this non-rotating universe. According to Newton's laws the universe could have been a rotating one – its angular momentum could have been non-zero and it is a happy coincidence that it does not rotate. We have shown elsewhere (Lynden-Bell 1992, 1996; Lynden-Bell & Katz 1995) that a 'classical' mechanics without absolute space predicts this result.

Newton's mechanics was so successful in the solar system that early questioning of his basic concepts soon seemed irrelevant. More than two centuries later the physicist-philosopher Mach (1883) trenchantly criticized Newton's concepts. He pointed out that Newton did *not* perform his globes experiment! No-one is capable of removing the rest of the universe, so no-one can claim to know the result of such an experiment.

Referring to Newton's bucket experiment he averred: "All this experiment shows is that the concavity of the water is not measurably affected by the rotation or non-rotation of the bucket. It does not show that rotation is absolute – it could be relative to the masses in the rest of the Universe". Prophetically he questioned, "Would the result have

† This is the current best estimate as updated by Barrow, Juskiewicz & Sonada (1985).

remained unchanged had the bucket been massive and many leagues thick?" Another Machian criticism of absolute space comes from Newton's third law that every action has an equal and opposite reaction. According to Newton, Absolute Space profoundly affects the relative motions of bodies; but there is no back-reaction; the motions of bodies have no effect on Absolute Space.

Both in the Prague meeting, (Prosser & Folta 1989), and at the recent Tübingen conference on Mach's Principle it was clear that different participants meant different things by it (Barbour & Pfister 1995). Here we adapt the clear statement in Bondi's book *Cosmology* (1952). All motions, velocities, accelerations, rotations are relative. Local inertial frames are determined through the distributions of energy and momentum in the universe by some weighted average of the apparent motions. The first indication that Mach might be right came from an early version of General Relativity. Einstein (1913) wrote to Mach from Zürich on 25 June 1913:

If a spherical shell S rotates about an axis passing through its centre relative to the fixed stars (Rest system), then a Coriolis field arises inside the shell, i.e. the plane of the Foucault pendulum is being carried along (though with a practically immeasurably small velocity). In the same letter he pointed out that if this theory was confirmed by experiment then – "your brilliant investigations on the foundations of mechanics will have received a splendid confirmation. For it follows of necessity that *inertia* has its origin in some kind of *interaction* of the bodies exactly in accordance with your argument about Newton's bucket experiment."

Notice that as soon as Einstein realises that some inertia comes from other masses he extrapolates and assumes with Mach that it all does. This is in keeping with his later statement: The postulate of the *relativity of inertia* cannot be satisfied by merely demonstrating that mass-energy influences inertia; *the mass-energy must entirely cause it!*

3. The Electromagnetic – Gravomagnetic Analogy

The excellent chapter 4 on Inertia in Bondi's book *Cosmology* (1952) first aroused my interest in the subject. While still an undergraduate I wrote an essay on the subject for my supervisor now Prof. W.F. Vinen (Physics, Birmingham). I had just learnt how electricity and special relativity inevitably lead to magnetism. Consider a charged particle fixed at the origin acting on a second charged particle at $\mathbf{r}$ moving with velocity $\mathbf{v}$. The force is given by $\mathbf{F} = q_1 q_2 \hat{\mathbf{r}}/r^2$ and the rate of working of that force is $\mathbf{F} \cdot \mathbf{v} = q_1 q_2 \hat{\mathbf{r}} \cdot \mathbf{v}/r^2$. In special relativity the rate of change of the second's momentum per unit proper time is the 4-force $(1 - v^2/c^2)^{-1/2}(\mathbf{F}, \mathbf{F} \cdot \mathbf{v}/c)$ and this is also a 4-vector. Thus we may discover the components of the 4-force in a moving frame by merely Lorentz transforming the vector. When that is done we shall have the 4-force in a frame in which both particles move. We find that this force consists of two parts: an "electric" part almost independent of the velocities of the particles and a "magnetic part" dependent on $\mathbf{v}_1'$ and $\mathbf{v}_2'$. We regard $q_1 \mathbf{v}_1'$ as a current which generates the magnetic field and $q_2 \mathbf{v}_2'$ as the current that receives it. In this way Coulomb's inverse square law plus special relativity lead to the concept of a magnetic field and the magnetic force on a moving particle $q(\mathbf{v}/c) \times \mathbf{B}$. However, Coulomb's law is not the only inverse square law known in classical physics. There is a striking correspondence to Newton's law of gravitation which may be treated by the same argument to deduce

Naively when I wrote this essay I thought I had discovered something new but I was much disappointed when I calculated its effects on solar system dynamics which were always negligible. In desperation I clutched at the idea that inertia should be relative.

$q_1 q_2 \hat{r}/r^2$	$\longleftrightarrow$	$-Gm_1 m_2 \hat{\mathbf{r}}/r^2$
Magnetic force on a moving charge $F = q(\mathbf{v}/c) \times \mathbf{B}$ The magnetic field $\mathbf{B}$ is generated by moving charges, i.e. currents of electricity		Gravomagnetic force on a moving mass $F = m(\mathbf{v}/c) \times \mathbf{B}_g$ The Gravomagnetic field $\mathbf{B}_g$ is generated by moving masses, i.e. by currents of mass-energy

Taking axes fixed on Earth would inevitably mean that the whole universe formed a vast matter current circulating around the Earth once in each sidereal day. Surely such a current must produce a significant gravomagnetic field at the Earth? Motion across such a field should produce a measurable force and indeed calculation shows that $\mathbf{F} = m(\mathbf{v}/c) \times \mathbf{B}_g = -2\boldsymbol{\Omega} \times \mathbf{v}mf$ where f is a positive dimensionless factor dependent on the structure of the universe which is very crudely given by

$$f = \frac{4\pi G}{c^2} \int_0^{c/H} \rho R \, dR \approx (3/4)\rho/\rho_c \tag{3.1}$$

where ρ is the density of the universe and ρ_c is the critical density given in terms of Hubble's constant by $H^2 = (8/3)\pi G \rho_c$. We do indeed see such a force in axes fixed in the Earth, $-2\boldsymbol{\Omega} \times \mathbf{v}m$, it is a major factor in cyclones and anticyclones and is commonly called Coriolis force. Notice that it has the same $\boldsymbol{\Omega}$ dependence, the same $\mathbf{v}$ dependence, the same m dependence, and the same sign as the force we have just deduced. However there is a major difference. Our force depends on how much density there is in the universe whereas Coriolis force does not. Can these forces nevertheless be identified? Are all the fictitious forces merely due to our neglect of distant matter? I thus arrived at a Machian view of Newton's inertial frames in a manner very similar to that found by Sciama a few years earlier. Any axes can be used for dynamics but forces due to distant matter must then be included. Inertial axes are useful because in them the forces due to distant matter balance and so may be omitted. But to achieve this new viewpoint we need the aforementioned dimensionless factor $f \sim 1$ that is $\tilde{\Omega} \equiv \rho/\rho_c \approx 1$.

4. General Relativity and the Unification of Gravity and Inertia

In Special Relativity light cones only appear straight in unaccelerated non-rotating axes. Hence inertial axes are part of space-time independent of its mass content. This is incompatible with Mach's principle that the distribution and motion of mass determine the inertial axes. One possible space-time in General Relativity is Minkowski space. Hence neither Special nor General Relativity are Machian.

Einstein realised that this trouble came not from his equations but from the 'flat at infinity' boundary condition that is often imposed on them. He introduced his Λ term in the hope that space would then be closed and so spatial boundary conditions would be eliminated. But de Sitter found his solution that was also empty and so disobeyed Mach's principle.

Within General Relativity certain space-times have profound implications concerning Mach's principle. Lense and Thirring (1918) solved the problem of a slowly rotating spherical shell that weakly perturbs flat space; their result $\omega_{\text{drag}} = (4/3)(GM/ac^2)\,\Omega_s$ demonstrates that Newton's experiments conducted inside such a shell would find inertial axes that rotate relative to the inertial axes at infinity. Thus, contrary to Newton, rotation is not Absolute. The shell affects inertia. Here it is interesting to interpose two statements of Einstein's (1917):

1. There is no inertia of mass against space but only inertia of mass against mass.
2. The postulate of the relativity of inertia cannot be satisfied by merely demonstrating that mass-energy influences inertia. The mass-energy must entirely cause it.

Thirring and Lense had demonstrated that mass motions affect inertial axes but do they, as Einstein implies, determine them? Certainly not in the Lense-Thirring solution for most of the inertial effect comes from flat space at infinity and there is only a little dragging by the sphere. To get a stronger effect a more strongly curved space-time was needed.

Brill and Cohen (1966) provided this by considering a slowly rotating massive sphere small enough that the space outside was perturbed Schwarzschild space with $2GM/ac^2$ no longer small. Indeed when that parameter reached one it appeared that the inertial axes in the flat space within the sphere were dragged along synchronously with it. However, infinite stresses appear in the shell before that condition is reached. To alleviate this problem, Lindblom and Brill (1974) considered a slowly rotating massive shell falling together freely under its own gravity. They demonstrated the surprising result that within the shell the dragging was instantaneous. There was no time delay between observers just inside the shell and those at the centre; the internal inertial frames all rotated together. However, because the space-time outside the shell was now time dependent, it was difficult to decide whether or not Mach's principle was being perfectly fulfilled.

In Lynden-Bell, Katz & Bičák (1995) the Lindblom & Brill problem was generalised to consider general 'vector' perturbations of Tolman universes and Friedmann-Robertson-Walker universes. We show how *all* the inertia arises from mass-energy provided that the Universe is closed. Many of the important points of this investigation are well illustrated by the example in which each spherical shell about an observer at $r = 0$ falls and slowly rotates at its own angular velocity. The perturbed metric is

$$ds^2 = e^{2\nu(r,t)}dt^2 - e^{2\lambda(r,t)}dr^2 - r^2[d\hat{\mathbf{r}} - \omega(r,t) \times \hat{\mathbf{r}}dt]^2 \tag{4.1}$$

The unperturbed metric is given by setting $\omega \equiv 0$ and the perturbed solution is given as the solution of the instantaneous equation

$$(\partial\omega/\partial r)_t = -6Gc^{-2}e^{\lambda+\nu}\mathbf{J}(< r)/r^4 \tag{4.2}$$

where $\mathbf{J}(< r)$ is the total angular momentum inside r

$$\mathbf{J}(< r) = \int_0^r \frac{8}{3}\pi\rho e^{\lambda-\nu}(\boldsymbol{\Omega} - \omega)\left(1 - \frac{v^2}{c^2}\right)^{-1/2} r^4 dr \tag{4.3}$$

$\boldsymbol{\Omega}(r,t)$ is the angular velocity of the shell at r. It is natural to integrate (4.2) and to impose the boundary condition that ω is zero at infinity which gives

$$\omega(r,t) = 2Gc^{-2}\left[W\mathbf{J}/r^3 + \int_r^\infty W(r',t)\partial\mathbf{J}/\partial r'(r')^{-3}dr'\right] \tag{4.4}$$

where the first term is a contribution to the dragging potential from the inside and the second is the contribution from all $r' > r$. The positive weight function W is unity for a flat space but in general it is given by

$$W(r,t) = r^3 \int_r^\infty 3e^{\lambda+\nu}r^{-4}dr\,. \tag{4.5}$$

At the origin the dragging potential ω is the dragging of the inertial frame itself (see 4.1). However, the value of ω there is as much determined by the boundary condition that ω

must be zero at infinity as it is by the distribution of angular momentum. Thus part of inertia comes from a spatial boundary condition and part from mass. This disagrees with Einstein's relativity of inertia principle. However, application of (4.2) in a closed universe gives a much more interesting result. Here the condition that the solution is regular at the second point where $r = 0$ around the back of the universe can only be satisfied when the total angular momentum is zero. This result, which reminds us of our condition for a classical relative mechanics, can also be proved far more generally for arbitrarily large perturbations of universes of hyperspherical topology. It follows from the method of superpotentials which shows that the angular momentum within a surface can be expressed as a surface integral. For a closed universe the bounding surface of an increasing volume eventually vanishes around the back of the universe leaving the whole universe 'inside' a zero surface area. This proves the result! In integrating (4.2) for a closed universe there is no zero point for ω so all that can be obtained are differences $\omega(r_1, t) - \omega(r_2, t)$. There is no absolute rotation. Although ω can not be determined from $\mathbf{J}$ it is interesting that $\mathbf{J}$ itself is absolute since it involves $\mathbf{\Omega} - \omega$ (see 4.3). A somewhat different problem is the determination of the dragging potential ω, not for a given distribution of angular momentum, but for given apparent rotations $\mathbf{\Omega}$ of the heavenly spheres. Interestingly this has a unique solution, so in a closed universe ω is wholly determined by ρ and $\mathbf{\Omega}$ whereas in an open one it is partly determined by them and partly by the spatial boundary condition at ∞. We deduce that these space-times behave Machianly for all closed universes but disobey Mach's principle for open universes. This conclusion is strongly fortified by the limiting cases considered next.

5. Space-Time does not exist without mass-energy

Inertia in an empty space was a conundrum to Mach, Einstein (1917) and Bondi. Here we show how the condition that the Universe must be closed with hyperspherical topology beautifully eliminates the conundrum. In any closed Robertson-Walker model the mass-energy within hyperspherical 'radius' χ is given by $M(<\chi) = (4/3)\pi a^3 \rho \sin^3 \chi$.

This rises to a maximum on the universe's equator $\chi = \pi/2$ and there $2GM/c^2 a_{\rm max} = 1$ when the radius of the universe is at maximum. Now perform the thought experiment of comparing different universes with successively smaller M. Evidently $a_{\rm max}$ is proportional to M so as M decreases $a_{\rm max}$ decreases. Furthermore, the mean density is proportional to $M/a^3_{\rm max}$ so it increases like M^{-2} and the time between the Big Bang and the Big Crunch decreases like $a_{\rm max}/c \propto M$. As $M \to 0$ closed universes shrivel up in space and time getting denser and denser. Small M universes do not approach flat Minkowski space at all; rather they get more and more highly curved. *In the limit $M = 0$ there is no space, no time and no inertia.* This beautifully obeys Mach's principle. Contrast this with the behaviour of open universes as we compare those with less and less mass. In them the period of gravitational deceleration after the Big Bang gets shorter and shorter as mass is removed. In the limit the deceleration disappears and space-time approaches the empty Milne model, which is a re-interpretation of (part of) Minkowski space. There all inertia comes from space and Mach's principle is certainly disobeyed. The $\widetilde{\Omega} \equiv \rho/\rho_c = 1$ universe is not closed; its topology is that of the open universes so it is likewise unacceptable. In all these universes inertia arises partly from mass but partly from space via a boundary condition at ∞. Only in the closed universes do the forces of gravity and inertia arise from mass; only in closed universes are they fully unified.

In Newtonian mechanics space and time are absolute and form an arena independent of the masses therein. In Special Relativity space and time are not separately absolute but are unified into the space-time of Minkowski which forms an absolute arena independent of its content. In General Relativity the arena itself is modified and bends in conformity with its contents. With hyperspherical topology and the closed-space (no-boundary) condition *the arena itself only exists because of its energy content.* There is no space-time without energy. Herein is Mach's principle fulfilled.

REFERENCES

BARBOUR, J.B. & BERTOTTI, B. 1982 *Proc. R. Soc. London A* **382**, 295.

BARBOUR, J.B. & PFISTER, H. 1995 *Mach's Principle: From Newton's bucket to Quantum Gravity.* Birkhauser, Berne.

BARROW, J.D., JUSKIEWICZ, B.R. & SONADA, D.H. 1985 *MNRAS* **213**. 917.

BONDI, H. 1952 *Cosmology* Ch. IV, Cambridge University Press, Cambridge.

BRILL, D.R. & COHEN, J.M. 1966 *Phys. Rev.* **143**, 1011.

EINSTEIN, A. 1913 *The Collected Papers of Albert Einstein,* Vol. 5 §448, p.360. English translation Beck, A. & Howard, D., Princeton Univ. Press 1995.

EINSTEIN, A. 1917 *SB preuss Akad. Wiss.* **142**.

EINSTEIN, A. 1920 *Sidelights on Relativity.* Dover, New York (reprinted 1983).

HAWKING, S.W. 1969 *MNRAS* **162**, 307.

LEIBNIZ, G.W. 1712 *The Leibniz-Clarke Correspondence* (ed. Alexander, H.) 1984. Barnes & Noble, New York.

LENSE, J. & THIRRING, H. 1918 *Phys. Z.* **19**, 156.

LINDBLOM, L. & BRILL, D.R. 1974 *Phys. Rev. D* **10**, 3151.

LYNDEN-BELL, D. 1992 In *ASP Conf. Ser. on Variable Stars and Galaxies* (ed. Warner, B.), Vol. 30, p.1. Astron. Soc. Pac. San Francisco.

LYNDEN-BELL, D. & KATZ, J. 1995 *Phys. Rev. D.*

LYNDEN-BELL, D., KATZ, J. & BIČÁK 1995 *MNRAS* **272**, 150.

LYNDEN-BELL, D. 1996 in *Gravitational Dynamics* (ed. Lahav, O. & Terlevich, R.J.), Cambridge University Press.

MACH, E. 1883, *The Science of Mechanics,* Open Court 1942, Nelson, London 1957.

NEWTON, I. 1687 *Principia.* (ed. Cajori F.), UCLA Press 1934, p.12.

PROSSER, V. & FOLTA, J. 1989 *Ernst Mach and the Development of Physics.* Charles University, Prague.

SCIAMA, D.W. 1953 *MNRAS* **113**, 34.

Internal Structure of Astrophysical Black Holes

By WERNER ISRAEL

Canadian Institute for Advanced Research Cosmology Program, Avadh Bhatia Physics Laboratory, University of Alberta, Edmonton, Alberta, Canada T6G 2J1

This is a brief review of recent progress in analyzing the structure of the inner-horizon singularity of a generic (rotating) black hole formed in a gravitational collapse, and of a geometrical formalism (covariant double-null dynamics) recently developed for this and other purposes.

1. Introduction

Igor Novikov's interests have ranged over every facet of cosmology and relativistic astrophysics, the latter a subject which he and Zel'dovich virtually created with their original researches and their famous Uspekhi review articles of the 1960's, later amplified into books (Zel'dovich and Novikov 1971, 1983).

While the task of confronting theory with observation has always been at the forefront of his endeavours, Igor has not shrunk from pursuing fundamental (some might say even philosophical) issues. The nature of the internal processes in black holes has been a persistent thread running through his work from the earliest days (Doroshkevich and Novikov 1978, Gürsel *et al.* 1979, Novikov and Starobinsky 1980, Frolov and Novikov 1993). Figure 1 reproduces the title page of an investigation carried out before the appearance of Kruskal's paper, and was among the first to provide insight into the physical nature of the Schwarzschild black hole interior.

It is perhaps appropriate here to recall an even earlier contribution by the Irish mathematician J. L. Synge, who passed away on 29 March 1995 at the age of 98. Figure 2 shows Synge's remarkable anticipation (in 1950) of the Kruskal diagram. Synge was the first to introduce a pair of lightlike co-ordinates for the description of a black hole's interior. However, because his co-ordinates were inverse trigonometric functions of Kruskal's, he was led to a lattice-like picture of the interior, in which a particle which has fallen into the hole oscillates like a shuttlecock. Up to the first of the $r = 0$ singularities, this manifold is, of course, isometric to Kruskal's. This work attracted almost no attention, perhaps partly because it appeared in the Proceedings of the Royal Irish Academy, a journal not delivered with the morning paper.

2. Singularity at the Cauchy horizon

I shall report very briefly on some recent progress in an area which Igor, with various collaborators, studied quite intensively between 1978 and 1980 in the papers already cited.

The general background of this problem is explained in recent reviews (Droz *et al.* 1996, Bonanno *et al.* 1994). The settling of a black hole toward a stationary, hairless condition in the wake of a collapse is attended by a tail of gravitational waves decaying as an inverse power of external time. Some of this radiation falls into the hole. Penrose (1968) noted that the Cauchy (inner) horizon of a rotating or charged hole—a lightlike hypersurface characterized by infinite advanced time v, but which infalling particles reach

NOTE ON THE SPACE-TIME METRIC INSIDE THE SCHWARZSCHILD SINGULAR SPHERE

I. D. Novikov

P. K. Shternberg State Astronomical Institute
Translated from Astronomicheskii Zhurnal, Vol. 38, No. 3, pp. 564-565, May-June, 1961
Original article submitted December 1, 1960

A physical interpretation is offered for the Schwarzschild solution of Einstein's equations for a vacuum inside a singlular sphere. The properties of the accompanying space for the corresponding frame of reference are investigated.

In astronomical applications of the general theory of relativity, considerable significance is attached to the discussion of solutions of equations of gravitation describing a centrosymmetric field in a vacuum (the gravitational field of the sun and other celestial bodies and of systems possessing spherical symmetry other than the former, and some cosmological applications). The familiar astronomical verifications of general relativity theory are based precisely on that solution.

As Birkhoff showed [1], any solution of Einstein's equations for a vacuum:

$$G_{\mu\nu} = 0, (\mu, \nu = 0, 1, 2, 3), \qquad (1)$$

exhibiting spatial spherical symmetry may be reduced, by a transformation of the coordinates, to the form:

FIGURE 1. Title page of Novikov (1961)

in finite proper time—has the property of blueshifting any incoming flux by a factor $e^{2\kappa_0 v}$, where κ_0 is the inner surface gravity. At the Cauchy horizon the density of inflowing radiation thus tends exponentially to infinity.

This effect was confirmed in a series of detailed investigations in the 1970's. But these were perturbative studies assuming a fixed geometrical background, and did not take account of the back-reaction on the geometry. The general guess was that, in accordance with the strong cosmic censorship hypothesis, the Cauchy horizon becomes singular and spacelike.

More recently, Eric Poisson (1990) and I examined the back-reaction for a spherical model. We found that the geometry indeed becomes singular at the Cauchy horizon, but this singular hypersurface remains lightlike. Amos Ori (1991) subsequently pointed out that the singularity is locally mild in the sense that tidal forces do not diverge fast enough to disrupt infalling objects up to the moment they reach the Cauchy horizon. Beginning from an asymptotically stationary configuration in its infinite past, this lightlike surface steadily contracts as transverse (scattered) radiation focuses its generators, finally tapering to a crushing spacelike singularity. In effect, the Cauchy horizon serves as a lightlike bridge, linking the asymptotically stationary and hairless outer layers of the hole (at late advanced times) to the crushing singularity associated with its formation (at early advanced times) (Fig. 3).

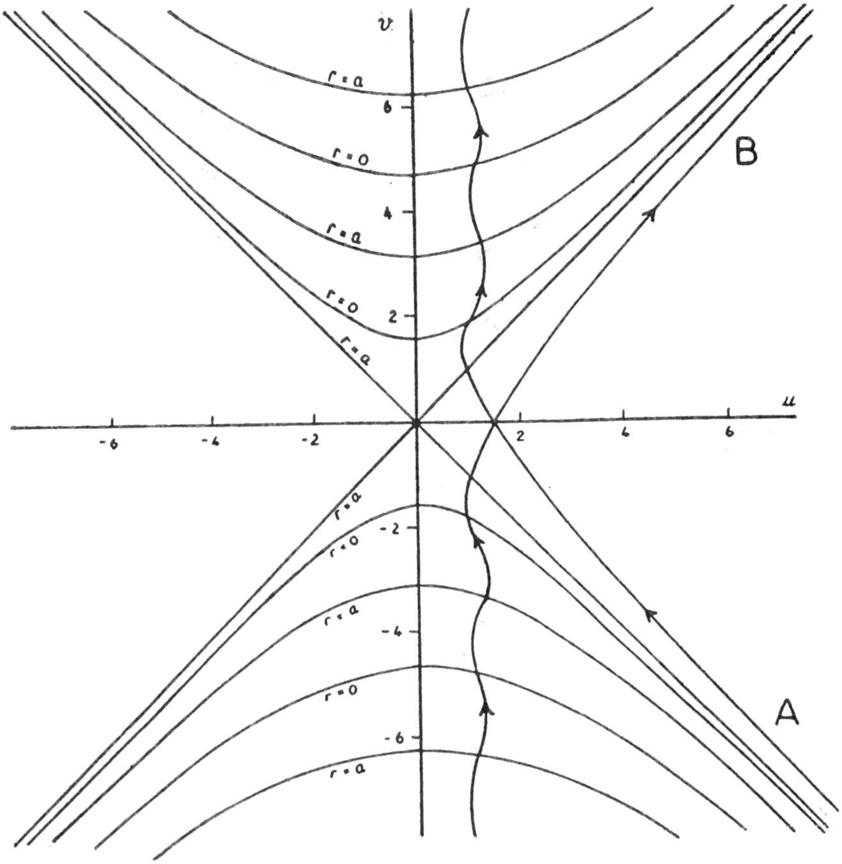

FIG. 14.—Histories of particles projected radially in from $r = \infty$ (*A*), and projected radially out to $r = \infty$ (*B*), with zero velocity at $r = \infty$ in each case.

FIGURE 2. Reproduced from Synge (1950)

Although doubts were initially expressed about the genericity of this picture (e.g. Gnedin and Gnedin 1993), it may now be considered established, at least for spherical holes (Bonanno *et al.* 1995, Brady and Smith 1995). The nonspherical case is more difficult, but also here there has been recent progress which I shall report on in the following sections.

It is easy to explain the main features of the spherical case. Any spherisymmetric metric can be written in the form

$$dz^2 = g_{AB} dx^A dx^B + r^2(x^A) d\Omega^2$$

where x^A $(A, B, \ldots = 0, 1)$ are any two co-ordinates labeling the 2-spheres of radius $r(x^A)$. The four-dimensional Einstein field equations can be reduced to two-dimensionally covariant equations, using g_{AB} as the 2-metric. The Schwarzschild mass function $M(x^A)$ is defined in terms of the gradient of r:

$$1 - 2M(x^A)/r = g^{AB}(\partial_A r)(\partial_B r). \tag{2.1}$$

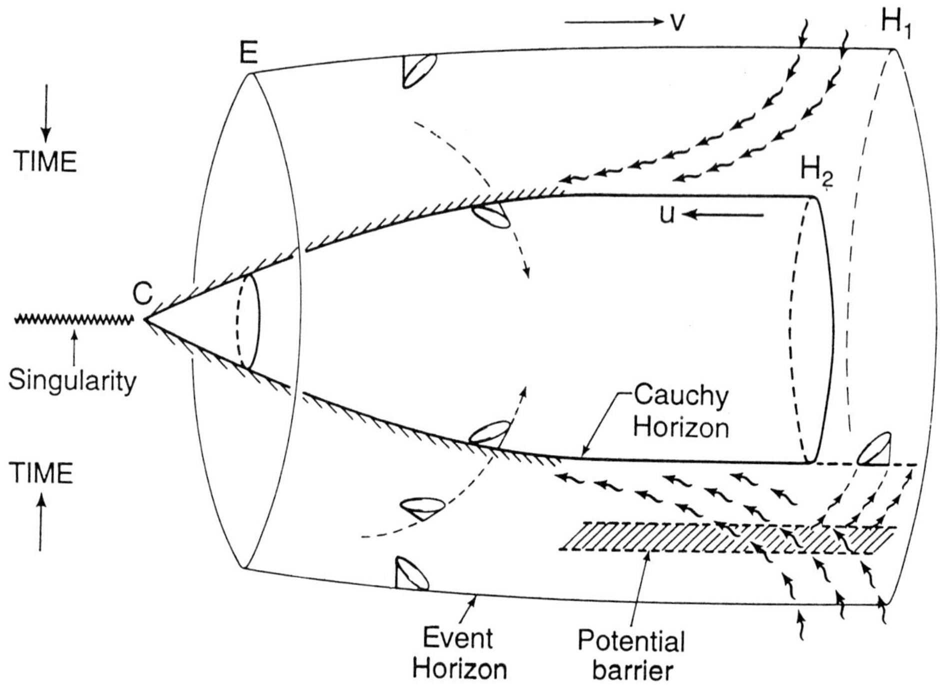

FIGURE 3. Interior of a charged, spherical black hole (one angular coordinate suppressed) showing infalling radiation partly backscattered from the potential barrier between horizons, and partly transmitted to accumulate along the Cauchy horizon.

Then it can be shown that

$$\partial_A M = 4\pi r^2 (T_A{}^B - \delta_A^B T_D^D)\partial_B r \tag{2.2}$$

where $T_A{}^B$ are components of the stress-energy tensor.

¿From (2.1) and (2.2) one can derive a (1 + 1)-dimensional wave equation for M. Leaving aside inessential terms, this is

$$\Box M = -(4\pi)^2 r^3 T^{AB} T_{AB}, \tag{2.3}$$

a form which brings out the essential nonlinearity of the Einstein equations with particular clarity.

The blueshifted influx of radiation near the Cauchy horizon, together with a nominal outflux, makes the right-hand side of (2.3) into a diverging source term for the mass function M. (For a pure influx, T_{AB} would be lightlike and its square would vanish. But some backscatter of the ingoing waves is unavoidable.) The detailed calculation shows that M—equivalently, the Weyl scalar $(C_{\alpha\beta\gamma\delta}C^{\alpha\beta\gamma\delta})^{\frac{1}{2}}$—grows near the Cauchy horizon like

$$M \sim |uv|^{-p} e^{\kappa_0(u+v)} \tag{2.4}$$

if the externally measured flux decays like v^{-p}. Here, u is an internal retarded time, calibrated so as to increase from $-\infty$ at the event horizon. Externally, M stays nearly constant. Further discussion of this "mass inflation" phenomenon can be found in the papers and reviews cited.

3. Double-null dynamics

To extend this analysis to the nonspherical case and examine the structure of the singularity at a generic Cauchy horizon, it is most convenient to slice spacetime into lightlike hypersurfaces parallel to this horizon. In this section I shall explain the elements of a compact formalism designed for this purpose (Brady *et al.* 1996).

The classic analysis of Arnowitt, Deser and Misner formulates gravitational dynamics in terms of the evolution of a spatial 3-geometry. The geometrical framework is the embedding formalism of Gauss and Codazzi for the foliation of spacetime by spacelike hypersurfaces.

Quite often, however — as in the analysis of Cauchy horizon singularities, but also in other contexts—one encounters circumstances where a lightlike foliation is especially suitable. Because of various degeneracies that arise in the lightlike case, the embedding relations are quite different and the situation not quite as familiar and under control. To bypass the degeneracies, one is forced to fall back to a foliation of co-dimension 2, by spacelike 2-surfaces.

Several such $(2+2)$-formalisms are extant, notably Geroch, Held and Penrose (1973), d'Inverno and Smallwood (1980) and Hayward (1993). Of course, all have basically the same content, but they take very different forms.

The distinguishing feature of the approach I shall present here is that it maintains manifest two-dimensional covariance and operates directly with geometrically significant objects. Two-dimensional covariance permits reduction of the Einstein field equations to an especially concise and geometrically transparent form: the ten Ricci components are embraced in a set of just three compact, two-dimensionally covariant expressions.

I now sketch the elements of the formalism.

We consider a double foliation of spacetime by a net of two intersecting families of lightlike hypersurfaces Σ^0 (with equations $u^0 = \text{const.}$) and Σ^1 (given by $u^1 = \text{const.}$), where $u^A(x^\alpha)$ $(A, B, \ldots, = 0, 1; \alpha, \beta, \ldots, = 1, \ldots, 4)$ are a given pair of scalar fields over spacetime, with lightlike gradients:

$$\nabla u^A \cdot \nabla u^B = e^{-\lambda} \eta^{AB},$$

and where the matrix

$$\eta^{AB} = \eta_{AB} = \begin{pmatrix} 0 & -1 \\ -1 & 0 \end{pmatrix}$$

will be used to raise and lower upper-case Latin indices, and $\lambda(x^\alpha)$ is a scalar function. The generators $\ell^{(A)}_\alpha$ of Σ^A are conveniently defined as

$$\ell^{(A)} = e^\lambda \nabla u^A.$$

Two hypersurfaces Σ^0 and Σ^1 intersect in a 2-surface S, with parametric equations

$$x^\alpha = x^\alpha(u^A, \theta^a) \qquad (a, b, \ldots = 2, 3)$$

where (θ^2, θ^3) are intrinsic co-ordinates of S. Both generators $\ell^{(A)}$ are orthogonal to S.

Holonomic basis vectors $e_{(a)}$ and the intrinsic metric of S may now be defined:

$$e^\alpha_{(a)} = \frac{\partial x^\alpha}{\partial \theta^a}, \qquad g_{ab} = e_{(a)} \cdot e_{(b)}.$$

The matrix g_{ab} and its inverse g^{ab} are used to lower and raise lower-case Latin indices, so that $e^{(a)} = g^{ab} e_{(b)}$ are the dual basis vectors tangent to S.

Two-dimensional shift vectors s^a_A are defined by

$$s^a_A = \frac{\partial x^\alpha}{\partial u^A} e^{(a)}_\alpha = -\ell^\alpha_{(A)} \frac{\partial \theta^a}{\partial x^\alpha}.$$

As in the Arnowitt-Deser-Misner formalism, the shift vector s^a_A measures how much one has to deviate from the normal direction $\ell_{(A)}$ to connect points on different 2-surfaces having the same intrinsic co-ordinates θ^a. An infinitesimal four-dimensional displacement dx^α can be decomposed as

$$dx^\alpha = \ell^\alpha_{(A)} du^A + e^\alpha_{(a)}(d\theta^a + s^a_A du^A).$$

Together with the completeness relation

$$g_{\alpha\beta} = e^{-\lambda}\eta_{AB}\ell^{(A)}_\alpha \ell^{(B)}_\beta + g_{ab}e^{(a)}_\alpha e^{(b)}_\beta$$

for the basis $(\ell^{(A)}, e_{(a)})$, this implies that the spacetime metric is decomposable as

$$g_{\alpha\beta}dx^\alpha\, dx^\beta = e^\lambda \eta_{AB} du^A\, du^B + g_{ab}(d\theta^a + s^a_A du^A)(d\theta^b + s^b_B\, du^B).$$

Associated with its two normals $\ell_{(A)}$, a 2-surface S has two extrinsic curvatures, defined by

$$K_{Aab} = (\nabla_\beta \ell_{(A)\alpha})e^\alpha_{(a)}\, e^\beta_{(b)}$$

and easily shown to be symmetric in a, b. (Since we are free to rescale the null vectors $\ell_{(A)}$, a certain scale-arbitrariness is inherent in this definition.)

A further basic geometrical property of the double foliation is given by the Lie bracket of $\ell_{(0)}$ and $\ell_{(1)}$. One finds

$$[\ell_{(B)}, \ell_{(A)}] = \epsilon_{AB}\omega^a e_{(a)}$$

where

$$\omega^a = \epsilon^{AB}(\partial_B s^a_A - s^b_B s^a_{A;b}),$$

the semicolon indicates two-dimensional covariant differentiation associated with metric g_{ab}, and ϵ_{AB} is the two-dimensional permutation symbol.

The geometrical significance of the "twist" ω^a can be read off from (2.4): the curves tangent to the generators $\ell_{(0)}$, $\ell_{(1)}$ mesh together to form 2-surfaces (orthogonal to the surfaces S) if and only if $\omega^a = 0$. In this case, it would be consistent to allow the co-ordinates θ^a to be dragged along both sets of generators, and thus to gauge both shift vectors to zero.

I shall denote by D_A the two-dimensionally invariant operator associated with differentiation along the normal direction $\ell_{(A)}$. Acting on any two-dimensional geometrical object $X^{a\cdots}_{b\ldots}$, D_A is formally defined by

$$D_A X^{a\ldots}_{b\ldots} = (\partial_A - \mathcal{L}_{s^d_A})X^{a\ldots}_{b\ldots}.$$

Here, ∂_A is the partial derivative with respect to u^A and $\mathcal{L}_{s^d_A}$ the Lie derivative with respect to the 2-vector s^d_A. As an example:

$$D_A g_{ab} = \partial_A g_{ab} - 2s_{A(a;b)} = 2K_{Aab}.$$

Geometrically, $D_A X^{a\cdots}_{b\ldots}$ is the projection onto S of the Lie derivative with respect to $\ell_{(A)}$ of the equivalent tangential 4-tensor $X^{\alpha\cdots}_{\beta\ldots}$.

The objects K_{Aab}, ω^a and D_A are all simple projections onto S of four-dimensional geometrical objects. Consequently, they transform very simply under two-dimensional co-ordinate transformations. Under the arbitrary reparametrization

$$\theta^a \to \theta^{a'} = f^a(\theta^b, u^A) \tag{3.5}$$

(which leaves u^A and hence the surfaces Σ^A and S unchanged), D_A is invariant, and ω_a and K_{Aab} transform cogradiently with

$$e_{(a)} \to e'_{(a)} = e_{(b)}\partial\theta^b/\partial\theta^{a'}.$$

By contrast, the shift vectors s^a_A undergo a more complicated gauge-like transformation, arising from the u-dependence in (3.5).

This geometrical groundwork is already sufficient to allow me to display the simple form that the Ricci components take in this formalism. (Notation for the tetrad components is typified by $R_{aA} = R_{\alpha\beta} e^{\alpha}_{(a)} \ell^{\beta}_{(A)}$.)

The results are

$$\begin{aligned}
{}^{(4)}R_{ab} &= \frac{1}{2}\,{}^{(2)}R g_{ab} - e^{-\lambda}(D_A + K_A) K^A_{ab} \\
&\quad + 2e^{-\lambda} K^{\ d}_{A(a} K^A_{b)d} - \frac{1}{2} e^{-2\lambda} \omega_a \omega_b - \lambda_{;ab} - \frac{1}{2}\lambda_{,a}\lambda_{,b} \\
R_{AB} &= -D_{(A} K_{B)} - K_{Aab} K_B^{\ ab} + K_{(A} D_{B)} \lambda \\
&\quad - \frac{1}{2}\eta_{AB}\big[(D^E + K^E) D_E \lambda - e^{-\lambda}\omega^a \omega_a + (e^{\lambda})^{;a}{}_{a}\big] \\
R_{Aa} &= K^b_{Aa;b} - \partial_a K_A - \frac{1}{2}\partial_a D_A \lambda + \frac{1}{2} K_A \partial_a \lambda \\
&\quad + \frac{1}{2}\epsilon_{AB} e^{-\lambda}\big[(D^B + K^B)\omega_a - \omega_a D^B \lambda\big],
\end{aligned}$$

where ${}^{(2)}R$ is the curvature scalar associated with the 2-metric g_{ab}, and $K_A \equiv K^{\ a}_{Aa}$.

It should be clear that this provides a remarkably economical and geometrically transparent generalization of the $(2+2)$ spherical formalism, whose application to spherical black holes was outlined in Sec. 2.

4. Further developments and heuristics. The generic Cauchy horizon

Double-null dynamics provides an efficient and even intuitively appealing way of extending the spherical arguments sketched in Sec. 2. As an example, one can write a nonspherical generalization of equation (2.2):

$$D_A M = 4\pi r^2 e^{-\lambda}\{(T_A^{\ B} + \tau_A^{\ B}) D_B r - T^B_B D_A r\} + f_A .$$

Here, r^2 is $({}^{(2)}g)^{\frac{1}{2}}$ (modulo a constant factor); the "mass" function M is given by

$$1 - 2M/r = e^{-\lambda} D^A r \cdot D_A r;$$

and the object

$$\tau_A^{\ B} = \sigma_{Aab}\sigma^{Bab} - \frac{1}{2}\delta^B_A\,\sigma_{Dab}\,\sigma^{Dab}$$

which is quadratic in the shearing rates

$$\sigma^{\ b}_{Aa} = K^{\ b}_{Aa} - \frac{1}{2}\delta^b_a\,K_A,$$

may conveniently be called "gravitational-wave stress-energy." Other nonspherical terms (involving $\partial_a\lambda$, $\partial_a r$ and ω_a) are swept into the correction term f_A (Morsink and Israel 1996).

Of course, these are just words. No operational prescription exists for defining quasi-local "mass" or gravitational "stress-energy," except sometimes in limiting cases (high frequency waves, etc.). But words can be helpful as heuristic guides to formal calculations, especially so in situations where "fluff" like f_A is relatively small or under control.

The systematic analysis of the characteristic initial-value problem in a neighbourhood of the Cauchy horizon has recently been carried by Brady, Droz and Morsink (1996). [Other studies are due to Ori (1992), Brady and Chambers (1995) and Flanagan and Ori (1996).] The conclusion which eventually emerges is that the spherical picture remains qualitatively unchanged. The singularity remains lightlike and locally mild. Now, however, the "mass-inflation" is overlaid by a gravitational shock wave imploding along the Cauchy horizon.

5. Concluding remarks

Because of the difficulties (such as breakdown of predictability) associated with the formation of timelike singularities, Penrose postulated in his strong cosmic censorship hypothesis that generic singularities inside a black hole are always spacelike. From the analyses reported here, a slightly more complex picture emerges, which modifies the conjecture in detail, but not in spirit. It appears that there is a milder and *lightlike* precursor to the final, spacelike crushing singularity, which extends indefinitely along the inner horizon back into the past. The generic picture is quite similar to the spherical one illustrated in Fig. 3.

I owe a great debt to my fellow black-hole speleologists Alfio Bonanno, Pat Brady, Serge Droz, Sharon Morsink and Eric Poisson, and would like to thank them for an exciting collaboration. My thanks to the organizers for arranging such a stimulating conference.

This contribution is dedicated to Igor Novikov in affection and respect.

The work was supported by NSERC of Canada and the Canadian Institute for Advanced Research.

REFERENCES

Bonanno, A., Droz, S., Israel, W. & Morsink, S. 1994 The internal evolution of black holes. *Can. J. Phys.* **72**, 755–759. (erratum: *Can. J. Phys.* **73**, 251 (1995)).

Bonanno, A., Droz, S. Israel, W. & Morsink, S. 1995 Structure of the charged spherical black hole interior. *Proc. Roy. Soc.* **A450**, 553–567.

Brady, P. R., Droz, S., Israel, W. & Morsink, S. 1996 Covariant double-null dynamics. *Class. Quantum Grav.* (in press).

Brady, P. R., Droz, S. & Morsink, S. 1996 The generic black hole singularity. (Submitted for publication).

Brady, p. R. & Chambers, C. M. 1995 Nonlinear instability of Kerr-type Cauchy horizons. *Phys. Rev.* **D51**, 4177–4186.

Doroshkevich, A. G. & Novikov, I. D. 1978 Spacetime and physical fields inside a black hole. *Soviet Physics JETP* **47**, 1.

Droz, S., Israel, W. & Morsink, S. 1996 Black holes: the inside story. *Physics World* **9** (January), 34–37.

Flanagan, E. & Ori, A. 1996 How generic are null spacetime singularities? *Phys. Rev.* **D53**, R1754–1758.

Frolov, V. P. & Novikov, I. D. 1993 Wormhole as a device for studying a black hole's interior. *Phys. Rev.* **D48**, 1607–1615.

Geroch, R., Held, A. & Penrose, R. 1973 A space-time calculus based on pairs of null directions. *J. Math. Phys.* **14**, 874–881.

Gnedin, M. L. & Gnedin, N. Y. 1993 Destruction of the Cauchy horizon inside a charged spherical black hole. *Class. Quantum Grav.* **10**, 1083–1102.

Gürsel, Y., Novikov, I. D., Sandberg, V. D. & Starobinsky, A. A. 1979 Final state of the evolution of the interior of a charged black hole. *Phys. Rev.* **D20**, 1260–.

Hayward, S. A. 1993 Dual null-dynamics of the gravitational field. *Class. Quantum Grav.* **10**, 779–790.

D'INVERNO, R. A. & SMALLWOOD, J. 1980 Covariant 2 + 2 formulation of the initial-value problem in general relativity. *Phys. Rev.* **D22**, 1233–1247.

MORSINK, S. & ISRAEL, W. 1996 Paper in preparation.

NOVIKOV, I. D. 1961 Note on the space-time metric inside the Schwarzschild singular sphere. *Soviet Astronomy AJ* **5**, 423–424.

NOVIKOV, I. D. & STAROBINSKY, A. A. 1980 Quantum-electrodynamic effects inside a charged black hole and the problem of Cauchy horizons. *Soviet Physics JETP* **51**, 1.

ORI, A. 1991 Inner structure of a charged black hole: an exact mass-inflation solution. *Phys. Rev. Letters* **67**, 789–792.

ORI, A. 1992 Structure of the singularity inside a realistic rotating black hole. *Phys. Rev. Letters* **68**, 2117–2120.

PENROSE, R. 1968 In *Battelle Rencontres* (ed. C. M. DeWitt and J. A. Wheeler), p. 222. W. A. Benjamin.

POISSON, E. & ISRAEL, W. 1990 Internal structure of black holes. *Phys. Rev.* **D41**, 1796–1809.

SYNGE, J. L. 1950 The gravitational field of a particle. *Proc. Roy. Irish Acad.* **53 A6**.

ZEL'DOVICH, YA. B. & NOVIKOV, I. D. 1971, 1983 *Relativistic Astrophysics* I & II, Univ. of Chicago Press.

Black Hole Entropy: External Façade and Internal Reality

By Valeri Frolov

CIAR Cosmology Program; Theoretical Physics Institute,
University of Alberta, Edmonton, Canada T6G 2J1;
P.N.Lebedev Physics Institute, Moscow, Russia

The problem of the black hole entropy has many different aspects (faces). We discuss the dynamical origin of the entropy of a black hole and relations between what one might observe (the *thermodynamical entropy* S^{TD}) and what one might calculate by counting dynamical degrees of freedom of a black hole (the *statistical-mechanical* entropy S^{SM}). In the general case these quantities are different. We explain why this is so and establish the relation between S^{TD} and S^{SM}. The situation is qualitatively different if the Einstein gravity itself is an effective low-energy theory induced by one-loop effects generated by fundamental constituent fields. In particular in the Sakharov's induced gravity model the Bekenstein-Hawking entropy of a black hole can be naturally related to the entropy of the constituents that generate the gravitational field.

1. Black-Hole Entropy

My interest in the problem of black hole entropy was greatly stimulated by long discussions with Igor Novikov during my stay in Copenhagen in 1992-93. A lot of new important and interesting results in this field were obtained during the last three years. It is virtually impossible to cover all or most of them in the brief paper. My purpose is much more modest. It happens that many of the questions and puzzles discussed with Igor have been solved and understood recently, sometimes in quite unexpected ways.

In this talk I shall give brief review of some of the aspects of the problem as I see them now. According to the thermodynamical analogy in black hole physics, the entropy of a black hole in the Einstein theory of gravity is

$$S^{BH} = A_H/(4l_{\rm P}^2), \tag{1.1}$$

where A_H is the area of a black hole surface and $l_{\rm P} = (\hbar G/c^3)^{1/2}$ is the Planck length (Bekenstein 1972, 1973, 1974; Hawking 1975). In black hole physics the Bekenstein-Hawking entropy S^{BH} plays essentially the same role as in the usual thermodynamics. In particular it allows to estimate what part of the internal energy of a black hole can be transformed into work. Four laws of black hole physics that form the basis in the thermodynamical analogy were formulated by Bardeen, Carter, & Hawking (1973). The generalized second law (Bekenstein 1972, 1973, 1974; see also Novikov & Frolov 1989; Wald 1992; Frolov & Page 1993; and references therein) implies that when a black hole is a part of a thermodynamical system, the total entropy (i.e. the sum of the entropy of a black hole and the entropy of the surrounding matter) does not decrease.

Thermodynamical entropy S^{TD} is defined by the response of the free energy F of a system to the change of its temperature:

$$dF = -S^{TD} dT. \tag{1.2}$$

This definition applied to a black hole determines its Bekenstein-Hawking entropy. This entropy is directly related to the quantities that can be measured in (at least gedanken) experiments. For example, one might consider a spherical (of radius r_0) mirror-like boundary heated up to the temperature T. If the temperature is high such a cavity will contain a black hole in the thermal equilibrium with thermal radiation.

Such an experiment (which might require for its realization supernatural forces) is shown in the Figure to the left. By heating the boundary one changes the temperature of radiation and the parameters (such as mass, gravitational radius, and so on) of the black hole. The change of the energy of the system dE is directly related to the change of its entropy $dE = TdS^{TD}$. The four laws of black hole thermodynamics allow one to predict the result of this and similar experiments.

A similar experiment can be performed if one replaces a black hole by a usual black body. In the latter case we know that such a body has internal degrees of freedom. Their excitation requires energy, and the number of excited internal degrees of freedom can be connected with the contribution of the black body to the total entropy of the system. Is the analogy between a black hole and a black body complete? Do black holes also possess internal degrees of freedom? The success of the thermodynamical analogy in black hole physics allows one to hope that the analogy is really deep and that it is possible to develop a statistical-mechanical foundation of black hole thermodynamics.

Before discussing the possibility of introduction of internal degrees of freedom of a black hole, it is worthwhile to remember that the thermodynamical and statistical-mechanical definitions of the entropy are logically different. *Statistical-mechanical entropy* S^{SM} is defined as

$$S^{SM} = -\mathrm{Tr}(\hat{\rho} \ln \hat{\rho}), \tag{1.3}$$

where $\hat{\rho}$ is the density matrix describing the internal state of the system under consideration. For a black body in the thermal equilibrium with radiation the density matrix is $\hat{\rho} \sim \exp(-\hat{H}/T)$, where $\hat{H}$ is the Hamiltonian of the internal degrees of freedom and T is the temperature. By using this representation and the expression for the free energy $F = -T \ln \mathrm{Tr}[\exp(-\hat{H}/T)]$ it is possible to prove that for a usual body S^{TD} defined by Eq. (1.2) coincides with S^{SM} defined by Eq. (1.3). As we shall see, this is not generally true for a black hole.

The statistical-mechanical foundations of black hole thermodynamics must include the solution of the following three problems: (1) definition of internal degrees of freedom of a black hole, (2) a calculation of the hole's statistical-mechanical entropy S^{SM} by counting the dynamical degrees of freedom described by the black hole density matrix $\hat{\rho}^H$ and (3) establishment of a relation between the statistical-mechanical S^{SM} and the thermodynamical S^{TD} entropies.

2. Dynamical Degrees of Freedom of a Black Hole

The problem of the dynamical origin of the black hole entropy has been much discussed lately. One of the proposed basic ideas is to relate the dynamical degrees of freedom of a black hole to its quantum excitations. This idea has different realizations.

In the framework of the *membrane paradigm* the dynamical degrees of freedom are identified with different possible states of thermal atmosphere of a black hole, while the entropy of a black hole is identified with the amount of information about the state of thermal atmosphere, which is lost by stretching the horizon (Thorne & Zurek 1985, 1986).

In his *brick wall model* 't Hooft (1985) proposed to consider a mirror-like boundary, located outside a black hole at the close distance to its horizon. He assumed that outside the boundary there exist thermal radiation with temperature equal to the Hawking temperature. He showed that the entropy of such thermal radiation is of the same order of magnitude as the Bekenstein-Hawking entropy, provided the distance of the mirror-like boundary from the horizon is chosen to be of order of Planck's length.

Bombelli et al (1986) called attention to the fact that even in a flat spacetime in Minkowski vacuum state one can obtain a non-vanishing entropy if one restricts his observations to a spatially bounded part of the space. The corresponding *entanglement entropy* arises due to the presence of correlations of zero-point fluctuations of those modes that are propagating in the vicinity of the boundary of the chosen spatial region. It was proposed to relate black hole entropy with the entanglement entropy associated with the presence of the horizon (Bombelli et al. 1986, Srednicki 1993).

We (with Igor Novikov) arrived at a similar idea independently by trying to answer the question of where the entropy of a black hole is located. In principle, there might be three different possibilities: (i) the entropy is connected with the black-hole surface physics, (ii) the physical effects responsible for the entropy takes place in the space of black hole interior, and (iii) it is the property of a singularity hidden inside the black hole. In order to distinguish between these possibilities we consider the gedanken experiment proposed in our earlier paper (Frolov & Novikov 1993a). Namely, we assume that there exist a traversable wormhole, and its mouths are freely falling into a black hole.

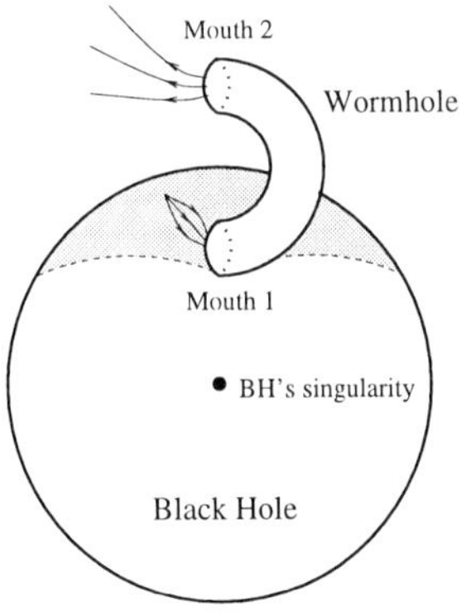

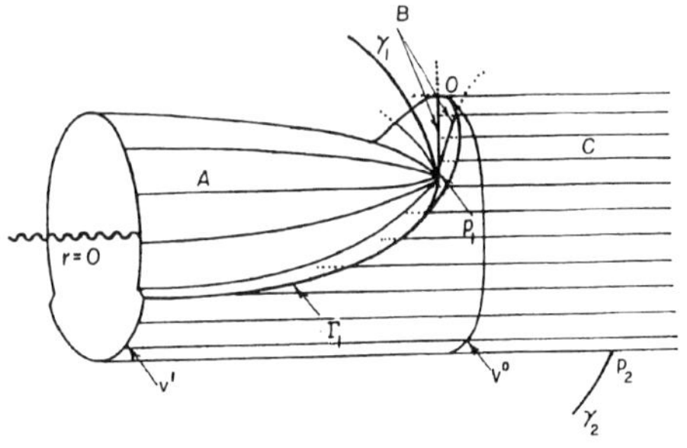

If one of the mouths (say mouth 1) crosses the gravitational radius earlier than the other (mouth 2), then rays passing through the first mouth can escape from the region lying inside the gravitational radius. Such rays would go through the wormhole and enter the outside region through the second mouth. This situation is schematically shown in the left figure above. As a result, the shadowed region inside the gravitational radius becomes visible to an external observer. The position of the event horizon is changed. The above right figure shows the position of the event horizon as the function of advanced time (in the figure the time is directed to the right). The worldlines of the mouths 1 and 2 are represented by lines γ_1 and γ_2. During the period of time when the first of the mouths is inside and the other mouth is outside the gravitational radius the surface area of the horizon decreases.

The device (wormhole) described above can be used to extract information from the regions located inside the gravitational radius (apparent horizon). The remarkable feature of this gedanken experiment is that the change of the area of the event horizon remains finite even in the limit when the cross-section of the wormhole becomes negligibly small. If we assume that the black hole entropy is related to the surface area of a black hole, then the only possibility to escape contradictions with the second law is to assume that this entropy is really connected with physics of narrow region in the vicinity of the event horizon so that during the experiment the decrease of entropy is related to the possibility of getting access to some new information concerning black hole internal states located near the horizon.

At first sight it looks like a puzzle. We know that (at least in the classical General Relativity) a black hole at late time is completely specified by finite number of parameters. For a non-rotating uncharged black hole one needs to know only one parameter (mass M) to describe all its properties. It is true not only for the exterior, where this property is the consequence of no-hair theorems, but also for the black hole interior [Doroshkevich & Novikov (1978)].

The reason why it is impossible for an isolated black hole at late time to have non-trivial classical states in its interior in the vicinity of the horizon is basically the same as for the exterior regions. These states might be excited only if a collapsing body emits the pulse of fields or particles immediately after it crosses the event horizon. Due to the red-shift effect, the energy of the emitted pulse must be exponentially large in order to reach the late time region with any reasonable energy. Short time after the formation of a black hole (say $100M$) it is virtually impossible because it requires the energy of emission much greater than the black hole mass M.

In quantum physics the situation is completely different due to the presence of zero-point fluctuations. To analyze states of a quantum field it is convenient to use its decomposition into modes. Besides the positive-frequency modes which have positive energy, there also exist positive-frequency modes with negative energy. In a non-rotating uncharged black holes such modes can propagate only inside the horizon where the Killing vector used to define the energy is spacelike. It is possible to show that at late time for any regular initial state of the field these states are thermally excited and the corresponding temperature coincides with the black hole temperature $T_H = (8\pi M)^{-1}$. In the framework of this *dynamical-black-hole-interior* model we proposed to identify the dynamical degrees of freedom of a black hole with the internal modes of all physical fields. The set of the fields must include the gravitational one.

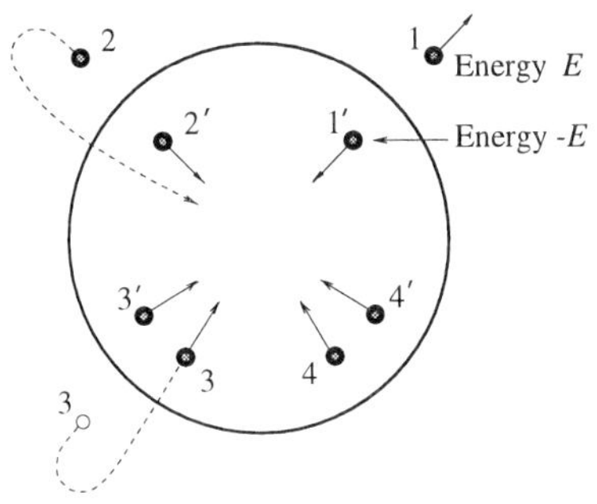

The Figure to the left gives a schematic picture of a black hole at some 'instant of time'. Quanta of different fields are created in the Hawking process. The quanta are created in pairs. The energy conservation does not allow for the both of the components of the pair to be created outside the black hole. Different possibilities are shown in Figure. A particle 1 has enough energy to penetrate the potential barrier and to escape to infinity. Such particles form the flux of Hawking radiation of the black hole. Particle 2 has not enough energy to escape to infinity and will fall down into a black hole later, but at the moment of consideration it is located outside the black hole. Particle 3 was created earlier outside the horizon but after reflection from the potential barrier it fell down into the black hole. Particle 4 as well as its companion $4'$ are created inside the black hole.

After averaging of the given initial state over the states of the particles located outside the black hole at the given moment of time one obtains the density matrix describing the state of the black hole interior. In the model of the dynamical interior this density matrix is identified with the density matrix of the black hole itself. Pairs 3-$3'$ and 4-$4'$ do not contribute to the entropy because both of the components are located inside, and their state is pure. It means that only pairs 1-$1'$ and 2-$2'$ are responsible for the entropy of the black hole. A particle belongs to the type 1 if its angular momentum l is small: $l \lesssim \sqrt{27}ME$. Since particles with energies greater that $T_H = (8\pi M)^{-1}$ practically are not created, one can conclude that particles of type 2 (trapped by the potential barrier) give the main contribution to the statistical-mechanical entropy of the black hole. [For more details see, e.g., Frolov & Novikov (1993b).]

In accordance with our definition of the internal degrees of freedom the density matrix of a black hole is

$$\hat{\rho}_H = \underbrace{Tr}_{\text{'visible'}} (\hat{\rho}_{total}) , \tag{2.4}$$

where $\hat{\rho}_{total}$ is the total density matrix for quantum fields, and the trace operation is performed over the modes located outside the horizon (that are 'visible' by a distant observer). The corresponding statistical-mechanical entropy is

$$S^{SM} = - \underbrace{Tr}_{\text{'invisible'}} (\hat{\rho}_H \ln \hat{\rho}_H) , \tag{2.5}$$

where the trace operation is performed over the modes located inside the horizon (that are 'invisible' by a distant observer). By using the mode decomposition and the fact, that the internal modes are thermally populated with respect to the Killing energy, one gets

$$S^{SM} = \underbrace{\Sigma}_{\text{'visible' } J} s_J . \tag{2.6}$$

Here

$$s_J = \frac{\beta\omega}{\exp(\beta\omega) - 1} - \ln[1 - \exp(-\beta\omega)], \tag{2.7}$$

is the entropy of a quantum oscillator of frequency ω at the temperature $T = \beta^{-1}$. It is important to stress that in our model the dynamical degrees of freedom of a black hole are connected with 'invisible' modes, nevertheless the summation in Eq. (2.6) is performed over the 'visible' states. The reason is simple: only those pairs, for which one of the components is 'visible' do contribute to the entropy†.

It can be shown that for any chosen field the number of the modes infinitely grows as one considers the regions located closer and closer to the horizon. For this reason the contribution of a field to the statistical-mechanical entropy of a black hole, calculated by counting the hole's internal modes, is formally divergent. In order to make it finite one might restrict oneself by considering only those modes which are located at the proper distance from the horizon greater than some chosen value l. For this choice of the cut-off the contribution of a field to the statistical-mechanical entropy of the black hole is

$$S^{SM} = \alpha \frac{A}{l^2}, \tag{2.8}$$

where A is the hole's surface area and the dimensionless parameter α depends on the type of the field.

The statistical-mechanical entropy S^{SM} in this as well as other 'dynamical' approaches possesses the following main properties: (1) $S^{SM} \sim A$, where A is the surface area of the hole; (2) S^{SM} is divergent and requires regularization $S^{SM} \sim A/l^2$, where l is the cut-off parameter; (3) S^{SM} depends on the number of fields which exist in Nature; (4) for $l \sim l_{Pl}$ one has $S^{SM} \sim S^{BH}$.

It might be tempting to use a special choice of the cut-off and to identify S^{SM} with S^{BH}. Unfortunately the situation is not so simple. A black hole really is a thermodynamical system, but it also possesses a very special property which singles it out: namely its size and temperature are not independent parameters.

One of the consequences is that S^{TD} and S^{SM} for the black hole are different (Frolov 1995). In order to explain why it happens we need to discuss a more accurate derivation of the thermodynamical properties of black holes.

3. Renormalized Effective Action and Free Energy

The complete information concerning the canonical ensemble of a black hole with a given inverse temperature β at the boundary is contained in the partition function $Z(\beta)$ given by the Euclidean path integral (Hawking 1979)

$$Z(\beta) = \int D[\phi] \exp(-I_E[\phi]). \tag{3.9}$$

Here the integration is taken over all fields including the gravitational one that are real on the Euclidean section and are periodic in the imaginary time coordinate τ with period β. The quantity ϕ is understood as the collective variable describing the fields. In particular it contains the gravitational field. Here $D[\varphi]$ is the measure of the space of fields ϕ and I_E is the Euclidean action of the field configuration. The action I_E includes the Euclidean Einstein action. The state of the system is determined by the choice of the boundary conditions on the metrics that one integrates over. For the canonical ensemble of the gravitational fields inside a spherical box of radius r_0 at temperature T one must integrate over all the metrics inside r_0 which are periodically identified in the imaginary time direction with period $\beta = T^{-1}$. Such a partition function must describe in particular the thermal ensemble of a black hole. The partition function Z is related to the effective action $\Gamma = -\ln Z$ and the free energy $F = \beta^{-1}\Gamma = -\beta^{-1} \ln Z$.

† There is an elegant way to describe the internal degrees of freedom introduced above in the framework of the so called no-boundary wave function of the black hole introduced by Barvinsky, Frolov, & Zelnikov (1995).

By using the stationary-phase approximation one gets

$$\beta F \equiv -\ln Z = I_E[\phi_0] - \ln Z_1 + \dots . \qquad (3.10)$$

Here ϕ_0 is a (generally speaking, complex) solution of classical field equations for action $I[\phi]$ obeying the required periodicity and boundary conditions. Besides the tree-level contribution $I_E[\phi_0]$, the expression (3.10) includes also one-loop corrections $\ln Z_1$, connected with the contributions of the fields perturbations on the background ϕ_0, as well as higher order terms in loops expansion, denoted by $(\dots)$. The one-loop contribution of a field ϕ can be written as $\ln Z_1 = -\frac{1}{2}\mathrm{Tr}\ln(-D)$, where D is the field operator for the field ϕ inside the box r_0.

The one-loop contribution contains divergences and requires the renormalization. In order to be able to absorb these divergences in the renormalization of the coefficients of the initial classical action we chose the latter in the form

$$I_{cl} = \int d^4x\sqrt{g}L, \qquad (3.11)$$

$$L = -\frac{\Lambda_B}{8\pi G_B} - \frac{R}{16\pi G_B} + c_B^1 R^2 + c_B^2 R_{\mu\nu}^2 + c_B^3 R_{\alpha\beta\mu\nu}^2 . \qquad (3.12)$$

By using heat-kernel representation for $\ln Z_1$ one can write

$$-\frac{1}{2}\ln\det(-D) = \frac{1}{2}\int_{\delta^2}^{\infty}\frac{ds}{s}\mathrm{Tr}K(s), \qquad (3.13)$$

where $K(s)$ is the heat-kernel of the operator D which has the following Schwinger-DeWitt expansion

$$K(s) = e^{-sD} = \frac{1}{16\pi^2 s^2}\sum a_n s^n, \qquad s \to 0. \qquad (3.14)$$

For the particular case of a scalar massless field

$$a_0 = 1, \qquad a_1 = (1/6 - \xi)R, \qquad (3.15)$$

$$a_2 = \frac{1}{180}R_{\alpha\beta\mu\nu}^2 - \frac{1}{180}R_{\mu\nu}^2 - \frac{1}{6}\left(\frac{1}{5} - \xi\right)\Box R + \frac{1}{2}\left(\frac{1}{6} - \xi\right)^2 R^2 . \qquad (3.16)$$

By substituting this expansion into (3.13) one can conclude that the one-loop contribution Γ_1 to the effective action can be written in the form

$$\Gamma_1 = \Gamma_1^{div} + \Gamma_1^{fin} . \qquad (3.17)$$

where

$$\Gamma_1^{div} = -\frac{1}{32\pi^2}\int d^4x\sqrt{g}\left[\frac{a_0}{2\delta^4} + \frac{a_1}{\delta^2} - 2a_2\ln(\delta)\right] . \qquad (3.18)$$

The divergent part of the one-loop effective action has the same structure as the initial classical action (3.11) and hence one can write

$$\Gamma = \Gamma_{cl}^{ren} + \Gamma_1^{ren}, \qquad (3.19)$$

$$\Gamma_1^{ren} = \Gamma_1 - \Gamma_1^{div} = \Gamma_1^{fin} . \qquad (3.20)$$

Here Γ_{cl}^{ren} is identical to the initial classical action with the only change that all the bare coefficients Λ_B, G_B, and c_B^i are substituted by their renormalized versions Λ_{ren}, G_{ren}, and c_{ren}^i

$$\frac{\Lambda_{ren}}{G_{ren}} = \frac{\Lambda_B}{G_B} + \frac{1}{8\pi\delta^4} + \dots , \qquad (3.21)$$

$$\frac{1}{G_{ren}} = \frac{1}{G_B} + \frac{1}{2\pi\delta^2}\left(\frac{1}{6} - \xi\right) + \dots , \tag{3.22}$$

$$c^i_{ren} = c^i_B + \alpha^i \ln \delta + \dots , \tag{3.23}$$

whete dots denote possible less divergent terms. We shall refer to (3.19) as the loop expansion of the renormalized effective action. After multiplying the renormalized effective action by β^{-1} we get the expansion for the renormalized free energy.

The effective action Γ contains the complete information about the system under consideration. In particular the variation of Γ with respect to the metric provides one with the equations for the quantum average metric $\bar{g} = \langle g \rangle$:

$$\frac{\delta \Gamma}{\delta \bar{g}} = 0. \tag{3.24}$$

One usually assumes that quantum corrections are small and solves this equation perturbatively:

$$\bar{g} = g_{cl} + \delta g, \tag{3.25}$$

where g_{cl} is a solution of the classical equations. At this point we need to make an important remark. In principle, there exist two possibilities: either to begin with the solution of the classical equations for the action (3.11), or its renormalized version Γ^{ren}_{cl}, which is written in terms of the renormalized constants. One usually assumes that the renormalized values of Λ_{ren} and c^i_{ren} vanish $\Lambda_{ren} = c^i_{ren} = 0$. In the general case, however, their initial values are not vanishing unless one is dealing with some special type of theory (e.g. assuming sypersymmetry). In other words, the global properties of the solutions for I_{cl} and Γ^{ren}_{cl} are generally different. Thus, in order to provide that δg in (3.25) be small, one begins with the metric g_{cl} that is an extremum of Γ^{ren}_{cl}

$$\frac{\delta \Gamma^{ren}_{cl}}{\delta g_{cl}} = 0. \tag{3.26}$$

We assume that the renormalization of the coupling constants in the classical action is made from the very beginning and we shall assume that the 'classical' field g_{cl} is a solution of the equation (3.26). In our case g_{cl} is the Euclidean black hole metric, while the metric $\bar{g}$ describes the Gibbons-Hawking instanton deformed due to the presence of quantum corrections to the metric. The quantity $\Gamma[\bar{g}]$, being expressed as the function of boundary conditions (β and r_0), specifies the thermodynamical properties of a black hole.

4. Thermodynamical Entropy

For the canonical ensemble of gravitational fields described above the leading tree-level contribution to the renormalized effective action is given by the Euclidean gravitational action of the Euclidean black hole solution (the so-called Gibbons-Hawking instanton).

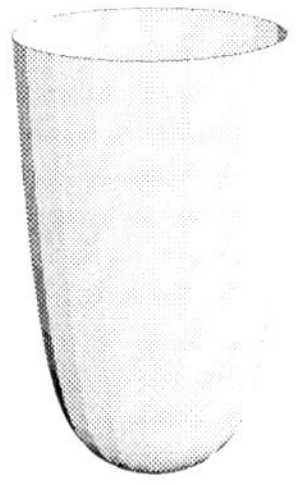

Under the assumption that $\Lambda_{ren} = 0$ and $c^i_{ren} = 0$ the tree-level renormalized effective action coincides with the standard Einstein action. The Euclidean vacuum solution is $ds^2 = (1 - r_+/r)d\tau^2 + (1 - r_+/r)^{-1}dr^2 + r^2 d\omega^2$. Here $r_+ = 2G_{ren}M$ is the gravitational radius of a black hole of mass M, and $d\omega^2 = d\theta^2 + \sin^2\theta d\phi^2$ is a line element on a unit sphere. The condition of regularity at the Euclidean horizon requires that τ is periodic with the period $4\pi r_+$. The embedding diagram of the 2D (τ, r)– section for such a regular instanton is shown in the Figure.

The tree-level contribution to the renormalized free energy of the black hole is (York 1986, Braden et al. 1990)

$$F_0^{ren} \equiv \beta^{-1}\Gamma_{cl}^{ren}[g_{cl}] = r_0\left(1 - \sqrt{1 - r_+/r_0}\right) - \pi r_+^2 \beta^{-1}. \tag{4.27}$$

For a given temperature β^{-1} at the boundary, r_+ is defined by the relation

$$\beta = 4\pi r_+(1 - r_+/r_0)^{1/2}. \tag{4.28}$$

The inverse Hawking temperature $\beta_H = 4\pi r_+$ is also determined by this equation as the function of the inverse temperature at the boundary β. The relation (4.28) can be also rewritten as $\beta_\infty = \beta_H$, where $\beta_\infty = \beta(1 - r_+/r_0)^{-1/2}$.

According to the definition the thermodynamical entropy of a black hole S_0^{TD} is determined by the response of the free energy of a system including a black hole to the change of the temperature. One can easily verify that

$$S_0^{TD} = -\frac{dF_0^{ren}}{dT} \equiv \beta^2 \frac{dF_0^{ren}}{d\beta} = \frac{A_H}{4l_\mathrm{P}^2}, \tag{4.29}$$

and hence the tree-level S^{TD} coincides with the Bekenstein-Hawking expression S^{BH}. (It is assumed that r_+ in F_0^{ren} is expressed in terms of β and r_0 before differentiation with respect to β.) One-loop contribution in Eq.(3.10) describes quantum correction to the entropy of a black hole as well as the entropy of thermal radiation in its exterior. The latter evidently depends on the radius r_0 of the boundary. Since the Euclidean black hole background is regular, the corresponding contribution F_1^{ren} is finite. For this reason the quantum corrections to the Bekenstein-Hawking entropy $4\pi M^2$ are small unless the mass of a black hole M is comparable with the Planckian mass (Anderson et al. 1994). Due to the presence of the conformal anomalies one might expect that the leading corrections are of the order $\ln M$ [see, e.g., Fursaev (1995)].

5. Statistical-Mechanical Entropy

Derivation of the thermodynamical entropy of a black hole requires *on-shell* calculations. It means that one uses only a regular Euclidean metrics that are the solutions of the field equations (3.24). The discussion of the relation of the thermodynamical and statistical-mechanical entropy of a black hole requires *off-shell* calculations (Frolov, Fursaev & Zelnikov 1996ab).

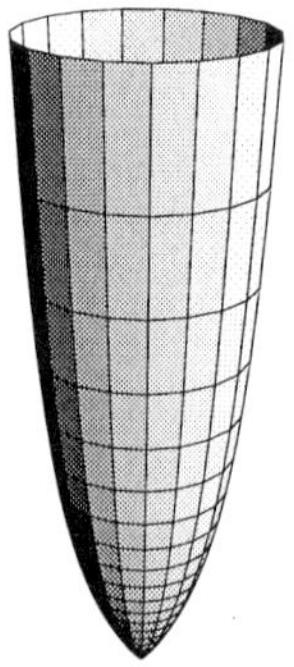

The words *off-shell* mean that in the calculations one must consider the inverse temperature β and the black hole gravitational radius r_+ as independent parameters. In other words we must consider situations where the black hole is not in equilibrium with the surrounding thermal radiation. Strictly speaking, for $\beta_\infty \neq \beta_H \equiv 8\pi M$ there are no regular vacuum Euclidean solutions with the Euclidean black-hole topology $R^2 \times S^2$. Such solutions can be obtained only if one excludes a horizon sphere S^2. The alternative is to consider a non-vacuum solution with the cone-like singularity shown in the Figure. This is the so called *singular instanton.* For $\beta_\infty = \beta_H$ the singularity disappears.

The new feature arises when $\beta_\infty \neq \beta_H$: the corresponding renormalized one-loop corrections are divergent. In order to make the answer finite one might introduce spatial cut-off. We shall use the so called *volume cut-off method*, where the volume integrals that enter the expressions for the free energy and entropy are restricted to the regions lying outside the proper distance l from the cone-singularity. In the model with 'dynamical black-hole-interior' the presence of such a cut-off was connected with the quantum fluctuations of the horizon (Frolov & Novikov 1993b). There are other off-shell methods and corresponding cut-off procedures. For example, in the 'brick wall model' ('t Hooft 1985) one introduces an internal mirror-like boundary which does not allow quanta to penetrate the region near the horizon. The details of the calculations depend on the particular off-shell procedure, but physically important conclusions remain similar.

The renormalized one-loop contribution F_1^{ren} to the free energy is of the form

$$F_1^{ren} = F_1^{ren}[\beta, \beta_H, \varepsilon], \tag{5.30}$$

where $\varepsilon = [l/(2G_{ren}M)]^2$ is the dimensionless cut-off parameter. For $\varepsilon \to 0$ and $\beta_\infty \neq \beta_H$ the one-loop free energy F_1^{ren} is divergent $F_1^{ren} \sim \varepsilon^{-1} f(\beta, \beta_H)$. For $\beta_\infty = \beta_H$ the divergence disappears, so that F_1^{ren} is finite. This quantity is directly related to quantum (one-loop) corrections to the thermodynamical entropy of a black hole

$$S_1^{TD} = \beta^2 \frac{d}{d\beta} F_1^{ren}[\beta, \beta_H(\beta), 0] \equiv \left[\beta^2 \frac{\partial F_1^{ren}}{\partial \beta} + \beta^2 \frac{\partial \beta_H}{\partial \beta} \frac{\partial F_1^{ren}}{\partial \beta_H}\right]_{\beta_\infty = \beta_H}. \tag{5.31}$$

The expression (5.30) allows one also to get the statistical-mechanical entropy S^{SM}. Dowker & Kennedy (1978) and Allen (1986) made an important observation that

$$F_1^{ren} = F_{vac}^{ren} + F_{therm}^{ren}, \qquad \frac{\partial F_{vac}^{ren}}{\partial \beta} = 0, \tag{5.32}$$

$$F_{therm}^{ren} = -\beta^{-1} \ln \mathrm{Tr}\left[e^{-\beta \hat{H}}\right] = \ln\left[\sum \exp(-\beta E_n)\right], \tag{5.33}$$

where E_n is the energy (eigenvalue of the Hamiltonian $\hat{H}$ of the field φ). By using the expansion in eigenfunctions one can obtain

$$F_{therm}^{ren} = \sum_\lambda f(\beta\omega_\lambda) = \int d\omega N(\omega|\beta_H, \varepsilon) f(\beta\omega). \tag{5.34}$$

$f(\beta\omega) = \beta^{-1} \ln[1 - \exp(-\beta\omega)]$ is the free energy of an oscillator of frequency ω at inverse temperature β, and $N(\omega|\beta_H)$ is the density of the number of states at the given energy ω in a spacetime containing a black hole of mass M. This density of the number of states diverges. In order to make it finite we have introduced the cut-off ε; thus the presence of ε as an argument of N. The expression similar to (5.34) is usually a starting point for the 'brick wall' model.

The statistical-mechanical entropy S^{SM} is

$$S^{SM} = \left[\frac{\partial F_1^{ren}}{\partial \beta}\right]_{\beta_H} = \left[\frac{\partial F_{therm}^{ren}}{\partial \beta}\right]_{\beta_H} \tag{5.35}$$

$$= \int d\omega N(\omega|\beta_H, \varepsilon) s(\beta\omega) \tag{5.36}$$

Here $s(\beta\omega) = \beta\omega/(e^{\beta\omega} - 1) - \ln(1 - e^{-\beta\omega})$ is the entropy of a quantum oscillator of frequency ω with inverse temperature β. S^{SM} is divergent in the limit $\varepsilon \to 0$. The divergence is directly related to the divergence of the density of the number of states located in the narrow region in the vicinity of the horizon.

By comparing the expressions (5.31) and (5.35) we can conclude that S^{TD} and S^{SM} differ from one another. This happens for the following two reasons: (1) Vacuum polarization (F_{vac}^{ren}) depends on M and hence on β_H; (2) $d/d\beta$ does not commute with Tr-operation. In the general case one gets (Frolov 1995)

$$S_1^{TD} = S^{SM} + \Delta S. \tag{5.37}$$

In the limit $\varepsilon \to 0$, $\Delta S \neq 0$ is also divergent, but S_1^{TD} remains finite and (for $M \gg \mu_{Pl}$) small.

The above general formulas can be illustrated by explicit calculations in a simplified two-dimensional model (Frolov, Fursaev & Zelnikov 1996ab). In this model one considers a special version of 2D dilaton gravity, that can be obtained from the Einstein action by its reduction to spherical spacetimes, and then uses a massless quantum field to describe the corresponding quantum degrees of freedom. The main result is that the entropy of a black hole can be written in the form

$$S^{TD} = \bar{S}^{BH} + S^{SM} - S^{SM}_{Rindler}. \tag{5.38}$$

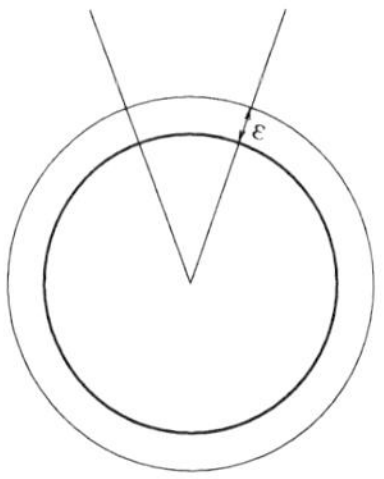

Black Hole

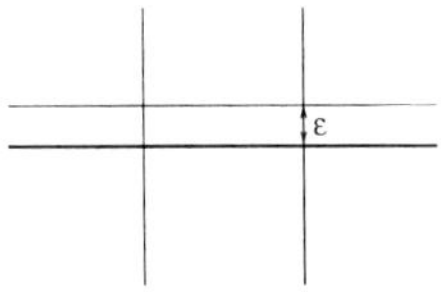

Rindler Space

The quantities that enter this relation have the following meaning. $\bar{S}^{BH}$ is the Bekenstein-Hawking entropy with the only change that for its calculation one must use instead of the classical gravitational radius r_+ its quantum corrected version. S^{SM} is the statistical-mechanical entropy calculated either with the volume cut-off or by the brick-wall method. $S^{SM}_{Rindler}$ is the analogous quantity calculated by the same method in the Rindler space (that is, in flat spacetime). Formula (5.38) holds provided in both calculations (for the black hole and in the Rindler space) one uses the identical values of the proper-distance cut-off parameter ε. Taken together, relations (5.37) and (5.38) can be viewed as an explanation of the entropy subtraction procedure proposed earlier by Zurek & Thorne (1985, 1986).

6. Black Hole Entropy in Sakharov's Induced Gravity

As shown above, the attempts to explain the Bekenstein-Hawking entropy S^{BH} by counting quantum excitations of a black hole meet an evident difficulty because S^{BH} arises at the tree-level, while the entropy of quantum excitations is a one-loop quantity.

It was pointed out by Susskind & Uglum (1994) and Callan & Wilczek (1995) that the renormalization of the Newton coupling constant might play an important role. But as soon as the initial bare parameter G_B^{-1} does not vanish there always would be an inevitable non-statistical (geometrical) contribution to the entropy. This difficulty can be overcome if the gravity itself arises as a result of quantum effects, so that the metric $g_{\mu\nu}$ is a collective variable and the general relativity is a low-energy effective theory.

One of the possible examples of such a theory is Sakharov's induced gravity (Sakharov 1968, 1975). According to Sakharov's idea the background fundamental theory is described by the action $I[\varphi_i, g_{\mu\nu}]$ of fields φ_i propagating in the external geometry with the metric $g_{\mu\nu}$. The metric originally is not dynamical so that the original gravitational action is simply put equal to zero.

By averaging over the constituent fields φ_i one gets the dynamical effective action for the metric $g_{\mu\nu}$

$$\exp(-W[g_{\mu\nu}]) = \int \mathcal{D}\varphi_i \exp(-I[\varphi_i, g_{\mu\nu}])\,. \tag{6.39}$$

An observable gravitational field is an extremum of the effective action $W[g_{\mu\nu}]$. The idea is to relate the Bekenstein-Hawking entropy to the statistical mechanics of ultraheavy constituents that induce the gravity in the low energy limit of the theory (Jacobson 1994).

This idea was recently realized by Frolov, Fursaev, & Zelnikov (1996c). The model proposed in this paper consists of Dirac fermion and scalar boson field constituents. It was demonstrated that one could obtain a vanishing induced cosmological constant and a finite induced Newton coupling constant G for a special choice of the spectrum of constituents' masses, provided at least some of the scalar fields were non-minimally coupled. The Newton constant G is the function of the parameters of the constituents

$$\frac{1}{G} = \frac{1}{12\pi}\left(\sum_s (1 - 6\xi_s)\, m_s^2 \ln m_s^2 + 2\sum_d m_d^2 \ln m_d^2\right), \tag{6.40}$$

where ξ is the parameter of the non-minimal coupling $\xi R\varphi^2$ in the action. The non-minimality of the scalar fields plays an important role since it allows one to make finite both G and the entropy of constituents (Frolov, Fursaev & Zelnikov 1996c, Frolov & Fursaev 1997). The value of the induced Newton constant is dominated by the masses of the heaviest constituents and at least some of the constituents must have Planckian mass $m \sim m_{Pl}$.

Relation (6.39) allows one to express the black hole entropy as the functional integral over the configurations of the constituent fields in a given gravitational background. Considering the states of constituents near the horizon one basically returns to the original model discussed by Frolov & Novikov (1993b). Pairs of constituents are created in the vicinity of the horizon, at least one of the component of the pair being created inside the black hole. The only (but very important) difference is that instead of counting the degrees of freedom of light (physically observable) particles one repeats similar calculations for the heavy (unobservable at low energies) constituents. The heavy constituents can effectively be excited only in the narrow region of the Planckian size close to the horizon. The remarkable fact is that, while the entropy of the black hole is generated by the same fields that generate the low energy gravity itself, the statistical-mechanical entropy of constituents in such models always coincides with the Bekenstein-Hawking entropy.

The proposed class of the induced gravity models certainly cannot be considered as the complete quantum gravity theory. They are simply phenomenological models that reproduce correctly the low energy behavior of the gravitational field. But since the results do not depend on the particular model from this class, one might hope that the same property, namely independence of the entropy of macroscopic black holes from the particular realization of the fundamental theory at high energies, is of more general origin. This point of view is supported by recent calculations of the black hole entropy in the superstring theory [see e.g. review by Horowitz (1996)]. Frolov, Fursaev, & Zelnikov (1996c) proposed to consider this consistency of the statistical mechanics of constituents in fundamental theory with the standard low energy gravitational calculations as a general principle, that we called the *low-energy censorship conjecture.*

This work was supported by the Natural Sciences and Engineering Research Council of Canada. I would like to thank Dmitri Fursaev for his help with preparation of pictures.

REFERENCES

ALLEN, B. 1986 *Phys.Rev.* **D33**, 3640.

ANDERSON, P. R., HISCOCK, W. A., WHITESELL, J., & YORK, J. W. 1994 *Phys.Rev.* **D50**, 6427.

BARDEEN, J. M., CARTER, B. & HAWKING, S. W. 1973 *Comm.Math.Phys.* **31**, 181.

BARVINSKY, A. I., FROLOV, V. P. & ZELNIKOV, A. I. 1995 *Phys.Rev.* **D51**, 1741.

BEKENSTEIN, J. D. 1972 *Nuov.Cim.Lett.* **4**, 737; 1973 *Phys.Rev.* **D7**, 2333; 1974 **D9**, 3292.

BOMBELLI, L., KOUL, R.K., LEE, J., & SORKIN, R. 1986 *Phys.Rev.* **D34**, 373.

BRADEN, H. W., BROWN, J. D., WHITING, B. F. & JORK, J. W. 1990 *Phys.Rev.* **D42**, 3376.

CALLAN, C. & WILCZEK, F. (1995) *Phys. Lett.* **B333** 55.

DOROSHKEVICH, A. G. & NOVIKOV, I. D. 1993 *Soviet Phys. JETP.* **63**, 1538.

DOWKER, J.S. & KENNEDY, G. 1978 *J.Phys.* **A 11**, 895.

FROLOV, V. 1995 *Phys.Rev.Lett.* **74**, 3319.

FROLOV, V. & FURSAEV, D. 1997 *Black Hole Entropy in Induced Gravity and Non-Minimal Coupling* (paper under preparation).

FROLOV, V., FURSAEV, D., & ZELNIKOV, A. 1996a *Phys.Rev* **D54**, 2711; 1996b *Phys.Lett.* **B382**, 220.

FROLOV, V., FURSAEV, D. , & ZELNIKOV, A. 1996c *Statistical Origin of Black Hole Entropy in Induced Gravity* Preprint Alberta-Thy 20-96, hep-th/9607104; to appear in *Nucl.Phys.* **B**.

FROLOV, V. & NOVIKOV, I. 1993a *Phys.Rev.* **D48**, 1607.

FROLOV, V. & NOVIKOV, I. 1993b *Phys.Rev.***D48**, 4545.

FROLOV, V. & PAGE, D. N. 1993 *Phys.Rev.Lett.* **71**, 3902.

FURSAEV, D. V. 1995 *Phys.Rev.* **D51**, 5352.

GIBBONS, G. W. & HAWKING, S. W. 1976 *Phys.Rev.* **D15**, 2752.

HAWKING, S. W. 1975 *Comm.Math.Phys.* **43**, 199.

HAWKING, S. W. 1979 in *General Relativity: An Einstein Centenary Survey.* (eds. S. W. Hawking and W. Israel), Cambridge Univ.Press, Cambridge.

HOROWITZ, G.T. 1996 *The Origin of Black Hole Entropy in String Theory*, Preprint UCSBTH-96-07, gr-qc/9604051.

JACOBSON, T. 1994 *Black Hole Entropy and Induced Gravity*, Preprint gr-qc/9404039.

NOVIKOV, I. & FROLOV, V. 1989 *Physics of Black Holes.* (Kluwer Academic Publ., Dordrech-Boston-London).

SAKHAROV, A.D. 1968 *Sov. Phys. Doklady* **12** 1040; 1975 *Theor.Math.Phys. (USSR)* **23**, 435.

SOLODUKHIN, S. N. 1995 *Phys.Rev.* **D51**, 609.

SREDNICKI, M. 1993 *Phys.Rev.Lett.* **71**, 666.

SUSSKIND, L. & UGLUM, J. 1994 *Phys. Rev.* **D50** 2700.

'T HOOFT, G. 1985 *Nucl.Phys.* **B256**, 727.

WALD, R. W. 1992 In:*Black Hole Physics* (Eds.V. DeSabbata and Z. Zhang), (Kluwer Academic Publ., Dordrech-Boston-London), 1992).

YORK, J. W. 1986 *Phys.Rev.* **D33**, 2092.

ZUREK, W. H. & THORNE, K. S. 1985 *Phys.Rev.Lett.* **54**, 2171; THORNE, K. S., ZUREK, W. H. & PRICE, R. H. 1986 in *Black Holes: The Membrane Paradigm* , edited by K. S. Thorne, R. H. Price, and D. A. MacDonald (Yale University Press, New Haven), p.280.

Accretion disks around black holes

By MAREK A. ABRAMOWICZ[1,2,3]

[1]Department of Astronomy & Astrophysics, Göteborg University, and Chalmers University of Technology, S-412 96 Göteborg, Sweden

[2]NORDITA, Blegdamsvej 17, DK-2100 Copenhagen Ø, Denmark

[3]International Centre for Theoretical Physics, I-34011 Trieste, Italy

Recent progress in understanding the basic physics of accretion disks around black holes demonstrated that the standard THIN model, developed in the early 1970s by Lynden-Bell, Pringle, Rees, Shakura, Sunyaev, Novikov and Thorne, is fundamentally inadequate in describing some most relevant accretion processes. In particular, it was found that thermal equilibria of hot, optically thin accretion disks are dominated by advective cooling. The advective cooling, neglected in the standard model, is not an extra technical complication that one may, or may not, include in astrophysical models of accretion disks. Indeed, it is one of the most fundamental features of the black hole accretion: all the black hole accretion disks are cooled by advection in their innermost parts, and for hot disks a strong advective cooling occurs everywhere. In addition, contrary to what was assumed in the standard model, real hot disks could be substantially non-Keplerian and have a significant radial pressure gradient. Recently, there was also a significant progress in understanding the origin and properties of viscosity in accretion disks achieved in fundamentally important works of Balbus and Hawley, Brandenburg, Papaloizou, Pringle, Dubrulle, Kato and their collaborators. These new developments call not only for a substantial revision of the standard model, but indeed for a totally reconstructed, new paradigm for the theory of black hole accretion. New understanding of thermal equilibria is a part of this new paradigm. It emerges from recent works of Abramowicz, Chen, Kato, Lasota, Narayan, Yi and others, and it is rooted in the SLIM accretion disk model developed earlier by Abramowicz, Lasota and their co-workers. This review describes the recent rapid progress in understanding of the role of advective cooling in accretion disks.

1. Introduction: the standard model

One of the goals of the theory of accretion disks around black holes is to explain quasars: the most powerful non-explosive energy sources in the Universe. In a relatively small volume, comparable to the sphere of influence of a single star, a typical quasar generates power exceeding that of some hundreds of billions of stars. This enormous energy output is powered by release of gravitational energy of matter infalling onto a very massive black hole located right in the galactic center. According to a widely accepted view, this occurs via an accretion disk which very rapidly rotates around the central black hole. In accretion disks viscous torques remove angular momentum which weakens the centrifugal support and causes matter orbiting the hole to fall in. Viscous friction converts orbital energy of rotating matter into heat which is radiated away as the quasar's radio, optical and X-ray emission.

In the past twenty years the paradigm for understanding accretion disks around black holes has been provided by the standard model developed by Lynden-Bell (1969),Pringle & Rees (1972), Shakura (1972), Shakura & Sunyaev (1973), Novikov & Thorne (1973), and others. The main assumptions of the standard model are: (1) The vertical thickness $H(R)$ of the disk is very small, i.e. at a distance R from the central black hole, $H(R)/R \ll 1$. All the physical quantities are replaced by their vertical averages. For example, the density $\rho(R, Z)$ is replaced by the surface density $\Sigma(R)$,

$$\Sigma(R) = \int_{-H(R)}^{+H(R)} \rho(R, Z) dZ \approx H(R)\rho(R, 0). \tag{1.1}$$

(2) The angular velocity of rotating matter $\Omega(R)$ equals the Keplerian one, $\Omega(R) = \Omega_K(R)$. (3) The heat generated by viscous friction at a given place $Q^+(R)$ is radiated away through the disk surface at the same place, $Q^+(R) = Q^-_{\rm rad}(R)$. (4) The only non-vanishing component of the viscous torque is $t_{\varphi R}$, and the kinematic viscosity coefficient equals

$$\nu = \alpha H c_S, \tag{1.2}$$

where c_S is the sound speed, and α is a constant free parameter of the theory. (6) At the inner edge of the disk $R = R_{\rm in}$ the viscous torque vanishes, $t_{\varphi R}(R_{\rm in}) = 0$. This no-torque inner boundary condition introduces the radial function

$$f(R) = 1 - \frac{\ell_{\rm in}}{\ell_K(R)}, \tag{1.3}$$

where $\ell_K(R)$ corresponds to the specific Keplerian angular momentum and $\ell_{\rm in} = \ell_K(R_{\rm in})$. Asymptotically, for R going to infinity f goes to one. In the Newtonian theory $f(R) = 1 - (R_{\rm in}/R)^{1/2}$.

The standard model gives the physical quantities, for example the surface density Σ, in terms of the model's free parameters: mass of the central black hole M, accretion rate $\dot{M}$, viscosity parameter α, and the distance from the centre R. It is convenient to express the mass in the solar mass units, $m = M/M_\odot$, the accretion rate in the Eddington units, $\dot{m} = \dot{M}/\dot{M}_{\rm Edd}$ ($\dot{M}_{\rm Edd} = L_{\rm Edd}/c^2$, $L_{\rm Edd} = 10^{38} m$ [erg/sec] is the Eddington luminosity), and distances in the units of the gravitational radius of the black hole, $r = R/R_G$ ($R_G = 3 \times 10^5 m$ [cm]).

At a fixed distance from the center, one considers a sequence of standard accretion disk models corresponding to a fixed mass of the central black hole and a fixed value of the viscosity parameter. Different solutions on this sequence are characterised by different accretion rates, and each physical quantity X varies along the sequence according to the linear relation $\log X = A + B \log \dot{m}$. The straight lines on the $\log \dot{m}$ vs. $\log \Sigma$ plane are called branches.

1.1. *Thermal and viscous stability of the branches*

Piran (1978) introduced four convenient parameters,

$$\mathcal{K} = \left(\frac{\partial \ln Q^-}{\partial \ln H}\right), \ \mathcal{L} = \left(\frac{\partial \ln Q^-}{\partial \ln \Sigma}\right), \ \mathcal{M} = \left(\frac{\partial \ln Q^+}{\partial \ln H}\right), \ \mathcal{N} = \left(\frac{\partial \ln Q^+}{\partial \ln \Sigma}\right). \tag{1.4}$$

The necessary conditions for thermal and viscous stability are,

$$\mathcal{K} - \mathcal{M} > 0, \quad \frac{\mathcal{N}\mathcal{K} - \mathcal{M}\mathcal{L}}{\mathcal{K} - \mathcal{M}} > 0. \tag{1.5}$$

The criterion for thermal stability was first derived by Pringle, Rees & Pacholczyk (1973), (see also Shakura & Sunyaev 1976), while the condition for viscous stability was derived by Lightman (1974) and Lightman & Eardley (1974).

1.2. *Optically thick branches: Shakura & Sunyaev*

In the two extreme cases when the total pressure $P = P_{\rm gas} + P_{\rm rad}$ is dominated either by the gas pressure $P = P_{\rm gas}$ or by the radiation pressure $P = P_{\rm rad}$, logarithm of each physical quantity $\log X$ may be written as a linear combination of logarithms of the model parameters and the function f. For example, the surface density in the case $P = P_{\rm gas}$ is given by

$$\log \Sigma[{\rm g/cm^2}] = 4.85 + \frac{1}{5}\log m + \frac{3}{5}\log \dot{m} - \frac{4}{5}\log \alpha - \frac{3}{5}\log r + \frac{3}{5}\log f, \qquad (1.6)$$

while in the case $P = P_{\rm rad}$,

$$\log \Sigma[{\rm g/cm^2}] = 0.63 - \log \dot{m} - \log \alpha - \log f. \qquad (1.7)$$

Note, that because $f = 0$ at the inner edge of the standard disk, the above standard solutions are singular there. This demonstrates that the standard model is not consistent near the inner edge.

In the optically thick standard disk model, assuming that kinematic viscosity is proportional to the *total* pressure, $\mathcal{K} = 4(1+\beta)/B$, $\mathcal{L} = -\beta/B$, $\mathcal{M} = 2$, $\mathcal{N} = 1$. Here $\beta = P_{\rm gas}/P$ is the ratio of gas to total pressure and $B = 4 - 3\beta$. Thus, the condition for thermal and viscous stability is $\beta > 2/5$. Therefore, in the standard model, optically thick, radiation pressure supported parts (i.e. with $\beta \sim 0$) are unstable. This is a problem, because the innermost parts of the standard optically thick black hole accretion disks *are* radiation pressure supported.

1.3. *Optically thin branch: Shapiro, Lightman & Eardley*

Thorne & Price (1975) proposed that the instability may swell the inner part into much hotter, gas pressure dominated, optically thin state. This idea was further developed in the classic paper by Shapiro, Lightman & Eardley (1976) who constructed a hot, optically thin, two-temperature model of accretion disk. The model was attractive, because it could solve the problem with an insufficient temperature of the standard optically thick disks. Indeed, the maximal effective temperature of optically thick disks is, according to the standard model,

$$T_{\rm max} \approx 1 \times 10^7 m^{-1/4} \dot{m}^{1/4}\ [^\circ{\rm K}]. \qquad (1.8)$$

For the supermassive black holes relevant for quasars and sub-Eddington accretion rates this gives $T_{\rm max} \lesssim 10^5$ [°K]. Thus, an optically thick disk is too cold to emit the X-ray radiation observed in the spectra of quasars. In Shapiro, Lightman & Eardley (1976) hot, optically thin, two-temperature accretion disks, the electron temperature is of the order of $T_{\rm e} \sim 10^9 - 10^{10}[^\circ{\rm K}]$, and the proton temperatures may reach the virial regime, $T_{\rm p} \sim 10^{10} - 10^{12}[^\circ{\rm K}]$. This would be sufficient to explain the presence of X-ray photons in Quasars' spectra.

The instability problem, was not solved, however. The two-temperature hot disks were still violently thermally unstable.

2. The need of a new paradigm

It was always known that due to the unphysical singularity, the standard model cannot correctly describe the physics at the innermost part of accretion disks. This was not, however, considered as a serious problem, because it was believed that the part of the standard solution affected by the singularity was very small and therefore not relevant in astrophysical applications.

More concern was raised recently when it was realized that the advective cooling (ignored in the standard model) is not an extra technical complication that one may, or may not, include, but that instead it is one of the most fundamental features of the black hole accretion disks. Indeed, all black hole accretion disks are cooled by advection in their innermost parts, and for the hot, optically thin disks a strong advective cooling occur *everywhere*, also very far from the black hole.

Non-standard optically thick accretion disk models with a significant advective cooling were constructed already more than ten years ago by Abramowicz, Lasota and their collaborators Abramowicz, Lasota & Xu (1986), Abramowicz, Czerny, Lasota & Szuszkiewicz (1988), but these models never belonged to the accepted part of the mainstream of the accretion theory.

Here, I would like to discuss the importance of advection by discussing a few relatively simple issues. For pedagogical convenience, I use a Newtonian model for the gravitational field of non-rotating black hole proposed by Paczyński & Wiita (1980). It is not conceptually more difficult to discuss the same problems for Schwarzschild or Kerr black holes using the full apparatus of Einstein's relativity (see *eg.* Jaroszyński, Abramowicz & Paczyński, 1980), but this would bring irrelevant and purely formal complications to the present review.

Paczyński's black hole is a Newtonian object with the gravitational potential $\Phi(R)$ given by,

$$\Phi = \frac{GM}{R - R_G}, \quad R_G = \frac{2GM}{c^2}, \tag{2.9}$$

where M is the mass of the black hole. I use spherical coordinates R, θ, ϕ centered at the black hole.

Free particles orbiting the Paczyński black hole on circular orbits have 'Keplerian' angular momentum $\ell_K(R) = R^{3/2}\sqrt{|d\Phi/dR|} = \sqrt{GM}R^{3/2}/(R - R_G)$. The function $\ell_K(R)$ has a minimum at $R = 3R_G \equiv R_{MS}$. Keplerian orbits with $R > R_{MS}$ are stable, and those with $R < R_{MS}$ are unstable. The unstable orbit with $R = 2R_G \equiv R_{MB}$ has zero binding energy, and unstable orbits with $R < R_{MB}$ are unbound, *i.e.* they have negative binding energy. The same general properties of Keplerian motion around non-rotating black holes, *and* numerical values of R_{MS} and R_{MB} follow from exact calculations in general relativity. The angular momentum corresponding to the marginally stable orbit at $R = R_{MS}$ is $\ell_K(R_{MS}) \equiv \ell_{MS} = (3/2)^{3/2}R_G c$.

I start with a short discussion of stationary and barytropic, $P = P(\rho)$, tori rotating around black holes. The fluid lines are assumed to be co-planar circles along the axis $R\sin\theta = 0$. These tori do not pretend to be models of any realistic accretion flows, but they are helpful in explaining concepts that are most relevant here: although accretion flows are never in the state of dynamical equilibrium (pressure, gravity and centrifugal force in perfect balance), some of the disks could be partially in a state very nearly dynamical equilibrium. The barytropic tori are models of these near equilibrium states. In dynamical equilibrium of a torus, only the azimuthal component of the velocity is non-zero, $v_\phi \neq 0$, and both the angular velocity $\Omega = v_\phi/R\sin\theta$, and the angular momentum $\ell(R\sin\theta) = v_\phi R\sin\theta = \Omega(R\sin\theta)^2$, depend only on the distance $R\sin\theta$ from the axis (von Zeipel's theorem). From the condition for the balance of gravitational, centrifugal and pressure forces, it follows that in a stationary equilibrium, surfaces of constant pressure P (and density ρ) must coincide with the 'equipotential' surfaces of $W(R,\theta) = \Phi(R) - \int \ell^2(R\sin\theta)(R\sin\theta)^{-3}d(R\sin\theta)$. Obviously, any stationary and finite distribution of rotating matter around a black hole must be bounded by a closed and finite $P = 0$ surface. Thus, such a stationary distribution could exist if and only if there exist a closed $W(R,\theta) =$const equipotential surface around the black hole.

Whether the closed equipotential surface does exist depends on the angular momentum distribution of rotating matter, i.e. on the function $\ell = \ell(R\sin\theta)$. I shall now discuss the simplest case $\ell(R\sin\theta) = \ell_0 =$ const, because it captures all the relevant features of the most general situation. In this case the equipotential surfaces are given by,

$$W(R,\theta) = -\frac{GM}{(R - R_G)} + \frac{1}{2}\frac{\ell_0^2}{R^2 \sin^2\theta}. \tag{2.10}$$

Suppose that there exist a closed surface $W(R,\theta)$ =const. This surface must cross the equatorial plane $\theta = \pi/2$ at least twice, at two different radii R_{in} and R_{out}, say. Let us consider the function $W_0(R) = W(R, \pi/2)$. (NB: this function is identical to the effective potential for radial motion of a particle with the angular momentum ℓ_0.) Obviously, $W_0(R_{in}) = W_0(R_{out})$ which means that $(dW_0/dR)_C = 0$ for a particular $R = R_C$ located between R_{in} and R_{out}. It is easy to see that from $(dW_0/dR)_C = 0$ it follows that $\ell_0 = \ell_K(R_C)$. However, we know that the minimum value of $\ell_K(R)$ is ℓ_{MS}. Thus, in the simplest case $\ell = \ell_0$ = const, the necessary and sufficient condition for the existence of a closed equipotential surface — and therefore the necessary condition for the existence of a stationary and finite distribution of rotating matter around a black hole is $\ell_0 > \ell_{MS}$, otherwise it would be not possible to have $\ell_0 = \ell_K$. The above condition captures the essential point here. Indeed, for an *arbitrary* angular momentum distribution the general condition is quite similar: the angular momentum of matter should be greater than the Keplerian one, $\ell(R) > \ell_K(R)$, for at least some range of R.

Thus, we have two possibilities here: either the flow is everywhere sub-Keplerian, or it is super-Keplerian somewhere. This two possibilities correspond to quite different types of accretion flows, and the very fact of the existence of these two types is one of the most general features of the black hole accretion. This was first recognized by Abramowicz & Zurek (1981) who have called the first type 'Bondi-like' and the second one 'disk-like' accretion. They found that there is no smooth transition between these two types, because 'intermediate' solutions have globally unacceptable topology of the sonic point (Lu 1985, Lu & Abramowicz 1988). Abramowicz & Zurek suggested that when one of the flow parameters experiences a secular change that leads to the change of the flow type, a shock may develop. Muchotrzeb (1983) was the first who pointed out that the question which of the two types types is chosen, depends mostly on the value of the viscosity coefficient α. Above the critical value, $\alpha_{\rm crit} \approx 0.1$, the Bondi type is chosen, and below $\alpha_{\rm crit}$, the disk type. The parameter space considered by Muchotrzeb corresponded to small accretion rates and fixed, almost Keplerian outer boundary conditions. Her results have been expanded in great details by Kato and his collaborators: Matsumoto, Kato, Fukue & Okazaki (1984), Okazaki, Kato & Fukue (1987), and Abramowicz & Kato (1989). Later, several other authors discussed detailed differences between the two types of black hole accretion flows in specific astrophysical situations, and examined the question of critical α. In particular, Fukue (1987) and Abramowicz & Chakrabarti (1990) have presented a toy model for the shock formation based on the Bondi-disk (and disk-disk) jumps. The model employed free parameters that assumed *ad hoc* how much energy and entropy was lost by radiation emerging vertically from the shock. In a realistic model, the amounts of energy and entropy that are lost should be calculated from equations governing the internal shock structure, and not just assumed. Calculating the shock structure would be a very difficult, but also a very interesting task. Unfortunately, a significant part of the recent literature on shocks in accretion disks follows a different path. It is neither difficult, nor interesting. Numerous different 'solutions' are constructed, based on numerous different *ad hoc* choices of the free parameters (see Chakrabarti 1996 for a review). In my opinion, these works are mostly a mathematical curiosity, and they add nothing new to the physical insights achieved in the early results of Abramowicz, Zurek, Muchotrzeb and Kato.

2.1. *The Bondi-like type.*

If the condition $\ell(R) > \ell_K(R)$ is not obeyed, then, typically, the flow is everywhere far from dynamical equilibrium. Radial and vertical components of velocity are not negligibly small, and non-local dissipation is dominant: a substantial part of locally dissipated energy, entropy and angular momentum is carried along (advected) with the matter and lost in the central black hole. Therefore, the Bondi-like accretion has a low radiative efficiency. For *exactly* spherical accretion with high (near Eddington) rates, this was first noticed by Katz (1977). He pointed out that radiation diffuses outward at a speed $v_{\rm diff} = c/\tau$, where τ is the optical depth. With v being the velocity of matter, $v_{\rm diff}/v = \dot{M}_{\rm Edd}/2\dot{M}$. Thus, as $\dot{M}$ approaches $\dot{M}_{\rm Edd}$, radiation is trapped within the accreting matter and cannot escape: $\dot{M}$ may be very large without L exceeding $L_{\rm Edd}$. This idea of the trapping radius was later worked out in more details by Rees (1978) and Begelman (1978, 1979). Rees (1982) and Rees *et al.*(1982) have suggested that at low (very sub Eddington) accretion rates, starved black holes have low radiative efficiency. They argued that this is relevant for radio galaxies, the $\sim 10^6 M_\odot$ black hole in the Galactic Center and for some stellar mass black holes in our Galaxy.

2.2. *The disk-like type.*

If the condition $\ell(R) > \ell_K(R)$ is obeyed, then a part of the flow could be in a state *very close* to dynamical equilibrium. In this case, pressure gradient, gravitational, and centrifugal forces are almost in balance, and radial and vertical components of velocity are very small, $v_R \ll v_\phi$, $v_\theta \ll v_\phi$. For the flow in nearly dynamical equilibrium, there could be a substantial local reprocessing (e.g. by radiative losses) of locally dissipated energy, entropy and angular momentum because fluid elements stay at the same radius during a large number ($\sim v_\phi/v_R \gg 1$) of dynamical periods. For disk-like accretion, equation $dW_0/dR = 0$ has then two solutions in the equatorial plane $\theta = \pi/2$: the first one $R = R_R < R_{MS}$, which corresponds to a maximum, and the second one $R = R_C > R_{MS}$, which corresponds to a minimum of the effective potential $W_0(R)$. Let us define $W_R = W(R_R) = \text{const}$. The equipotential surface $W(R,\theta) = W_R$ is called the Roche lobe. It crosses itself along the circle $R = R_R$ which is called 'the cusp'. Close to the cusp, accretion flow is always fully determined by the Roche lobe overflow. Suppose that far away from the cusp, the surface of the flow ($P = 0$) coincides with the equipotential surface $W(R,\theta) = W_S = \text{const} > W_R$, *i.e.* matter distribution in the disk overflows the Roche lobe. This induces a mass loss from the disk through the cusp. Between the cusp and the black hole the flow is not in dynamical equilibrium: it is always of the Bondi-like type there. The existence of the self-crossing Roche lobe is a topological property of the system, and in a wide range of parameters does not depend on details. Matter distribution which overflows the Roche lobe cannot be in the equilibrium because it would touch the surface of the black hole (horizon), but no matter in hydrostatic equilibrium could be present at the horizon. The local accretion rate at the cusp depends very strongly on the amount of the Roche lobe overflow. This provides a self-regulatory stabilization mechanism against thermal and viscous perturbations Abramowicz (1981). To see how the mechanism works, consider a perturbation of a steady state accretion disk due to a local overheating at the cusp. The disk which is locally overheated will locally expand. This will increase the Roche lobe overflow and therefore also the local accretion rate will increase. In turn, the higher accretion rate will induce stronger advective cooling through the cusp, which will suppress the effect of initial overheating and bring both the overflow and the accretion rate back to their steady state values.

Therefore, in a steady state, the amount of the Roche lobe overflow self-regulates according to the steady accretion rate which is fixed by the outer boundary conditions.

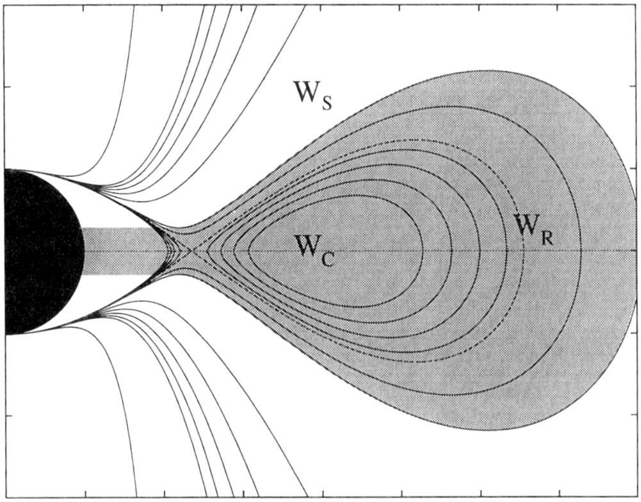

FIGURE 1. Equipotential surfaces $W(R,\theta)$ =const for the Schwarschild black hole. When the matter distribution (indicated by the dark shadow) overflows the Roche lobe $W(R,\theta) = W_R$, no static equilibrium is possible, and a mass loss from the disk occurs. For the situation that corresponds to a 'disk-like' case, the topology of equipotential surfaces shown in this picture is *independent* of the angular momentum distribution. Thus, *all accretion disks must behave this way in their innermost parts.* Abramowicz (1981) demonstrated that the Roche lobe overflow induces strong advective cooling that dominates the thermal equilibrium of the disk close to the cusp. Figure taken from Eriguchi, Lanza, Nishida & Abramowicz (1996).

One particular consequence of this is that, at the location of the cusp, the condition of thermal balance has the form,

$$Q^+ = Q^-_{\rm rad} + Q^-_{\rm adv}, \quad Q^-_{\rm adv} = \xi Q^+, \quad \xi \approx 1, \tag{2.11}$$

and that modifies the inner boundary condition function $f^*(r_0) = f(r_c) + \xi(r_c)$. This means that when one considers the advective cooling induced by the Roche lobe overflow, which is always the dominant cooling mechanism in the innermost part of the disk, the artificial $f = 0$ singularity of the standard model is removed.

3. Advective cooling and thermal equilibria

It was initially thought, *cf.* Wheeler (1981), Shapiro & Teukolsky (1983), that the effect of the Roche lobe overflow stabilization could be relevant only in the immediate vicinity of the cusp. Later, in the context of slim disks, Lasota and I realized that the reason for stabilization of thermal modes was not directly connected to the Roche lobe overflow, but rather to the fact that the thermal timescale of vertical expansion $\tau_{\rm th} \sim 1/(\alpha\Omega)$ was of the same order as the timescale of inward advection $\tau_{\rm adv} \sim r/v_R$. When this happens, the instability is advected inwards before it has a chance to grow, in an analogous way to what stabilizes the spherical accretion Moncrief (1980).

In standard thin accretion disks the timescale for radial advection is given by the viscous timescale $\tau_{\rm adv} \sim \tau_{\rm vis} \sim \tau_{\rm th}/(H/R)^2$. Thus, stabilization of the thermal modes by advective *cooling* should occur when $H \sim R$. This is consistent with the fact that in general

$$Q^-_{\rm adv} \sim \left(\frac{H}{R}\right)^2 Q^+, \tag{3.12}$$

and therefore when $H \sim R$, the cooling is dominated by advection: $Q^-_{\rm adv} \sim Q^+$, and radiative cooling is thus insignificant in the heat balance.

I shall turn now to a much more exciting, and new, possibility. In deriving the condition $H \sim R$ for the importance of advective cooling, we have not assumed any particular regime of optical depth. Thus, the condition should also be valid for *optically thin* disks. Indeed, the very hot, optically thin disks always have $H \sim R$ and thus their thermal equilibria must be dominated by advective cooling. This case have been recently intensely studied by several authors, mostly by Abramowicz, Björnsson, Chen, Kato, Lasota, Narayan, Yi and their collaborators.

The need for a re-discussion of the thermal equilibria of hot, optically thin accretion disks was recognized by Narayan and his collaborators at Harvard in 1994, and later that year (but independently) by Abramowicz, Lasota and Regev during their collaboration at Technion. A new type of non-standard, *stable*, hot, optically thin accretion disks was then found by these two groups. The new solutions had been anticipated in an earlier paper on boundary layers by Narayan & Popham (1994) and also in Narayan & Yi (1994, 1995a). The solutions have been explicitly constructed by Abramowicz, Chen, Kato, Lasota & Regev (1995), who assume that radiative cooling is given by the bremsstrahlung and pointed out that they exist only below a certain accretion rate (which is smaller than the Eddington one), and by Narayan & Yi (1995b) who consider a more realistic radiative cooling (with synchrotron radiation and Comptonization).

Works of Narayan and his collaborators (Narayan, McClintock & Yi 1996; Narayan & Yi 1994, 1995ab; Narayan, Yi & Mahadevan 1995) inspired most of the recent interest in the subject because of their direct applications to several astrophysical objects that are underluminous and very hot, like X-ray transient sources, some Seyfert galaxies and the Galactic Center. Although it was much earlier recognized by Rees and his collaborators that in these sources the radiative efficiency must be very low, the most recent progress in understanding the role of advective cooling offered a new, detailed and consistent physical explanation for the phenomenon of low radiative efficiency. Works by Kato and co-workers (Kato, Abramowicz & Chen 1996; Kato, Yamasaki, Abramowicz & Chen 1996), were pioneering in discussing stability properties of these disks, very much different from those in the standard case. It looks that, thanks to the combined efforts of these and other researchers, a rather complete picture of accretion disk thermal equilibria has gradually been taking shape. I shall now summarize the main elements of this picture.

3.1. *Type (2):* $\alpha > \alpha_{\rm crit}$, *$\tau$ varies,* $Q_{\rm adv}/Q_{\rm rad} \ll 1$.

It was found by Chen, Abramowicz, Lasota, Narayan & Yi (1995) and confirmed by Björnsson, Abramowicz, Chen & Lasota (1996) that at a given radius R, there exist exactly four physically distinct types of accretion disk thermal equilibria. Two of these types correspond to a low viscosity, $\alpha < \alpha_{\rm cr}(r)$, and the two others correspond to a high viscosity, $\alpha > \alpha_{\rm cr}(r)$. The critical viscosity $\alpha_{\rm cr}$ depends strongly on the radius, and weakly on the mass of the central accreting object. According to Chen, Abramowicz, Lasota, Narayan & Yi (1995), in the inner part of the disk $\alpha_{\rm cr} \sim 0.1$. This value was changed by the more accurate treatment of microphysics and inner boundary condition by Björnsson, Abramowicz, Chen & Lasota (1996) who found $\alpha_{\rm cr} > 1$. The two low viscosity type of flows are further differentiated by whether the flow is optically thin or optically thick, and the two high viscosity flow solutions by whether advection is

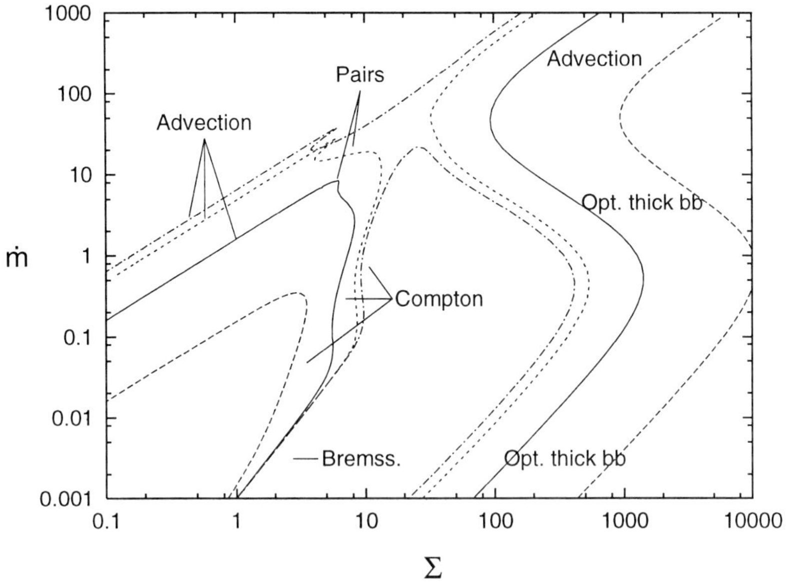

FIGURE 2. Dominant cooling processes for accretion disks around $10M_\odot$ black hole at a particular radius $r = r_G$ (from Björnsson, Abramowicz, Chen & Lasota 1996). Dash-dotted lines correspond to $\alpha > \alpha_{\rm crit}$, all other lines to $\alpha < \alpha_{\rm crit}$. All lines correspond to a fixed value of α: the long-dashed, solid, short dashed and dash-dotted correspond to $\alpha = 0.1$, 1.0, 3.0, 4.0 respectively. (Σ is in units g/cm^3.) Although these results have been obtained from a *local* model, they were fully confirmed by later *global* calculations (see Figure 3). In particular, the optically thin, very hot disks may exist *only* for accretion rate less than a critical one. The precise value of the critical rate is very sensitive to details (microphysics, mathematical approximations) and it is still rather poorly known.

negligible or dominant. I shall now describe these four types in details. One should remember, that this is a *local* classification, at a fixed radius. It would be convenient for the reader to consult Figure 2 and find the corresponding types and branches there.

3.2. *Type (1):* $\alpha < \alpha_{\rm crit}$, $\tau > 1$, $Q_{\rm adv}/Q_{\rm rad}$ *varies.*

This is an "S"-shaped sequence of three branches: lower, middle and upper. The *lower* branch corresponds to a gas-pressure dominated, radiatively cooled classical Shakura & Sunyaev solution. Disks here are cool, and optically thick. Radiation is of the black body or modified black body type. For low accretion rates, corresponding to those characteristic for dwarf novae, this branch itself bends into an S-shaped sequence of three sub-branches due to a strong opacity dependence on temperature. (There could be even more sub-branches depending on complications in heat transport, *e.g.* by a vertical convection.) For higher accretion rates the opacity is determined by electron scattering and the branch is both thermally and viscously stable. Most of what is firmly known about physics of accretion comes from a very detailed comparison of theory and observations for disks from the lower branch of this type. The *middle* branch is also a part of the classical Shakura & Sunyaev solution. It is optically thick, radiation pressure dominated, and cooled by radiation. It is thermally and viscously unstable. Thus, it has no astrophysical applications in modeling steady state disks. The *upper* branch was found by Abramowicz, Lasota & Xu. It is often called *slim* accretion disks branch. It is optically thick, radiation pressure supported, and cooled by radial advection of heat. It is both thermally and viscously stable. It exists at moderately super-Eddington accretion rates and has radiative efficiency up to factor of 10 smaller than the classical Shakura & Sunyaev disks. The (modified) black body spectra calculated for slim disks explain the Big Blue Bump in the observed spectra of active galactic nuclei *far more* accurately than those calculated for the standard Shakura & Sunyaev ones.

3.3. *Type (2):* $\alpha > \alpha_{\rm crit}$, τ *varies,* $Q_{\rm adv}/Q_{\rm rad} \ll 1$.

Again, this type consists of three branches, right, upper, and left, that together form a shape of a "$\bigcap$". The *right* branch is identical in properties (except that now the viscosity is higher) with the lower branch of the Type 1. The *upper* branch corresponds to disks that have $\tau \approx 1$ and $P_{\rm gas}/P_{\rm rad} \approx 1$. Because radiative transfer equation is very difficult to solve for plasma with $\tau \approx 1$, a phenomenological treatment ("bridging formulae") is used. It is firmly established, however, that the disks at this branch are thermally and viscously unstable. The *left* branch corresponds to the Shapiro, Lightman & Eardley models. It is optically thin, and very hot. Protons have much higher (approaching the virial regime) temperature than electrons. Radiative processes include bremsstrahlung, synchrotron radiation and Comptonization. Creation and annihilation of the e^+e^- pairs is also considered as an important contribution to thermal equilibrium. Advective cooling is unimportant (but advective *heating* may be present). These disks are thermally unstable.

Wandel \$ Liang (1991), and more recently Luo & Liang (1994) demonstrated that, within this type, one may construct models that for large radii are Shakura & Sunyaev type, *i.e.* optically thick, and cool, while for small radii they are Shapiro, Lightman & Eardley type *i.e.* optically thin, and hot. In the 'bridging' region they have optical depth close to unity. Thus, they experience optically thick — optically thin transitions at some intermediate radii. One may *a posteriori* check that, formally, in some range of radii, $|\xi| \gg 1$, so there is strong non-local advective heating in these disks. However, this makes *no physical difference* for the thermal equilibrium: such disks are predominantly cooled by radiation, and not by advection, and therefore they are violently thermally unstable. Although some authors are recently constructing models of these disks that formally include advection, it is difficult to see that these models could be seriously *physically different* from those already constructed by Liang and his collaborators.

3.4. *Type (3)* $\alpha < \alpha_{\rm crit}$, $\tau < 1$, $Q_{\rm adv}/Q_{\rm rad}$ *varies.*

This type consist of two branches, lower and upper, that together form a shape similar to an asymmetric ">". The *lower* branch is identical with the Shapiro, Lightman & Eardley, but now with a lower viscosity α. It is gas pressure dominated, very hot, and optically thin. *It is thermally and viscously stable.* Radiative efficiency is very low. The spectral characteristics of the disks are dominated by Comptonization of bremsstrahlung and synchrotron soft photons. Pairs are suppressed by strong advective cooling. Detail calculations show that these disks are excellent models for very hot, underluminous sources. Solutions of this type exist only *below* a certain critical rate. Its value depends strongly on detailed microphysics considered, but is typically sub Eddington.

There are several very interesting problems here that need to be solved: (i) In a range of accretion rates *two* stable solutions (type 1, lower branch) and (type 3, upper branch) are possible. Which one is chosen by the real flow? (ii) Why and how transitions between these two solutions occur in a non-stationary case? (iii) Are such transitions possible at some intermediate radii for (almost) stationary disks? (iv) Is there a simple physical reason which explains why the one-temperature hot disk models constructed by Esin, Narayan & Ostriker (1996) have negligible advective cooling?

3.5. *Type (4)* $\alpha > \alpha_{\rm crit}$, τ *varies*, $Q_{\rm adv}/Q_{\rm rad} \gg 1$.

There is one branch with the "/" shape. It is similar to that of the upper branches of the types (1) and (3) but now α is high.

Fully consistent global solution that obey realistic boundary conditions and smoothly pass through the sonic point have been constructed only very recently by Chen, Abramowicz & Lasota (1997) — see Figure 3 — Narayan, Honma & Kato (1997), Nakamura, Matsumoto, Kusunose & Kato (1997), and Abramowicz, Chen, Granath & Lasota (1997). In the last paper, fully relativistic equations in Kerr geometry have been solved.

I thank J.-P. Lasota and R. Narayan for their comments on a draft version of this report. Research reported here was supported by several grants from the Swedish NFR, Nordita's Nordic Project *Non-Linear Phenomena in Accretion Disks Orbiting Black Holes*, and by the SISSA-Chalmers/GU convention.

REFERENCES

ABRAMOWICZ M.A. 1981 *Nature* **294**, 235.

ABRAMOWICZ M.A. & CHAKRABARTI S. 1990 *Astrophys. J.* **350**, 281.

ABRAMOWICZ M.A., CHEN X., GRANATH M. & LASOTA J.-P. 1996, *Astrophys. J.*, in press.

ABRAMOWICZ M.A., CHEN X., KATO S., LASOTA J.-P. & REGEV O. 1995 *Astrophys. J.* **438**, L37.

ABRAMOWICZ M.A., CZERNY B., LASOTA J.-P. & SZUSZKIEWICZ E. 1988 *Astrophys. J.* **332**, 646.

ABRAMOWICZ M.A. & KATO S. 1989 *Astrophys. J.* **336**, 304.

ABRAMOWICZ M.A., LASOTA J.-P. & XU C. 1986 in *Quasars* (eds. G. Swarup & V.K. Kapachi), Reidel.

ABRAMOWICZ M.A. & LASOTA J.-P. 1995 *Comments Astrophys.* **18**, 141.

ABRAMOWICZ M.A. & ZUREK W.H. 1981 *Astrophys. J.* **264**, 314.–????

BEGELMAN M.C. 1978 *Mon. Not. Roy. Astr. Soc.* **184**, 53.

BEGELMAN M.C. 1979 *Mon. Not. Roy. Astr. Soc.* **187**, 237.

BJÖRNSSON G. ABRAMOWICZ M.A., CHEN X. & LASOTA J.-P. 1996 *Astrophys. J.*, in press.

CHAKRABARTI S. 1996 *Phys. Rep.* **266**, 229.

CHEN X., ABRAMOWICZ M.A. & LASOTA J.-P. 1997 *Astrophys. J.*, in press.

CHEN X., ABRAMOWICZ M.A., LASOTA J.-P., NARAYAN R. & YI I. 1995 *Astrophys. J.* **443**, L61.

ERIGUCHI Y., LANZA A., NISHIDA S. & ABRAMOWICZ M.A. 1996 *Mon. Not. Roy. Astr. Soc* **278,** L41.

ESIN A.A., NARAYAN R., OSTRIKER E. & YI I. 1996 *Astrophys. J.* **465**, 312.

FUKUE J. 1987 *Publ. Astr. Soc. Japan* **39**, 309.

HONMA F., MATSUMOTO R. & KATO S. 1991 *Publ. Astr. Soc. Japan* **43**, 147.

JAROSZYŃSKI M., ABRAMOWICZ M.A. & PACZYŃSKI 1980 *Acta Astr.* **30**, 1.

KATO S., ABRAMOWICZ M.A. & CHEN X. 1996a *Publ. Astr. Soc. Japan* **48**, 67.

KATO S. & HONMA F. 1991 *Publ. Astr. Soc. Japan* **43**, 95.

KATO S., HONMA F. & MATSUMOTO R. 1988 *Publ. Astr. Soc. Japan* **40**, 709.

KATO S., YAMASAKI, T., ABRAMOWICZ M.A. & CHEN X. 1996b *Publ. Astr. Soc. Japan*, in press.

KATZ J.I. 1977 *Astrophys. J.* **215** 265.

LASOTA J.-P., NARAYAN R. & YI I. 1996 *Astron. Astrophys.*, in press.

LIGHTMAN A.P. 1974 *Astrophys. J.* **194**, 429.

Lightman A.P. & Eardley D.M 1974 *Astrophys. J.* **187**, L1.

Lu J.F. 1985 *Ph.D. thesis, SISSA, Trieste.*

Lu J.F. & Abramowicz M.A. 1988 *Acta Astrophys. Sinica* **8**, 1.

Luo C., and Liang E.P. 1994 *Mon. Not. Roy. Astr. Soc.* **266**, 386.

Lynden-Bell D. 1969 *Nature* **223**, 690.

Matsumoto R., Kato S., Fukue J. & Okazaki A.T. 1984 Publ. Astr. Soc. Japan **36**, 71.

Moncrief V. 1980 *Astrophys. J.* **235**, 1038.

Muchotrzeb B. 1983 *Acta Astron.* **33**, 79.

Nakamura K.E., Matsumoto R., Kusunose M. & Kato S. 1996 *Pub. Astr. Soc. Japan*, in press.

Narayan R., Honma F. & Kato S. 1996 *Astrophys. J.*, in press.

Narayan R., McClintock J.E. & Yi I. 1996 *Astrophys. J.*, **457.** 821.

Narayan R. & Popham R. 1993 *Nature* **362**, 820.

Narayan R. & Yi I. 1994 *Astrophys. J.* **428**, L13.

Narayan R. & Yi I. 1995a *Astrophys. J.* **444**, 231.

Narayan R. & Yi I. 1995b *Astrophys. J.* **452**, 710.

Narayan R., Yi I. & Mahadevan R. 1995 *Nature* **374**, 623.

Novikov I.D. & Thorne K.S. 1973 in *Black Holes* (eds. C. DeWitt & B. DeWitt) Gordon and Breach.

Okazaki A.T., Kato S. & Fukue J. 1987 *Publ. Astr. Soc. Japan* **39**, 457.

Paczyński B. & Wiita P.J. 1980 *Astron. Astrophys.* **88**, 23.

Piran T. 1978 *Astrophys. J.* **221**, 652.

Pringle J.E. & Rees M.J. 1972 *Astron. Astrophys.* **21**, 1.

Pringle J.E., Rees M.J. & Pacholczyk 1973 *Astron. Astrophys.* **29**, 179.

Rees M.J. 1982 in *The Galactic Center*, eds. G. Riegler & R. Blandford (New York: AIP).

Rees M.J., Begelman M.C., Blandford R.D. & Phinney E.S. 1982 *Nature* **295**, 17.

Shakura N.I. 1972 *Astronom. Zh.* **49**, 921.

Shakura N.I. & Sunyaev R.A. 1973 *Astron. Astrophys.* **24**, 337.

Shakura N.I. & Sunyaev R.A. 1976 *Mon. Not. Roy. Astr. Soc.* **175**, 613.

Shapiro S.L., Lightman A.P. & Eardley D.M. 1976 *Astrophys. J.* **204**, 187.

Shapiro S.L. & Teukolsky S.A. 1983 *Black Holes, White Dwarfs and Neutron Stars*, Wiley

Thorne K.S. & Price R.H. 1975 *Astrophys. J.* **195**, L101.

Wandel A. & Liang E.P. 1991 *Astrophys. J.* **380**, 84.

Wheeler J.C. 1981 *Nature* **294**, 230.

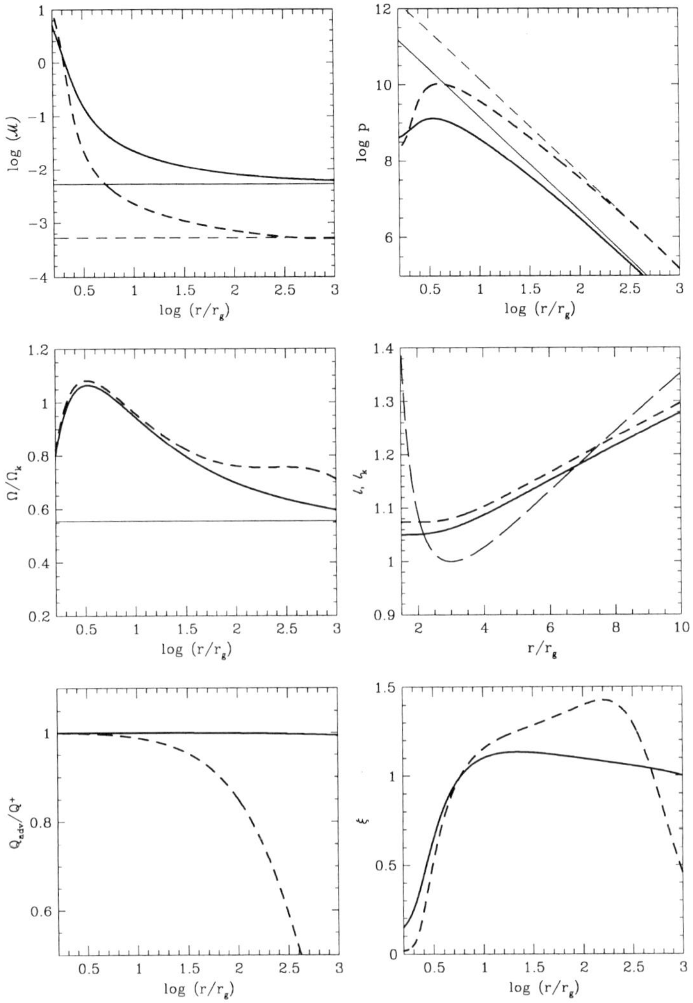

FIGURE 3. Global solutions for accretion disks around $10M_\odot$ black hole ($\alpha = 0.01$ and $\dot{m} = 10^{-5}$, solid lines) and ($\alpha = 0.001$ and $\dot{m} = 10^{-5}$, dashed lines). From Chen, Abramowicz & Lasota (1997). For comparison, straight lines show the corresponding approximate self-similar solutions with constant $\xi = d\ln S/d\ln r$ (S is the specific entropy) that have been constructed in earlier works. Upper: (left) the Mach number $\mathcal{M}$. One should note that the solutions pass smoothly the critical (sonic) point. Upper: (right) the pressure. Middle: (left) angular velocity, (right) angular momentum. One should note that the disk is non-Keplerian. Lower: (left) ratio of the advective cooling to viscous heating. Lower: (right) $\xi = \xi(R)$. Note, that ξ was not *assumed* (as in the earlier works on the subject), but self-consistently calculated, together with *all other* physical characteristics of the disk, from the *full set of differential equations* that describe the problem.

Black Hole X-ray Transients

By J. CRAIG WHEELER[1]

[1]Department of Astronomy, University of Texas, Austin, TX 78712, USA

The observations and theory of the exciting new class of galactic black hole X-ray transients is reviewed. Seven of these systems have measured mass functions or mass estimates in excess of stable neutron stars, making them excellent black hole candidates. Two of them have revealed "superluminal" radio jets. Study of the hard and soft radiation from these sources has given tight constraints on the physics of the viscosity of the accretion disk and promises firm proof that these systems contain black holes. This will allow us to search for black holes of more moderate mass and apply the knowledge of these systems to suspected supermassive black holes in AGN's. The most plausible mechanism for triggering the outburst of black hole candidate X-ray transients is the ionization thermal instability. The disk instability models can give the deduced mass flow in quiescence, but not the X-ray spectrum. Advection models that can account for the quiescent X-ray spectrum are difficult to match with the non-steady state, quiescent Keplerian disks. Self-irradiation of the disk in outburst may not lead to X-ray reprocessing as the dominant source of optical light, but may play a role in the "reflare." The hard power-law spectrum and radio bursts may be non-thermal processes driven by the flow of pair–rich plasma from the disk at early times and due to the formation of a pair–rich plasma corona at late times. The repeated outbursts in systems like GRO J0422+32 suggest some sort of clock, but it is unlikely that it has anything to do with a simple X-ray heating of the companion star. These systems typically have low mass secondaries and their evolutionary origin is still mysterious.

---

1. Introduction

In the past, Igor Novikov helped us to think about the inside of black holes and the methods to search for black holes. These subjects are today an active part of his legacy. In pursuing the astrophysics of black holes it is useful to remind ourselves about the subject of our quest lest we become too blasé or conclude the quest is finished.

Some black holes may arise from stars, but it is critical to remember that black holes are not ordinary stars, even exotic ones such as white dwarfs or neutron stars. They may come from matter, but they are not matter. Black holes represent the ultimate construct of strong gravity, of curved space time. Even though black holes harbor their secrets within the shroud of the event horizon, they are our conceptual link to the ultimate physics problem of singularities, fundamental notions of information as they evaporate, and to related concepts such as wormholes and closed time-like curves. It is thus of great importance to establish beyond any doubt that our Universe contains black holes and then proceed to study them, probe them, learn from them. Does our Universe contain black holes?

How does one go about the search for black holes? There are perhaps 10^{11} stars in the Galaxy, around 10^8 neutron stars, and maybe a million black holes left from the deaths of massive stars throughout the history of the Galaxy. The vast majority of these black holes will be isolated and thus hidden from any search. Logic suggests that because stars tend to congregate with like stars and black holes are thought to arise from massive stars that we should restrict our search to bright, hot, massive stars that might orbit a previously deceased companion, now a black hole. Massive stars are, however, rare. If the black hole is the product of a rare, massive star and it is in orbit around another intrinsically rare massive star while that star lives its relatively brief life, then the likelihood of the combination is low. The result is that there may be only one such system in our Galaxy and we have found it: the venerable Cygnus X-1.

It was with rather great surprise, then, that the last several years have taught us that the most common form of binary black hole candidates in the Galaxy, perhaps 100 to 1000 of them (Tanaka 1990), are black holes orbiting very small, but intrinsically common, half-solar mass K dwarfs. These systems are proving to be far more common than those like Cygnus X-1, but also to provide a fascinating laboratory to study the astrophysics of black holes.

We are one hundred percent sure we have discovered neutron stars. Have we discovered black holes? The probability is $P = 1 - \epsilon$. Many may think that $\epsilon = 0$ given the great work of HST (Ford *et al.* 1994) and the very clever result based on OH maser emission by Miyoshi *et al.* (1995) showing that active galaxies probably harbor massive black holes, as long suspected. In addition, at least six of the black hole transients to be discussed below have mass functions and hence minimum masses in excess of $3\,M_\odot$, the neutron star limit. One might, however, argue that this evidence is still stubbornly circumstantial given that we simply know these objects have high gravity, large mass and no optical stellar counterpart. That might leave ϵ finite, perhaps a few percent. What do we need to provide incontrovertible *proof* that we have discovered black holes? The black hole X-ray transients may help to yield that proof.

Section 2 will give an overview of the observations. Some discussion of the relevant disk instability that probably triggers the outbursts and issues of self-irradiation of the disk and reprocessing is given in §§3 and 4. Section 5 describes the important lessons we have learned from the exponential decline about the physical nature of the disk viscosity. The curious "reflare" phenomenon is discussed in §6 and various aspects of the new advection solutions are given §7. The quasar-like hard power law is considered in §8. Section 9 describes some of the problems of progenitor evolution and related issues. A summary and outline of key issues is given in §10.

2. Observations

There are by now upward of 24 binary X-ray sources that have been suggested as black hole candidates (Tanaka & Lewin 1995; Chen *et al.* 1996). Six of these have measured mass functions that make them excellent candidates on that basis alone: AO 620-00 (Nova Mon 1975), $M_x > 3\,M_\odot$ (McClintock & Remillard 1986); H 1705-250 (Nova Oph 1977), $M_x > 4\,M_\odot$ (Remillard *et al.* 1996); GS 2000+25 (Nova Vul 1988), $M_x > 5\,M_\odot$ (Filippenko, Matheson, and Ho 1995); GS 2023+338 (V404 Cyg), $M_x > 6\,M_\odot$ (Casares & Charles 1994; Filippenko, Matheson, & Barth 1995); GS 1124-683 = GRS 1121-68 (Nova Muscae 1991), $M_x > 3\,M_\odot$ (Remillard, McClintock, & Bailyn 1992); GRO J1655-40 (Nova Sco 1994), $M_x > 3\,M_\odot$ (Bailyn *et al.* 1995). Other systems are placed in this category of black hole candidates because they share morphological similarities to these best candidates, for instance in terms of their light curves or spectra or have other confirming orbital data. Among the more prominent of these are LMC X-1, LMC X-3, GX 339-4, 1E1740.7, GRO J0422+32 (Nova Persei 1992), GRS 1716-249 (Nova Ophiuchi 1993), GRS 1009-45 (Nova Vel 1993) and GRS 1915+105. Note that after a hiatus between the discovery of the prototype A0620-00 and that of GS 2000+25 in 1988, candidate systems have been observed at the rate of about one per year. Since 1992 this has been in great part due to the effectiveness of the BATSE experiment on CGRO (Fishman *et al.* 1989; Harmon *et al.* 1994). This discovery distribution is a powerful argument for the maintenance of all-sky monitors in a variety of wavelengths.

Some of the black hole transients are known to have repeated, A0620-00 in 1916 and

1975, V404 Cyg in 1938, 1989, and perhaps in 1956, and all of them are suspected to have done so on time scales of one to a few decades. Some repeat in about a year, but with properties that may be different than the major outbursts. Nova Sco which was first discovered in 1994 is undergoing another outburst at this writing (May 1996).

Many of the outbursts are characteristically marked by a very rapid rise on the time scale of about a day and then a nearly exponential decline in soft X-rays with an e-folding time of 30–40 days. This behavior was clearly displayed by A0620-00, GS200+25 and Nova Muscae 1991. This form of the light curve is not universal, however. V404 Cyg had a very irregular light curve near maximum. Other potential candidates have shown other types of behavior, plateaus, triangles, etc (Chen *et al.* 1996). A caveat here is that the fast rise and 30–40 day decay is a signature of the soft X-rays. The hard X-rays, with a manifestly different physical origin, often display burst behavior, but with a rather different morphology than the soft X-rays. If one only monitored the hard X-rays, as is the recent circumstance with BATSE but without the regular coverage of Ginga, then the picture one gets is rather different. It is thus difficult to know whether or not the various light curve morphologies BATSE has observed belie a very different underlying structure of the object, accretion disk, outflows, coronae, etc. or whether or not one is seeing the tail, not the dog, because of energy response limited to $\gtrsim$ 20 keV.

Another common feature of the light curves of the black hole candidates is a secondary brightening sometimes called a "reflare" at 60–80 days after maximum. This feature is not universal, but has been observed in all the sources with the canonical fast rise and exponential decay. Such a feature has never been observed in a neutron star soft X-ray transient with a fast rise and an exponential decline (Aql X-1 or Cen X-4), so there is some suspicion that it, too, is a signature of a black hole. V404 Cyg did not show this clear "reflare" and GRO 0422+32 showed only a bare flattening at $\sim$50 days. On the other hand the "reflare" is a feature of the soft X-rays, presumably arising from the geometrically thin, optically thick inner accretion disk. Neither V404 Cyg nor GRO J0422+32 showed this soft X-ray component, but only the hard component which has some other physical origin. Some of the most interesting recent sources, the "super-luminal" sources GRS 1915+105 and GRO J1655-40 to be discussed below, showed irregular light curves in the BATSE band, with little correlating information.

Where observations have been sufficient, many systems have shown a "second maximum" in flux about 200 days after the primary outburst, among them A0620-00, Nova Muscae, and GRO J0422+32. This feature can be observed in both soft and hard X-rays, but the soft component is probably the low energy extension of the hard power law component and not at all from the accretion disk.

Two systems, GRO J0422+32 (Callanan *et al.* 1995) and Nova Vela 1993 (Bailyn & Orosz 1995; Della Valle, Benetti & Wheeler 1996), have shown yet more subsequent outbursts. These bursts have been detected primarily in the optical with very irregular coverage in other bands, soft and hard X-ray, radio. It is thus difficult to determine their true nature and the connection to previous flux peaks.

Many of the black hole candidates show a characteristic two-component spectral structure. One component is a power-law that extends to high energies >20 keV perhaps >400 keV. As mentioned above, this power-law presumably also extends to lower energies. Rising above this power law at low energies, a few keV, is an extra "ultra-soft" X-ray component that is thought to arise in the accretion disk. These two components do not vary in lock step. As the soft component declines more or less monotonically in the characteristic exponential decay, the hard component can vary non-monotonically. In some phases the behavior is complementary with the hard rising as the soft declines, reminiscent of the different "γ-states" of Cyg X-1 (Ling *et al.*1987; Crary *et al.* 1996).

In particular, it is important to note that while (one component of) the soft X-ray flux varies with the presumed mass flow rate in the inner disk, the power law component does not behave so simply. Its power may ultimately derive from the release of gravitational energy, but there remains at least some possibility that some of its energy derives from the black hole, perhaps in a Blandford-Znajek process (Blandford & Znajek 1977) .

As noted above there are exceptions to this characterization of the spectral behavior. The "ultra-soft" component was definitely absent in V404 Cyg and GRO J0422+32 and Nova Oph 1993. It is not clear whether this could be some geometrical effect obscuring emission from the disk or a true absence of the inner soft X-ray emitting accretion disk. Tanaka (1990) argues that the large amplitude noise of the emission from V404 Cyg near maximum, a characteristic of the power law that is suppressed when the soft X-rays dominate, is evidence that the inner disk component is absent. On the other hand GRO J0422+32 did not show the great variability of V404 Cyg. Nova Oph had a strange square light curve (Harmon *et al.* 1994), and it is not clear how it fits into the general scheme.

The short-term variability is also of great interest in these systems. Nova Muscae showed QPO's at 3, 5, and 8 Hz (Miyamoto *et al.* 1993) and GRO J0422+32 at 0.04 and 0.1 Hz (Kouveliotou *et al.* 1993). The physics behind this variation is not yet understood, but it is clear that black hole sources and neutron star sources can have rather similar behavior and so the phenomena can not be restricted to any disk/magnetosphere interaction. There is some suggestion that this sort of time variability should be used to help classify the physical state of the system (van der Klis 1994, 1996).

Radio emission is also a ubiquitous feature of these systems where again no obvious parallel has been observed in the analog neutron star transient systems. The radio emission tends to follow the primary outburst, but most sources are not spatially resolved or well sampled temporally, so one does not know exactly where or when the radio was generated. This emission has been interpreted as an expanding synchrotron bubble (Han and Hjellming 1992), but it may all be better interpreted as a jet (Hjellming 1996). The relativistic particles might be created at some "working surface" that is somehow energized by power from the underlying system, but if jets are the rule then the fast particles almost certainly arise in the disk. The manner in which they are created is still not clear (§8).

More dramatically, two sources, GRS 1915+105 (Mirabel & Rodriguez 1995) and Nova Sco 1994 (Harmon *et al.* 1995; Hjellming & Rupen 1995; Tingay *et al.* 1995) have shown well-collimated "super-luminal" outflow. If the notion that we are dealing with black holes that emit power-law hard radiation and radio synchrotron radiation were not already sufficient, the discovery of these superluminal sources has completed the "mini-quasar" analogy. The X-ray light curves of both of these objects are somewhat unorthodox, but a well sampled soft X-ray light curve is lacking. GRS 1915+105 showed a long slow rise and then a outburst in the radio. It has a classic symmetric double-sided structure. Nova Sco had a rapid rise, but an irregular structure in which the radio bursts seems to correlate with a decline in the hard X-ray. It had an irregular hard X-ray light curve that bears no simple relation to the fast rise and exponential decline format. The radio structure is more irregular, with asymmetric blob structure.

Nova Muscae was well-sampled by Ginga and Sigma/GRANAT which detected both the ultra-soft flux and sufficiently high energy photons to sample the separate behavior of the power law tail. Miyamoto *et al.* (1993) have provided a decomposition of the light curve that is both rich in import for the physics of this and related systems, but also illustrative of the need for good sampling in both soft and hard X-rays as well as other bands, e.g. optical and radio.

The total power of Nova Muscae in all Ginga bands rose to a peak of 10^{38} erg s^{-1} in about a day and then declined exponentially following the soft X-rays. The power went up at the "reflare" when the power law component was at a minimum, clearly establishing that the "reflare" is a phenomenon of the soft flux from the accretion disk. The "reflare" is also observed in optical and UV, so either the outer disk is involved or the mechanism of the inner disk can radiate in the optical. The total power went through a local minimum around day 170 and then another local maximum around day 200. This "second maximum" falls approximately on the extrapolation of the exponential. This may imply that the minimum is a decrease in flow rate or efficiency, rather than the 200 day maximum being an enhancement of either. On the other hand, in A0620-00 the "second maximum" at 200 days has an amplitude considerably in excess of an extrapolation of the exponential, so it may not be possible to draw general conclusions from this behavior.

Miyamoto *et al.* (1993) deconvolve the total power into the power law component, including an extrapolation to soft X-rays, and the component ascribed to the disk, the excess in soft X-rays above the power-law extrapolation. The soft X-ray flux ascribed to the accretion disk rose at the "reflare" as remarked above. In A0620-00 there is some suggestion that the soft X-ray flux declined with a steeper slope after the "reflare." The data might also be compatible with an increase of the soft X-ray flux to a higher level at the time of the "reflare" with the same decline rate immediately after the "reflare" followed by a drop in flux to the original trend at about 100 days. For Nova Muscae the soft X-ray light curve shows a rise at the "reflare" followed by a decay with a somewhat shallower slope before the decline at about 140 days that precedes the "second maximum" at 200 days. GS2000+25 might show a similar behavior, but the sparse data is open to considerable interpretation. In the optical, A0620-00 showed a flatter slope throughout the exponential decline with an e-fold time of about 68 days, much shallower than the 30 days of the X-ray. The optical "reflare" was basically contemporaneous with the soft X-ray "reflare," but the flux returned to the same trend as before the event within about 10 days. There was an unfortunate paucity of optical data on Nova Muscae. The origin of this "reflare" and the subsequent behavior of the light curve are not well understood, but may have something to do with the self-irradiation of the disk as will be discussed below (§4).

In Nova Muscae the disk flux began to plummet about 140 days after peak and faded to near the threshold of detectability by 170 days, corresponding to the minimum in the total power curve, a factor of over 100 in 30 days. There is no apparent contribution from the accretion disk to the "second maximum" at 200 days. This shows that this local maximum may be entirely ascribed to the power-law component and not to the disk, just the opposite of the "reflare."

By deconvolving the spectrum into a disk and power law components the parameters that describe the disk model can be constrained. In particular the inner edge of the disk seems to be remarkably constant in a number of sources (GS2000+25; Nova Muscae; LMC X-3) when the soft X-ray luminosity and hence presumably the mass flow rate through the inner disk varies by factors of 10–100. There is only one obvious characteristic radius of the inner disk that should remain constant independent of $\dot{M}$ and that is the last stable circular orbit that depends only on the mass of the black hole. Although somewhat uncertain, the absolute value of the derived inner radius is also consistent with the last stable orbit of a black hole of a few $M_\odot$(Miyamoto *et al.* 1993; Ebisawa *et al.* 1994). These observations are remarkably consistent with what one expects from a black hole and inconsistent with any kind of neutron star magnetosphere where the radius should adjust to the mass flow.

The hard component of the first outburst ascribed to the power-law source by Miyamoto *et al.* (1993) peaked before the soft flux in Nova Muscae and then declined rapidly, much faster than the disk flux. It went through a local minimum about five days after the power law peak, at about the time of the peak of the disk component as deconvolved by Miyamoto *et al.* and then rose again to begin a somewhat more slow decay. As the power law began this new trend the transient feature at 480 keV which has been ascribed to red-shifted annihilation radiation was detected (Sunyaev *et al.* 1992; Goldwurm *et al.* 1993). The power law flux began a more rapid decline again about 30 days after the peak and went through another sharp local minimum at about 40 days after peak. One wonders whether there was another annihilation event at that time, but there were no appropriate observations. The power law component went through a minimum from 70 to 100 days after peak and then began a rise to dominate the flux at the "second maximum" at 200 days.

Although the spectral index in the first peak and the "second maximum" were similar, it is possible that the physics of these two events were rather different. The first was associated with a radio burst and the apparent annihilation event. There are reasons to think that this event was associated with some form of strong dynamic outflow (Moscoso & Wheeler 1994; §8). The "second maximum" may be more similar to the classic coronae discussed in the literature which are assumed to be in hydrostatic equilibrium (Liang 1990). Either of these configurations could interdict the inner portion of the disk, thereby preventing it from reaching to small radii where soft X-ray ($\sim$ keV) can be radiated efficiently. This may be why the hard flux precedes the soft flux on the rise while the disk is "filling in" the inner region from which the outflow emerges and why the disk flux plummets before the "second maximum" as the "corona" swells to fill up the region previously occupied by the inner disk. Unfortunately, the observations of Nova Muscae were not sufficient to determine whether the radius of the inner edge of the disk increased as the "disk" component declined at 140 days (Ebisawa 1996).

3. Disk Instability

The most plausible origin of the outburst in these systems is thermal instability associated with the ionization of hydrogen and helium which is also thought to be the fundamental mechanism of dwarf novae where the central star is a white dwarf (Meyer & Meyer-Hofmeister 1981; Cannizzo, Ghosh, & Wheeler 1982; Smak 1982; Faulkner, Lin, & Papaloizou 1983; Mineshige & Osaki 1983; for a recent review see Cannizzo 1993a). This mechanism arises as matter accumulates in the disk and becomes denser and warmer. Hydrogen begins to ionize and this increases the opacity. In order to maintain thermal equilibrium between viscous heating and radiative losses the density must decrease to maintain the optical depth in the face of the rising opacity. This leads to the characteristic "S-shaped" curve in the plane of surface density and temperature. The surface density of the disk cannot decrease, however, since matter is accumulating. Instead the disk structure is driven out of thermal equilibrium. It heats until it attains a hot state where H is fully ionized and thermal equilibrium can be restored. The hot state has higher viscosity and this leads to a mass flow in excess of the transfer rate from the companion. The surface density and temperature decline until a critical density is reached where hydrogen begins to recombine. At this point a cooling instability ensues returning the disk to a cold state.

This cycle is imposed whenever the transfer rate from the companion would require a portion of the disk to be in the region of thermal instability. It depends quantitatively, but not qualitatively, on the nature of the viscosity. The instability in geometrically thin Keplerian accretion disks generates heating waves that can propagate throughout the disk raising it to the bright, hot state that represents the outburst. This outburst can start in the outermost disk if the time scale for mass accumulation there is shorter than the viscous time scale. This depends on the mass transfer rate and the viscosity prescription in models. In other circumstances, the outburst can start in the middle or inner portions of the disk. After the outburst, the surface density profile becomes more concentrated at smaller radii. This leads to a decrease in the surface density at large radii and the subsequent cooling instability and cooling wave is automatically started near the outer edge of the disk. The subsequent cycles of heating and cooling to quiescence account for the basic observations of dwarf novae (Cannizzo 1993b) and candidate black hole transients (Mineshige and Wheeler 1989; Cannizzo 1993b, Cannizzo 1994, Cannizzo, Chen, & Livio 1995).

It is important to note that in the steady state the emission from the disk surface does not depend on the viscosity prescription (Shakura & Sunyaev 1973), but only on the local mass flow rate $\dot{M}$ in the disk (aside from some boundary condition restrictions at the edges). In the time dependent case, the flux does depend on the viscosity prescription. This is a two-edged sword. It means the predictions of the models depend on the uncertain physics of the angular momentum transport, but it also means that observations coupled with models with particular viscosity prescriptions can constrain the physical nature of the viscosity. The success of this approach will be outlined below (§5). In general, the transfer rate into the disk will not be equal to the mass flow rate through any radius in the disk and in particular not to the mass flow rate through the inner disk that determines the X-ray luminosity in the black hole models. One must avoid analyzing time-dependent situations with convenient, but erroneous, steady-state assumptions.

There are various tests of the disk instability in the context of the black hole transients. The models naturally give a rapid rise and slower decline in the soft X-ray and optical. They can give an exponential decay as observed in some sources (Cannizzo, Chen, and Livio 1995; §5). The mass transfer from the companion in A0620-00 greatly exceeds that attributed to the soft X-rays from the inner disk (Marsh, Robinson & Wood 1994; McClintock, Horne, & Remillard 1995). This shows that the disks are not in steady state in quiescence, a principle prediction of the disk instability models. There has been difficulty in detecting the cooling wave that should attend the decline (Cheng *et al.* 1992), but this may be because coronal heating yields a different color temperature than the effective temperature predicted by basic models.

4. Disk Irradiation

The question of the effect of disk self-irradiation is an important one for any source that emits intense X-rays, that is any neutron star or black hole system. In cases like Cygnus X-1 irradiation from the companion may also play a role, but this has not yet been quantitatively investigated. Irradiation from the companion to the disk is not important in the systems with small mass companions, but irradiation of the companion stars by the disk may be.

Van Paradijs and McClintock (1994, 1995) argue that the optical luminosity in black hole transients may be, in analogy to low mass X-ray binaries, dominated by X-ray reprocessing in the disk. Van Paradijs (1996) has extended this argument to investigate whether the disk irradiation can alter the critical transfer rate below which the disk instability sets in. If the disk is heated by irradiation it can be maintained in a stable, ionized state even if the mass transfer rate alone is insufficient to keep the disk ionized. On the other hand, Cannizzo (1994) has argued that the irradiation in the disk is negligible based on his study of the viscosity parameter α and its relation to the decay time scales in dwarf novae and X-ray novae. Huang and Wheeler (1989) and Mineshige and Wheeler (1989) noted that disk instability models of black hole sources with no irradiation give ample optical luminosity.

A first step toward the understanding of irradiation in the context of time-dependent, irradiated disk models has been taken by Kim, Wheeler, and Mineshige (1996a,b). They adopt two simple models of the irradiation: direct irradiation from the innermost hot disk and irradiation as might be reflected by a corona or disk atmosphere or chromosphere above the disk. The X-ray luminosity of the irradiation is given by

$$L_X(t) = \epsilon \dot{M}_{in} c^2, \tag{1}$$

where the efficiency ϵ =0.057, and $\dot{\mathrm{M}}_{\mathrm{in}}$ is the mass accretion rate at the inner edge of the disk R_{in}, taken to be $3r_g$ for a Schwarzschild black hole. For the indirect irradiation the assumption is made that the luminosity of Eq. 1 effectively arises from a point at the center of the disk (R = 0). The indirect irradiation flux is then given by

$$F_i(t) = C_X \left(\frac{L_X(t)}{4\pi R^2}\right), \tag{2}$$

where C_X is a constant. More sophisticated models incorporating radiative transfer would give $C_X = C_X(R,t)$ (see discussion in Tuchman, Mineshige & Wheeler 1990), but such time-dependent radiative transfer is too difficult at this time. For the direct irradiation, the prescription of Fukue (1992) is followed. This model assumes that the irradiation from a hot, geometrically thin annulus near the inside of the disk can be approximated by that from an infinitesimally thin, filled, uniformly radiating surface centered on the black hole. For regions in the disk at distances much larger than the radius of the annulus, this leads to the following expression:

$$F_d(t) = (1 - A)\left(\frac{L_X(t)}{2\pi R}\right)\frac{d}{dR}\left(\frac{H(t)}{R}\right)^2. \tag{3}$$

For simplicity, we take the X-ray albedo, A= 0.5. The specific choice of albedo does not appreciably affect the disk structure since shadowing suppresses the direct irradiation (see below). Eq. (3) shows that the direct irradiation is a function of the gradient of the disk height and hence very sensitive to small variation in the disk profile. Once $\dot{M}_{in}$ and hence $F_i(t)$ and $F_d(t)$ are determined, the irradiated flux is added to the flux generated internally in the disk by viscous heating through an implicit numerical method, and hence other disk physical variables are corrected in each time step (see Tuchman *et al.* 1990 and Mineshige, Tuchman & Wheeler 1990 for details). In this formulation, the irradiation fluxes are parametrized by two constants, C_X and A, which are, by assumption, independent of radius and time.

Kim *et al.* (1996a,b) show that even relatively mild irradiation from the inner disk can affect the overall disk evolution. Irradiation enhances depletion of mass from the disk during outburst, which results in a longer quiescent interval. It can also affect the slope of the declining light curve. More specifically, irradiation can affect the limit cycle by causing portions of the disk to linger in the intermediate temperature metastable "stagnation state." Although the disks are in nearly steady state near maximum, the time dependent models show that this is not strictly so and this has important implications for the irradiation. There tends to be a local maximum in H/R at intermediate radii ($\sim 10^8$ cm) that causes the outer disk to fall in the shadow of the inner disk, even at the time of maximum X-ray flux. This strongly limits the effect of the direct irradiation. Direct irradiation does not play an important role in the outer disk during and after the maximum of the outburst, but can affect the middle of the disk during the outburst. Since it is by assumption not influenced by shadowing, the indirect irradiation affects the overall disk evolution, including the outer disk, throughout its evolution. The indirect irradiation is, unfortunately, very uncertain because it depends on the X-ray scattering in coronae of unknown physical nature which Kim *et al.* have not considered.

Kim *et al.* have not explored the regime of strong indirect irradiation (large C_X) in any detail on the grounds that the light curve of the non-irradiated or only mildly irradiated models match the observations reasonably well. Such strong irradiation is likely to destroy this agreement, although this is a regime that should be explored more thoroughly. It is true that strong irradiation can heat the outer disk and stabilize disks that would otherwise be unstable to the disk instability for a given disk size and mass transfer rate as argued by van Paradijs (1966). Van Paradijs adopts a steady-state model for direct irradiation and hence his criterion for stabilizing the disks by irradiation may not apply to the transient sources. The time-dependence tends to yield disks with non-monotonic height profiles and hence shadowing of the outer disk by the inner and middle disk even in outburst when the disk is approximately in steady state. This suggests that transient sources should still follow the criterion given by, e.g., Smak (1983) neglecting irradiation. Aql X-1 represents an interesting case in point since it lies very close to the irradiated instability line determined by van Paradijs. One might argue that as a transient it should be governed by Smak's criterion and be well below the minimum transfer rate for stability. Aql X-1 has a variety of interesting properties that make it difficult to straightforwardly interpret it in terms of a disk instability (Tanaka and Lewin 1995), but it is worthy of closer examination in this respect.

Further study is necessary to understand the correlation of the optical activity to X-ray and radio observations. The irradiated models presented by Kim *et al.* are based on parameters chosen to reproduce the basic behavior of the optical light curve of A0620-00. They show that the disk instability is initiated in the outer portions of the disk. This means that the optical should begin to rise before any activity in harder bands. Mass flow is, however, enhanced in the inner disk before optical maximum, so soft and hard X-rays might well precede optical maximum. These models give inner disk temperatures of several million degrees near maximum, but this is not hot enough to provide sufficient soft X-rays. Other choices of disk parameters might give larger maximum flow rates and hence higher disk temperatures.

5. The Significance of the Exponential Decay

One of the interesting features of the black hole X-ray novae is the tendency to show an exponential decline. Simple models in which one quickly reduces the transfer rate to a hot disk with constant viscosity parameter α generate geometrically declining, not exponential, light curves. Even models in which the decline is driven by the cooling wave of the disk instability tend to have geometrically declining light curves with constant α. Mineshige, Yamasaki, & Ishizaka (1993) have argued that to produce an exponential decline, the angular momentum of the inner disk must be removed at a rate proportional to the angular momentum. They note that this tends to be the behavior of disk instability models with $\alpha = \alpha_0(H/R)^n$ with n $\sim$ 1 - 2. Cannizzo (1994) has also addressed this argument by noting that both dwarf novae and the black hole transients have exponential declines. Cannizzo concluded that to reproduce the exponential one needs $\alpha \propto R^{\epsilon}$, with $\epsilon \sim 0.3 - 0.4$, which is consistent with Mineshige *et al.* Cannizzo carried the argument one step further, however, by making the case that the precise value of ϵ that leads to exponential decline is itself a function of other parameters of the problem such as the transfer rate and inner disk radius. From this he concluded that exponential decline requires some form of feedback to operate in the disk to give just this behavior. This may hint that the angular momentum transport process is non-local, as the theories where internal waves play a critical role imply (Vishniac & Diamond 1993).

These arguments have been extended significantly by Cannizzo, Chen, and Livio (1995). Cannizzo *et al.* have shown that finely zoned (non-irradiated) models give a cooling front width w $\propto \sqrt{HR}$ and that the light curve in decline can be exponential only if there are certain conditions on α. Cannizzo *et al.* express the speed of the cooling front as,

$$V_f = \frac{2}{3}\alpha_0 \left(\frac{H}{R}\right)^{n+1} \frac{c_s}{w} R, \tag{4}$$

where $H = c_s/\Omega$ is the vertical scale height at radius R, c_s is the sound speed, Ω the Keplerian angular velocity, w is the width of the cooling front and the viscosity parameter is given by $\alpha = \alpha_0(H/R)^n$. They argue that the relevant sound speed is that dictated by the temperature of recombination in the matter just preceding the cooling front which is essentially constant, giving $(H/R) \propto R^{1/2}$. Assuming the criterion for exponential decay to be $V_f \propto$ R the relation w$\propto \sqrt{HR}$ then yields the observed exponential decay only when n = 3/2.

Vishniac and Wheeler (1996) have amplified this issue by showing that the cooling front speed is set by the rarefaction wave (in surface density Σ) that precedes the cooling front and that, in turn, determines the cooling front width, rather than the other way around. They show that the condition $V_f \propto$ R does lead to an exponential decline in the mass in the inner hot region and an exponential decline in $\dot{M}$, but that the speed of the front depends on opacities. For Kramer's opacity, the value of n that leads to an exponential decline is closer to n = 1.65 rather than 1.5. This preferred value of n is consistent with the theory of angular momentum transport by an internal wave-generated dynamo driven by tidal instabilities at the outer edge of the disk (Vishniac, Jin, & Diamond 1990, Vishniac & Diamond 1992). Vishniac and Wheeler argue that the cooling wave propagation depends on the viscous flow in the hot state and is nearly independent of the actual cooling process and of the state of the disk in the cool, quiescent material that accumulates in the wake of the inward-propagating cooling wave. The speed of the cooling waves is determined by the rarefaction wave that precedes them and is approximately $\alpha_F c_F (H/R)^q$, where α_F is the dimensionless viscosity, c_F is the sound speed, and all quantities are evaluated at the cooling front. The scaling exponent q lies in the interval $[0, 1]$, depending on the slope of the (T, Σ) relation in the hot state; however, q will be close to 1/2 for most models of the disk hot state. These results are insensitive to the structure of the disk outside of the radius where rapid cooling sets in. Vishniac and Wheeler conclude that the exponential luminosity decay of cooling disks is probably compatible with the wave-driven dynamo model. It is not compatible with models with separate, constant values of α for the hot and cold states.

There are several conclusions to be drawn from this work of Cannizzo *et al.* and Vishniac and Wheeler. First, the whole analysis is predicated on the assumption that a cooling wave exists in the decline of the light curve of transient black hole candidates and related systems. While the evidence is indirect, one can thus regard the exponential decline as a strong argument that a cooling wave is the fundamental mechanism of the decline of these transients. Since the cooling wave is one of the principal aspects of the disk instability, the exponential decay adds to the evidence that the accretion disk ionization instability is the underlying physical cause of the transient outburst phenomenon.

Another important general conclusion is that the study of accretion disks with phenomenological models for the angular momentum transport have provided crucial constraints on the nature of the viscosity. The quantitative and even qualitative behavior of the black hole models depends on the prescription for α. In the case of a double valued, but radially constant prescription, the outburst will tend to occur in the inner disk, yielding a somewhat slower rise phase and more symmetric outbursts. A prescription in which $\alpha = \alpha_0(H/R)^n$ will give very small values of α in quiescence where H/R is found to decrease inward. This will yield a very long viscous time in the inner disk and promote outbursts that begin in the outer disk and propagate inward. This yields model outbursts with rapid rise and slower decline, in accord with the observations for the optical and soft X-ray light curves of the many of the X-ray novae. The conclusion of Vishniac and Wheeler that the exponential luminosity decay of cooling disks is not compatible with models with separate, constant values of α for the hot and cold states thus constrains the nature of the outburst and suggests that for the black hole cases, in particular, ignition of the heating on the outside (logarithmically) is to be expected.

More specifically, these analyses support the conclusion that the exponential decay of the luminosity of black hole disk systems following outbursts is consistent with a local law for the dimensionless disk viscosity $\alpha \propto (H/R)^n$ if, and only if, n is approximately 3/2. This scaling for α implies that disk systems in general should exhibit approximately exponential luminosity decay from peak luminosity whenever the hot state opacity follows a simple power law. This result is apparently compatible with the internal wave driven dynamo model for disk viscosity.

Cannizzo *et al.* (1995) have shown that the slope of the exponential is a function of M/α_0, where M is the mass of the central object and α_0 is the coefficient in the local viscosity law. Cannizzo *et al.* argue that $\alpha_0 \sim 50$ is a universal constant. Understanding why this coefficient differs so significantly from unity will be a challenge for the dynamo models. If Cannizzo *et al.* are correct, then the slope of the exponential gives a measure of the mass of the central objects. For the black hole sources, the result is compatible with $M \sim 5\,M_\odot$ and the conclusion that A0620-00, GS2000+25, and Nova Muscae all have very similar black hole masses (§9).

These new insights into the nature of the viscosity of the black hole sources also give a new perspective in which to ask questions of other astrophysical sources. The so-called "TOADS," tremendous outburst amplitude dwarf novae (Howell, Szkody & Cannizzo 1995) give evidence for especially low viscosity. Another internal wave-driven mechanism, the incoherent dynamo, (Vishniac & Brandenburg 1995) gives a minimum value for α_F. This may provide a floor opacity to the TOADS and they in turn may provide some direct constraints on the non-local nature of the wave-driven mechanisms.

Another problem of interest is that of disks in QSO's or AGN's. In that context there is no obvious companion to generate internal waves by tidal instability. Some other mechanism to generate internal waves, perhaps stellar collisions with the disk, must be sought.

In any case, the understanding of the import of the exponential decline of the black hole sources marks the end of a chapter in the study of accretion disks and opens up a new range of topics to be explored with some confidence that we do have the first rudiments of understanding of the viscosity of accretion disks.

6. The Reflare

The "reflare" phenomenon seems to be a common feature of the black hole transients that show the classic fast rise and exponential decline. This feature calls for a physical explanation in general and raises the expectation that it might prove to be an interesting physical diagnostic. There have been four studies of the phenomenon of the "reflare" observed in the decay of many black hole X-ray novae: Chen, Livio & Gehrels (1993), Augusteijn, Kuulkers & Shaham (1993), Mineshige (1994) and Kim, Wheeler & Mineshige (1996b).

The first two studies assume that X-ray irradiation can modulate the mass transfer rate from the companion by irradiating the companion during the rise, maximum and the early stage of the outburst decay to produce the "reflare" (mass transfer burst model). Chen *et al.* (1993) do not discuss in any detail the observable implications of their model such as optical or X-ray light curves. It is not clear, for instance, what the optical response of the disk would be, nor the degree to which even a sharp burst of added mass will be spread by the finite viscous response of the disk (especially when the outer parts of the disk are in the cold state) so that any later effect in the X-rays will be delayed with respect to the optical and very spread out in time. There are also questions of whether the disk blocking they invoke to account for the delay of the "reflare" is consistent with their estimates of mass transfer and energetics that depend on irradiating the companion. Similar issues arise in their model for the "second maximum."

Augusteijn *et al.* (1993) suggest an oscillation of the light curve in the decay in which each successive burst is a "reflection" of the previous burst that heats the companion and drives more mass transfer after some time delay. This model seems to be remarkably reminiscent of the recently discovered "mini-outbursts" in GRO J0422+32 (Callanan *et al.* 1995) and Nova Vela 1993 (Bailyn & Orosz 1995; Della Valle, Benetti, and Wheeler 1996). Augusteijn *et al.* (1993) even predicted bursts in Nova Per in August 1993 and December 1993, as observed. Augusteijn *et al.* (1993) deserve credit for drawing attention to the fact that there may be some "clock" underlying the bursts in Nova Per and perhaps other objects, but there are still open questions concerning their particular model. One point that requires clarification is that Augusteijn *et al.* (1993) did not clearly differentiate the "reflare" from the "second maximum" as we are defining them here. They first adjust parameters of their model to fit the "reflare" of GS 2000+25, but then calibrate the model on the "second maximum" of GRO J0422+32 in order to "predict" the later outbursts. It is not at all clear that the "reflare" and the "second maximum" involve similar physics since the "reflare" is a feature only of the soft X-rays and optical/UV and the "second maximum" is dominated by the power law hard X-rays, although it also has an optical component (§2). The later mini-outbursts may be related to the "second maximum," as Augusteijn *et al.* (1993) argued, but their spectral response is not yet well established. They may be purely optical phenomena. In addition, Augusteijn *et al.* (1993) predict bursts of decreasing amplitude whereas, as Bailyn & Orosz (1995) point out, the observed optical bursts have nearly constant amplitude. Augusteijn *et al.* (1993) also predict a burst on 21 April, 1993 for GRO J0422+32, and no such burst was observed. The models of Augusteijn *et al.* (1993) also do not consider the state of the disk, especially in its cool, quiescent, low-viscosity state, in a self-consistent way.

Unlike the pictures proposed by Chen *et al.* (1993) and Augusteijn *et al.* (1993), the models of Kim *et al.* (1996b) for the optical "reflare" require that the outer disk and the L_1 point are blocked by the inner disk from receiving any direct irradiation throughout the decay phase prior to the "reflare." This contradicts the hypothesis of the X-ray-irradiated mass transfer burst models, that the L_1 point be irradiated. Kim *et al.* (1996b) find the companion to intercept an angle of about 10° from the center of the disk. Near maximum light, the disk blocks all angles less than about 5°. Kim *et al.* estimate that the companion receives $\lesssim 10^9$ erg cm^{-2} s^{-1} of intercepted irradiation at maximum. Recent studies of the evolution of irradiated low mass companions show that the mass transfer rate will increase as the companion's atmosphere expands upon heating if the irradiation fluxes are $\gtrsim 10^{10}$ erg cm^{-2} s^{-1} (Podsiadlowski 1991) or $\gtrsim 10^{9}$ erg cm^{-2} s^{-1} (Hameury *et al.* 1993). Thus the disks always shadow the L1 point and the intercepted flux may be so low as to be unable to engender any appreciable structural changes in any case.

Mineshige (1994) suggests that the "reflare" is caused by a sudden increase of viscosity as the disk is irradiated and by a concurrent mass transfer burst instability. The origin of this sudden increase in viscosity by a factor of 2 to 4 in extended portions of the disk, is, however, not clear. In the irradiated models presented by Kim *et al.* (1996b), there is a sudden increase of the viscosity, but it is confined to the zone that undergoes the "reflare." The viscosity parameter α changes by a factor ~ 4.8 from ~ 10 days prior to the "reflare" to the maximum of the "reflare."

The irradiated disk instability models of Kim *et al.* can reproduce the optical "reflare" observed in A0620-00. In these models the optical "reflare" results from an intrinsic property of the disk instability, the "stagnation" phenomenon whereby the disk does not directly jump from the hot down to the cool state, but stays at an intermediate temperature due to an increase in the specific heat in the partially ionized matter. Portions of the disk can then jump from this "stagnation" state back to the hot state giving the optical reflare. In the irradiated accretion disk models, the stagnation is reinforced by the continuous heating from the irradiation reflected by the disk corona or chromosphere. This modulation of the outer disk gives no direct way to modulate the mass flow rate in the inner disk in order to produce a correlated soft X-ray "reflare" as commonly observed in these black hole candidates, a problem that plagues other models of the "reflare." Kim *et al.* have shown, at least, that the possibility of optical flaring in the outer disk should be considered as part of the relevant physical processes.

We do not yet have a complete understanding of the physical mechanism of the "reflare" (or subsequent flares). No model yet proposed can naturally account for why the "reflare" seems to coincide with the sudden drop in the hard X-ray flux in Nova Muscae (Miyamoto *et al.* 1993). Nevertheless, these irradiated models show that effects in the disk alone can give optical outbursts that may be related to the optical flares seen. They also give us a new perspective from which to consider questions of the irradiation of the companion.

A critical question remains as to the frequency of occurrence and the physical nature of the repeated flares of systems like GRO J0422+32 and Nova Vela. Since coverage of these systems at late times is spotty, these sorts of subsequent outbursts could be common. The spectral coverage is even more dismal. We have very little idea whether these bursts are primarily an optical phenomenon, or whether they have an associated soft or hard X-ray component as well. Such observational data are critical if we are to understand whether these repeated bursts are a product of the outer disk, the inner disk, and related to the disk instability, to a tidal instability, to a mass transfer instability, to a combination of these processes, or to some other process entirely.

7. Matching with Advective Solutions

The models of Kim *et al.* (1996a,b) give an inner region that remains in the hot ionized state even though the bulk of the disk drops into a cold, low-viscosity, quiescence. The cooling wave slows as the density is depleted in the inner disk and it does not reach the inner edge before the next eruption occurs. The mass accretion rates given by these models, $\sim 10^{11}$ g s^{-1} for non-irradiated models and $\sim 10^{9}$ g s^{-1} for irradiated models, agrees rather well with that deduced for A0620-00 in quiescence by McClintock *et al* (1995), but the models give too low a temperature which McClintock *et al* (1995) estimate to be $\sim 2 \times 10^6$ K. This problem has been discussed by Lasota (1995). The behavior of this slow inner cooling wave may be sensitive to the viscosity parameter and other model parameters in a manner that needs to be explored.

A possible solution to the problem of the X-ray emission is provided by hot flows in which quasi-radial advection of thermal energy is a key component as reviewed by Abramovicz in these proceedings (Narayan and Yi 1994; Abramovicz *et al.* 1995; Artemova *et al.* 1996). Narayan, McClintock, and Yi (1996) obtain a fit to both the optical and X-ray spectra of A0620-00 in quiescence by invoking a hot two-temperature advective disk solution in the inner disk matched to a steady state Keplerian disk in the outer portions that provides the optical luminosity. The advective solution, however, is of low efficiency and requires a mass flow rate of 4×10^{14}g s^{-1}, much higher than the estimates based on steady state, geometrically thin, optically thick disks by Marsh *et al.* (1994) and McClintock *et al.* (1995) and much higher than the quiescent flow rates obtained by Kim *et al.* (1996a,b). The steady-state disks appended to the advection solutions by Narayan, McClintock & Yi are not consistent with the quiescent state of the disk being modeled. The disk is almost surely not in steady state in quiescence. Note that, even if advection is a factor at some phases of quiescence, the disk instability associated with ionization of hydrogen and helium and its role during outburst evolution can not be avoided for the conditions that are thought to exist in quiescence.

Lasota, Narayan, and Yi (1996) address these issues by presenting models with a radially constant mass flow rate specified as a parameter. They explore conditions in which the outer cool disk can be in a globally stable steady state at low viscosity. They point out that such configurations can be thermally unstable depending on the inner boundary where the advection solution takes over. Alternatively, they might be triggered into disk instability by some enhancement of mass transfer from the companion star. They note that the disk is too cool, $\sim$2000 K, to provide the observed quiescent optical luminosity, an aspect common to all disk instability models for these systems (Huang & Wheeler 1989; Mineshige & Wheeler 1989; Mineshige, Kim & Wheeler 1990).

One obvious possibility is that the blue optical flux observed in quiescence of A0620-00 is that of the hot spot generated by the transfer stream that is, to first order, constant during the whole outburst process. A0620-00, GS 2000+25, and Nova Muscae are strikingly similar objects in terms of their light curves. They also have similar orbital period and mass function, which suggests they have similar accretion disks. Eclipse mapping of GS 2000+25 in quiescence shows evidence for a hot spot which is much brighter than the disk itself (Casares, Charles & Marsh 1995). The bright spot in GS 2000+25 clearly implies that there is continuous mass transfer in quiescence, the major assumption of the disk instability model. The fact that the bright spot dominates the disk luminosity is consistent with the cold disk we predict in quiescence. The possibility that the hot spot contributed substantially to the quiescent optical emission of A0620-00 was discussed by McClintock *et al.* (1983). The optical/UV observations reported by McClintock, Horne, and Remillard (1995) show that the bolometric flux detected by HST is consistent with that expected from the transfer stream. They fit the continuum spectrum of the "disk" with a reddened 9000 K black body and for a distance of 1 kpc and derive an emitting area of 1.2×10^{20} cm^2, 80 times smaller than the disk. The implied bolometric luminosity of this emission, assuming an optically thick sphere with this area and effective temperature, is 4.6×10^{31} erg s^{-1}. The rate of release of potential energy by the transfer stream striking the outer edge of the disk at radius R_d is given approximately by $1.3 \times 10^{30} M\dot{M}_{15}/R_{d,11}$ erg s^{-1} where M is the mass of the compact object in solar masses, $\dot{M}_{15}$ is the secondary transfer rate in 10^{15} g s^{-1} and $R_{d,11}$ is the outer disk radius in units of 10^{11} cm. Clearly, for $M\dot{M}_{15} \sim 30$, a plausible case could be made that all the optical luminosity arises in a hot spot and not in the disk proper. It is not clear whether ascribing all the optical luminosity to the hot spot is consistent with the lack of observed orbital modulation (McClintock, Horne & Remillard 1995).

There is still the question of the quiescent X-ray emission from A0620-00 which can be matched by the advection solutions with appropriate choices of parameters, but not with the present disk instability models. The advection solutions require some means of severely depleting the surface density of the inner portions of the disk as the disk approaches quiescence. It is difficult to see how such a solution matches physically in terms of the surface density and angular momentum with the outer thin quiescent disk. Perspective on this problem can be gained by considering some time scales inherent in the advection solution. The high temperature, low efficiency advection solutions require a very low density and hence a low mass in the advection region. The published solutions assume an *ad hoc* supply of mass from the Keplerian disk to the advection region in order to maintain this steady-state condition (Narayan, McClintock and Yi 1996; Lasota, Narayan, and Yi 1996). The disk instability models, on the other hand, give a much lower mass flow rate in quiescence. It is thus of interest to inquire how long the advective solution would survive if the mass flow were cut off. Using the expression for the density distribution given in Narayan, McClintock, and Yi and the parameters for their fit to A0620-00 ($\alpha = 0.1$, $M = 4.4 M_\odot$, $\dot{M}_{15} = 0.4$), the mass is $1.9 \times 10^2 R_2^{3/2}$ g. For the derived mass flow rate in the advective region, the time to consume the advective region in the absence of mass input is $M/\dot{M}$=14 s $R_{a,9}^{3/2}$, where $R_{a,9}$ is the boundary between the advective region and Keplerian disk in units of 10^9 cm. Clearly, the advection region is very fragile and must be constantly supplied at the high mass flow rates required by the low radiative efficiency in order to be sustained. This is not a trivial requirement.

The high mass flow rate necessary to maintain the advective region, $\dot{M}_{15} = 0.4$, does not arise naturally in the quiescent disk instability models. It must, rather, be supplied from the matter in the Keplerian disk by some sort of ill-understood ablation process. One can at least ask what the implications are of supplying mass at this rate. To determine this, we fit an approximate power law to the quiescent disk structure of the models of Kim *et al.* (1996a,b), recognizing that this will be different for different viscosity prescriptions. For the quiescent mass distribution corresponding to Figure 5, we find

$$\frac{dm}{dr} = \frac{1}{v_a}\frac{dm}{dt} \sim 10^{13}\, r_{10}^{1.4}\, g, \tag{5}$$

where $v_a = dR_a/dt$ is the speed with which the boundary of the advection region advances into the Keplerian disk and dm/dt is the rate of ablation necessary to feed the advection region. This implies a characteristic time scale to consume the disk of

$$\tau = \frac{R_a}{v_a} = 2.5 \times 10^8\, R_{a,10}^{2.4}\, s, \tag{6}$$

where $R_{a,10}$ is the outer boundary of the advection region in units of 10^{10} cm. The advection solution of Narayan, McClintock & Yi (1996) that reproduces the ROSAT X-ray observations cannot remain with an outer boundary of order 10^9 cm for times comparable to the recurrence time. If such an advection region is fed by ablation from the Keplerian disk 20 years after the outburst, the boundary must be in excess of 10^{10} cm. This seems to be a rather extreme criterion. An alternative is that the quiescent X-rays do not arise in the inner disk at all. The observed flux is low and it is possible that it arises in flare activity on the companion.

Clearly more work is needed to understand the state of the disk in both quiescence and outburst. The black hole X-ray transients are excellent laboratories for these studies and the disk instability models give a framework in which to pose relevant questions.

8. The Hard Power Law, Radio Outbursts, and Positrons

The hard power law component that is frequently observed in black hole X-ray transients is very similar to that observed in AGN as discussed in these proceedings by Svensson. Since at some level we know more about the transient systems – the definite existence of accretion disks, orbital parameters and, in the best cases, masses of the components – we may, with some hubris, expect that these transient systems may help us to understand the hard radiation mechanisms of AGN, rather than the other way around.

It is commonly assumed that the power law component represents a Comptonized thermal spectrum arising in a hot plasma that is itself part of a coronal structure in hydrostatic equilibrium. Such a simple model for the emission can fit some objects at some epochs, but that does not make it unique or correct. Such models ignore the obvious evidence for non-thermal particles and magnetic fields implied by the common radio outbursts that are frequently associated with the X-ray bursts (Han and Hjellming 1992). The recent super-luminal sources are only the most extreme example. It is most likely that the non-thermal particles and magnetic fields arise in the disk and hence must be incorporated into models of the hard power-law emission. Kusunose has shown that thermal models are not unique. Non thermal models can produce power law emission that has a constant slope independent of luminosity, as observed (Kusunose & Mineshige 1995). In addition, there is ample evidence for outflow at least at some epochs and so there is reason to question the assumption of hydrostatic equilibrium for the plasma.

The soft X-ray component that probably arose in the accretion disk peaked more slowly than the hard power law flux in Nova Muscae. This may mean that the inner disk was incomplete in quiescence or the early phase of the outburst. The radius of the geometrically thin, optically thick disk may have shrunk in response to increased mass flow attendant with the disk instability in the outer disk, thus giving rise to the delayed rise of the soft flux. It may be that in systems like V 404 Cyg and GRO J0422+32, the inner disk never properly forms to give soft X-rays, for reasons that are not currently understood.

The first flare of the hard flux in systems like Nova Muscae may be better associated, not with a static, thermal plasma, but with a non-thermal, magnetic, pair-rich outflow (Moscoso & Wheeler 1994). This phase shows QPO's, correlated radio synchrotron bursts, and at least in Nova Muscae, the line feature that is plausibly associated with annihilation. If this is the annihilation line, then it is much too narrow to represent annihilation in the region where positrons are created and hence implies flow of some kind from the location of the creation of the positrons to the locale of their annihilation.

The "second maximum," as defined here is primarily characterized by a hard power law spectrum and the virtual absence of a soft X-ray component from the inner disk. This feature may arise in a structure that more closely resembles a quasi-static corona of the sort frequently modeled in the literature (Liang 1990). The fact that the disk component of the soft X-ray flux declines as this late hard component comes up strongly suggests that the corona is displacing the geometrically thin disk. With a larger effective inner radius, the disk simply becomes too cool to emit soft X-rays.

There is some speculation that the "second maximum" may be related to an advection solution, but the problems associated with the time scales and the feeding of such a region by the outer thin disk outlined in the previous section may again apply. There is also some question of whether this feature could be related to a tidal instability that is triggered when the disk radius grows beyond the 3 to 1 resonance due to the onset of the disk instability. The long time delay, 200 days, might be accounted for by the slow linear growth time of the instability (Ichikawa, Mineshige, & Kato 1994). It is not clear that this picture can naturally account for the X-ray spectral characteristics of the "second maximum."

To better understand the possible orgin of the power law in dynamical environments and the nature of the "annihilation feature" in Nova Muscae, Moscoso, Kusunose, and Wheeler are constructing a model to explore the source of the outflow in the primary outburst. This model consists of an inner hot, pair-rich corona represented by a single zone. Above this corona, photon annihilation will generate electron/positron pairs and associated annihilation. The parallel component of the average momentum of the photons that produce pairs is assumed to represent the bulk outflow momentum of pairs. The remaining momentum is randomized to provide the thermal component of the pair energy. Account will be taken of both the isotropic and anisotropic Comptonization. This simple model will give an estimate of the typical flow time scales, speeds, and the optical depth so that annihilation line profiles can be estimated.

9. Evolution

Many fascinating questions arise with respect to the evolution of these black hole transients. These involve the large mass ratios, the suggestion that the black holes may be rather moderate in mass, and the significance of lithium in some of them.

With typical primary masses of $5\,M_\odot$and secondary masses of $0.5\,M_\odot$, the mass ratio of these systems, ~ 10, is very large. Most massive binaries are thought to have mass ratios closer to unity, but there are clearly strong selection effects that bias observations away from such large ratios. Nevertheless if this is merely an accident of birth, it seems odd that it is the future black hole systems that favor small mass companions.

It is not clear how much the secondary may have altered during the evolution of the system (De Kool, van den Heuvel & Plyser 1987). Many of the secondaries seem to be at least slightly evolved, the secondary in V404 Cyg definitely so (Casares, Charles & Naylor 1992). The atmospheres, where they can be observed in quiescence, are, with some small exceptions, characteristic of main sequence stars, so they can not be highly evolved. They may have lost mass not only to Roche lobe overflow, but also to the process of being blasted by X ray radiation at recurrent intervals since the black hole formed (Mineshige, Kim & Wheeler 1990). The small orbital periods that characterize these systems, suggest that they have been subject to common envelope evolution and ejection, but the low mass companions have little leverage on the envelope of the black hole progenitor and it is not clear how such systems could escape core coalescence. There is some speculation that the secondary was not primordial, but formed out of the envelope of the massive star by some adjunct process to creating the black hole (Podsiadlowski 1996).

There is increasing evidence that there are upper limits as well as lower limits to the masses of the black holes in these systems (see, eg, Chen, Shrader & Livio (1996) for a compilation). The upper mass limits are interestingly low, perhaps substantially less than $10\,M_\odot$in many cases. Rather tight limits have been derived for Nova Oph 1977 ($4.9\pm1.3\,M_\odot$; Remillard *et al.* 1996), GS 2000+25 (5.9 - $7.5\,M_\odot$; Filippenko, Matheson and Barth 1995), GRO J0422+32 (3.57 ± 0.34; Filippenko, Matheson, and Ho 1995), and Nova Sco (4.0 - $5.2\,M_\odot$; Bailyn *et al.* 1995). Nova Muscae has been estimated to be in the range 4 - $8\,M_\odot$(Chen *et al.* 1996). If a temporary eclipse by the disk was seen by Haswell *et al.* (1993) an upper limit of $5\,M_\odot$ can be assigned to A0620-00. V 404 Cyg is the sole current exception with a mass that might plausibly be as large as $12\,M_\odot$[Shahbaz *et al.* (1994); but see Sanwal *et al.* (1996) for an upper limit].

The possible clustering of masses around $5,M_\odot$would be entirely consistent with the notion that the decline rate is a measure of the mass of the compact object (divided by α_0; §5). In this interpretation, the commonly observed decline rate with an e-fold time of about 30 days would signify a common mass. An important caveat to this conclusion is the evidence that some neutron star systems also display a decline rate of nearly 30 days (Chen *et al.* 1996).

These masses suggest that the black holes did not simply form by the collapse of a massive star. Clearly the collapse of an entire massive star would give far larger masses. A more reasonable possibility is that the collapse only involved the core of the star, the envelope having been ejected in a wind or common envelope process. As a rule of thumb, the core of a massive star is about 1/3 the original main sequence mass. Since moderately massive stars are commonly presumed to give rise to supernovae and neutron stars, the progenitors of black holes might be expected to have main sequence masses in excess of $30\,M_\odot$(Shields and Wheeler 1973). This would make a core mass in excess of $10\,M_\odot$, still too massive to account for the low estimates given above. There may be a need for some other evolutionary process.

One possibility is that the object that is now a black hole began its compact life as a neutron star. If there was a stage of rapid mass transfer, then the neutron star might have been swamped so that an envelope formed around it. The resulting object with a neutron star in the center and a quasi-static envelope supported by energy from the gravitational energy or nuclear burning of accreted matter is known as a Thorne-Żytkow Object (Thorne and Żytkow 1977). In such a case the neutron star might continue to accrete until it surpassed the stable mass limit and then collapse to form a black hole and consume the remainder of the envelope. The mass of the resulting black hole would be the sum of the neutron star plus the envelope and could be rather modest. This scenario would demand that the secondary lose substantial mass. The secondary would have to be evolved, as seems to be frequently the case, although it could retain a thin H blanket. This possibility would have to be reconciled with the atmospheric observations. Forming a Thorne-Żytkow Object rather than burying both stars in a common envelope would also presumably require special conditions. The possibility that the original star is consumed, but the current secondary spun off to coalesce anew is a complication that might be worth considering.

One interesting connection of Thorne-Żytkow Objects with the current population of black holes is lithium. The two black hole candidates that have sufficiently bright quiescent luminosity to be examined, V 404 Cyg (Martin *et al.* 1992) and AO620-00 (Marsh *et al.* 1994), have both revealed excess of lithium in the atmospheres of the companions. This observation cannot be a unique signature of black holes systems because lithium has also been detected in the neutron star system Cen X-4 (Martin *et al.* 1994), but it is nevertheless intriguing. No such lithium enrichment has been observed in nova-like systems containing white dwarfs (Martin *et al.* 1995).

The lithium is short-lived in convective stellar atmospheres, so the strong presumption is that it must be freshly created in these systems, not primordial. The most likely origin is by spallation in the atmosphere of the secondary by fast particles accelerated by processes near the black hole (or neutron star). Another possibility is that the lithium is created in the disk, where spallation is, if anything, more likely, but then the problem exists of preserving and transporting the lithium to the secondary star where it is observed. Yet another possibility, more remote, but perhaps not impossible, is that the lithium is associated with a Thorne-Żytkow Object precursor phase. Under some circumstances, the deep convective burning in a Thorne-Żytkow Object can produce lithium (Cannon 1993; Biehle 1994; Podsiadlowski, Cannon, & Rees 1995). It is not clear that the modest mass Thorne-Żytkow Objects that would be the precursors of the modest mass black holes would have this property (this is perhaps more likely with the large mass black hole of V 404 Cyg), nor how the lithium could be transferred to and preserved in the atmosphere of the secondary.

The observation of the lithium excess also couples to one more conundrum in the black hole systems. The putative annihilation line in Nova Muscae and the corresponding broad feature that has appeared in 1E1740.7 (the "Einstein Source") are, within the errors, centered at about 480 Kev, not 511 Kev. This has been interpreted as requiring that the annihilation occur sufficiently deeply in a gravity well that the shift can be attributed to a gravitational redshift (Chen, Livio & Gehrels 1993; Moscoso & Wheeler 1994). As it happens, one of the principle γ-ray de-excitation lines of lithium formed in spallation is at 483 Kev. This coincidence has given rise to considerable speculation that the "annihilation feature" is actually de-excitation of the newly forming lithium observed in the companion (Martin *et al.* 1992, 1994; Martin, Spruit, and van Paradijs 1994). Although there are some numerical coincidences, no quantitative model of spallation of the lithium has been provided to account for the observations. One counter argument is that there should be other associated spallation γ-rays that are not observed and a counter-counter argument is that there may be some hint, in Nova Muscae, that the 480 Kev feature has a double peak. It will take much better signal to noise to resolve that issue. The 480 Kev feature in Nova Muscae was accompanied by another, weaker line at 175 Kev which can be kinematically accounted for by a blue-shifted back scattering of the annihilation line (Hua and Lingenfelter 1993). It is not clear that a lithium model can account for this "back-scatter" feature. Other objections to the lithium interpretation are that the 480 Kev feature, while too narrow to arise in the region where positrons are created, has a finite width in both Nova Muscae and the 1E source. This means that the feature is unlikely to arise in the atmosphere of the secondary even though that is where the optical line of lithium is seen. The accretion disk is a more likely locale to produce such broad de-excitation lines by thermal or Doppler broadening. One might then have the inverse problem of the annihilation line interpretation, because if the line is from lithium, then it can suffer very little or no gravitational redshift. Clearly the interpretation of this feature as either lithium or positronium requires more careful quantitative analysis. Yet another possibility is that circumstances contrive to give the "line" as a result of continuum scattering in an appropriately beamed flow (Skibo, Dermer & Ramaty 1994).

10. Conclusions

The basic picture in which a companion star feeds mass at nearly a constant rate into an accretion disk that is susceptible to an ionization limit-cycle instability is still the most plausible explanation for the black hole X-ray transients. The characteristic light curve with a fast rise and exponential decline is consistent (even demands) that the instability begins in the outer portion, logarithmically, of the disk. In such a case, the optical emission should rise before any hard emission. This is difficult to check with current search techniques, but something like an all sky optical monitor could provide critical constraints on this important qualitative prediction.

As the heating wave propagates inward and leads to a dramatic increase in the flow rate in the inner disk, one might expect outflows to be generated. These outflows could then, in turn, be related to the ubiquitous radio bursts including, in the most extreme cases, the superluminal sources, and the occasional observation of annihilation lines. If the outflow region interdicts the inner portion of the accretion disk so that the optically thick, geometrically thin disk can not, at first, extend down to the last stable circular orbit, then the disk may be too cool to emit X-rays. This may be why the hard flux arises earlier than the soft flux in Nova Muscae despite the prediction that the instability should begin in the outer disk.

In this picture, the hard power law flux associated with the onset of primary outburst in the fast-rise, exponential decline systems, should be associated with an outflow, probably pair rich and magnetic, not in a quasi-static corona, nor even in a quasi-radial advective region. An important task is to develop physical models that can help to discriminate these various possibilities. This is the stage at which systems like Nova Muscae display the strongest QPO's and hence one is invited to examine models for QPO's that directly involve not only the disk or a radial inflow, but a dynamic outflow as well.

After a period of order a few days after the rise of the hard flux, the disk may settle into a quasi-steady declining state where it can emit copious soft X-rays and the hard power law source, perhaps in the outflow, declines rather more sharply. The exponential decline of the soft X-rays is a critical constraint on the physical nature of the disk. It is completely consistent with the predictions of the disk instability model, demands that a cooling wave be propagating in the disk during the decline, and is consistent only with a tightly-constrained prescription of the disk viscosity. There is only one self-consistent model for disk viscosity in the literature that fulfils these constraints, that based on a magnetic dynamo driven by internal waves.

The common "reflare" phenomenon during the exponential decline may be related to the self-irradiation of the disk. The fact that it occurs in Nova Muscae as the hard component reaches a minimum may be a clue that requires pursuing. It is not at all clear that this feature is related to a modulation of the mass transfer from the companion.

The common "second maximum" is dominated by a hard power law source in Nova Muscae rather than the disk thermal radiation, and presumably also in other similar systems. This phase tends not to be accompanied by QPO's. The power law source in this phase may arise in a quasi-static corona of a more traditional sort that can arise as mass flow rates decline. Although the spectral index is similar, this phase may thus involve a very different physical environment than the first epoch of hard power law emission that is dominated by outflow. The formation of this coronal region in the inner disk may again interdict the inner optically thick disk so that the emission of soft X-rays is inhibited. This could account for the precipitous decline in the disk component of the soft X-rays in Nova Muscae as the hard power law source grows to dominate the "second maximum." The "corona" associated with the "second maximum" may be related to an advective flow, but the current advective models have been fit to the much later, currently observed quiescent phase of AO620-00 and it is not clear that they can naturally account for the transient behavior of the "second maximum." At any phase there are some difficult questions to be addressed as to how the relatively high mass flow rates of the advective solutions can be stably fed by the naturally low mass flow rates in the quiescent Keplerian disk.

The black hole X-ray transients provide a continuing source of stimulating physics problems worthy of Igor Novikov. Their time-dependence and multi-component spectra, while a complication, may yet yield the sort of direct irrefutable evidence that we are dealing with black holes that is one of the holy grails of modern astrophysics. As a practical matter, this would allow us to search for and identify black holes of stellar mass in a way that is not possible when one is restricted to argument by mass function that "it is too massive to be a neutron star." Finally, with the lessons learned in these marvelous laboratories, we will be more strongly armed to attack the stubborn problem of the nature of QSO's and AGN's and their supermassive black holes.

I would like to thank all my colleagues who have taught me so much about black hole systems and accretion disks, especially John Cannizzo, Pranab Ghosh, Min Huang, Shin Mineshige, Soon-Wook Kim, Michael Moscoso, Greg Shields, Rob Robinson, Ethan Vishniac, and Masaaki Kusunose, with whom I have had the pleasure of working at Texas on these issues. This work has supported in part by NASA through grants NAGW-2975 and NAG5-3079.

REFERENCES

ABRAMOWICZ, M. A., CHEN, X., KATO, S., LASOTA, J.-P., & REGEV, O. 1995, *ApJ*, **438**, L37.

ARTEMOVA, I. V., BISNOVATYI-KOGAN, G. S., BJÖRNSSON, G., & NOVIKOV, I. D. 1996, *ApJ*, in press.

AUGUSTEIJN, T., KUULKERS, E., & SHAHAM, J. 1993, *A&A*, **279**, L13.

BAILYN, C. D. *et al.* 1995, *Nature*, **374**, 701.

BAILYN, C. D. & OROSZ, J. A. 1995, *ApJ*, **440**, L73.

BIEHLE, G. T. 1994, *ApJ*, **420**, 364.

BLANDFORD, R. D., AND ZNAJEK, R. L. 1977, *MNRAS*, **179**, 433.

CALLANAN, P. J. *et al.* 1995, *ApJ*, **441**, 779.

CANNIZZO, J.K. 1993a, in *Accretion Disks in Compact Stellar Systems*, ed. J.C. Wheeler (World Scientific Press: Singapore), 6.

CANNIZZO, J.K. 1993b, *ApJ*, **419**, 318.

CANNIZZO, J.K. 1994, *ApJ*, **435**, 389.

CANNIZZO, J.K., CHEN, W., & LIVIO, M. 1995, *ApJ*, **454**, 880.

CANNIZZO, J. K., GHOSH, P., & WHEELER, J. C. 1982, *ApJ*, **260**, L83.

CANNON, R. C. 1993, *MNRAS*, **263**, 817.

CASARES, J. C. & CHARLES, P. A. 1994, *MNRAS*, **271**, L5.

CASARES, J. C., CHARLES, P. A., & NAYLOR, P. A. 1992, *Nature*, **355**, 614.

CHEN, W., LIVIO, M., & GEHRELS, N. 1993, *ApJ*, **408**, L5.

CHEN, W., SHRADER, C. R., & LIVIO, M. 1996, *ApJ*, in press.

CHENG, F. H., HORNE, K., PANAGIA, N., SHRADER, C. R., GILMOZZI, R., PARESCE, F., & LUND, N. 1992, *ApJ*, **396**, 664.

CRARY, D. J., KOUVELIOTOU, C., VAN PARADIJS, J., VAN DER HOOFT, F., SCOTT, D. M., PACIESAS, W. S., VAN DER KLIS, M., FINGER, M. H., HARMON, B. A., & LEWIN, W. H. G. 1996, *ApJ*, in press.

DE KOOL, M., VAN DEN HEUVEL, E. P. J., AND PYLYSER, E. 1987, *A&A*, **183**, 47.

DELLA VALLE, M., BENETTI, S. & WHEELER J. C. 1996, *A&A*, in press.

EBISAWA, K. *et al.* 1994, *PASJ*, **46**, 375

EBISAWA, K. 1996, private communication.

FAULKNER, J., LIN, D. N. C., & PAPALIOZOU, J. 1983, *MNRAS*, **205**, 359.

FILIPPENKO, A. V., MATHESON, T., & HO, L. C. 1995, *ApJ*, **455**, 614.

FILIPPENKO, A. V., MATHESON, T., & BARTH, A. J. 1995, *ApJ*, **455**, L139.

FISHMAN, G. J. *et al.* 1989, in *Gamma Ray Observatory Science Workshop*, Eds. C. R. Shrader, N. Gehrels, & B. Dennis, (CP-3137, NASA: Greenbelt, MD) p239.

FORD, H. C., HARMS, R. J., TSVETANOV, Z. I., HARTIG, G. F., DRESSEL, L. L., KRISS, G. A., BOHLIN, R. C., DAVIDSEN, A. F., MARGON, B, & KOCHHAR, A. K. 1994, *ApJ*, **435**,

L27.

Fukue, J. 1992, *PASJ*, **44**, 663.

Goldwurm, A. *et al.* 1992, *ApJ*, **389**, L79.

Han, X., & Hjellming, R. M. 1992, *ApJ*, **400**, 304.

Harmon, B. A. *et al.* 1994, *Second Compton Symposium*, Eds. C. E. Fichtel, N. Gehrels & J. P. Norris (AIP: New York), 210.

Harmon, B. A. *et al.* 1995, *Nature*, **374**, 703.

Haswell, C. A. *et al.* 1993, *ApJ*, **411**, 802.

Howell, S. B., Szkody, P., & Cannizzo *ApJ*, **389**, 337.

Huang, M., & Wheeler, J. C. 1989, *ApJ*, **343**, 229.

Hua, X.-M. & Lingenfelter, R. E. 1993, *ApJ*, **416**, L17.

Hjellming, R. M. 1996, private communication.

Hjellming, R. M. & Rupen, M. P. 1995, *Nature*, **375**, 464.

Ichikawa, S., Mineshige, S., & Kato, T. 1994, *ApJ*, **435**, 748.

Kim, S.-W., Wheeler, J. C., & Mineshige, S. 1996a, *ApJ*, submitted.

Kim, S.-W., Wheeler, J. C., & Mineshige, S. 1996b, *ApJ*, submitted.

Kouveliotou, C., Finger, M. H., Fishman, G. J., Meegan, C. A., Wilson, R. B., Paciesas, W. S., Minamitani, T., & van Paradijs, J 1993 in *Compton Gamma Ray Observatory*, Eds. M. Friedlander, N. Gehrels, & D. J. Macomb (AIP: New York), 319.

Kusunose, M., & Mineshige, S. 1995, *ApJ*, **440**, 100.

Lasota, J.-P. 1995 in *Compact Stars in Binaries; Proceedings of IAU Symposium 165*, Eds. E. P. J. van den Heuvel and J. van Paradijs (Kluwer: Dordrecht), in press.

Lasota, J.-P., Narayan, R., & Yi, I. 1996, submitted to *A&A*.

Liang, E. P. 1990, *A&A*, **227**, 447.

Ling, J. C., Mahoney, W. A., Wheaton, W. A., & Jacobson, A. S. 1987, *ApJ*, **321**, L117.

Marsh, T. R., Robinson, E. L., & Wood, J. H. 1994, *MNRAS*, **266**, 137.

Martin, E. L., Casares, J., Charles, P. A., & Rebolo, R. 1995, *A&A*, **303**, 785.

Martin, E. L., Rebolo, R., Casares, J. & Charles, P. A. 1992, *Nature*, **358**, 129.

Martin, E. L., Rebolo, R., Casares, J. & Charles, P. A. 1994, *ApJ*, **435**, 791.

Martin, E. L., Spruit, H. C., & van Paradijs, J. 1994, *A&A*, **291**, L43.

McClintock, J. E., Horne, K., & Remillard, R. A. 1995, *ApJ*, **442**, 358.

McClintock, J. E., Petro, L. D., Remillard, R. A., & Ricker, G. R. 1983, *ApJ*, **266**, L27.

McClintock, J. E., & Remillard, R. A. 1986, *ApJ*, **308**, 110.

Meyer, F. & Meyer-Hofmeister, E. 1981, *A&A*, **104**, L10.

Mineshige, S. 1994, *ApJ*, **431**, L99.

Mineshige S., Kim S.-W., & Wheeler, J. C. 1990, *ApJ*, **358**, L5.

Mineshige, S., & Osaki, Y. 1983, *PASJ*, **35**, 377.

Mineshige, S., Tuchman, Y., & Wheeler, J. C. 1990, *ApJ*, **359**, 176.

Mineshige, S., & Wheeler, J.C. 1989, *ApJ*, **343**, 241.

Mineshige, S., Yamasaki, T., & Ishizaka, C. 1993, *PASJ*, **45**, 707.

Mirabel, I. F., & Rodriguez, L. F. 1994, *Nature*, **371**, 46.

Miyamoto, S. *et al.* 1993, *ApJ*, **403**, L39.

Miyoshi, M., Moran, J., Herrnstein, J., Greenhill, L., Nakai, N., Diamond, P., & Inoue, M. 1995, *Nature*, **373**, 127.

Moscoso, M., & Wheeler, J. C. 1994, in *Interacting Binary Stars*, Astronomical Society of the Pacific Conference Series Vol. 56, ed. A. W. Shafter (San Francisco: Astronomical Society of the Pacific), 100.

NARAYAN, R., & YI, I. 1994, *ApJ*, **428**, L13.

NARAYAN, R., MCCLINTOCK, J. E., & YI, I. 1996, *ApJ*, **457**, 821.

PODSIADLOWSKI, PH. 1991, *Nature*, **350**, 136.

PODSIADLOWSKI, PH. 1996, private communication.

PODSIADLOWSKI, PH., CANNON, R. C., & REES, M. J. 1995, *MNRAS*, **274**, 485.

REMILLARD, R. A., MCCLINTOCK, J. E., & BAILYN, C. D. 1992, *ApJ*, **399**, L145.

REMILLARD, R. A., OROSZ, J. A., MCCLINTOCK, J. E., & BAILYN, C. D. 1996, *ApJ*, **459**, 226.

SANWAL, D., ROBINSON, E. L., ZHANG, E., COLOMÉ, C., HARVEY, P. M., RAMSEYER, T. F., HELLIER, C., & WOOD, J. H. 1996, *ApJ*, **460**, 437.

SHAHBAZ, T., RINGWALD, F. A, BUNN, J. C., NAYLOR, T., CHARLES, P. A., & CASARES, J. 1994, *MNRAS*, **271**, L10.

SHAKURA, N.I., & SUNYAEV, R.A. 1973, *A&A*, **24**, 337.

SHIELDS, G. A. & WHEELER, J. C. 1973, *Nature*, 259, 642.

SKIBO, J. G., DERMER, C. D., & RAMATY, R. 1994, *ApJ*, 431, L39.

SMAK, J. 1982, *Acta Astr*, **32**, 199.

SMAK, J. 1983, *ApJ*, **272**, 234.

SMAK, J. 1984, *PASP*, **96**, 1.

SUNYAEV, R. A. *et al.* 1992, *ApJ*, **389**, L75.

TANAKA, Y. 1990, in *the 23rd ESLAB Symposium on Two Topics in X-ray Astronomy*, Ed. N. E. White (ESA SP−296), p2.

TANAKA, Y., & LEWIN, W. H. G. 1995, in *X-Ray Binaries*, ed. W. H. G. Lewin *et al.* (Cambridge University Press: Cambridge), 126.

THORNE, K. S. & ŻYTKOW, A. N. 1977, *ApJ*, **212**, 832.

TINGAY, S. J. *et al.* 1995, *Nature*, **374**, 141.

TUCHMAN, Y., MINESHIGE, S., & WHEELER, J. C. 1990, *ApJ*, **359**, 164.

VAN DER KLIS, M. 1994, *A&A*, **283**, 469.

VAN DER KLIS, M. 1996, private communication.

VAN PARADIJS, J. 1996, *ApJ*, in press.

VAN PARADIJS, J. & MCCLINTOCK, J. E. 1994, *A&A*, **290**, 133.

VAN PARADIJS, J., & MCCLINTOCK, J. E. 1995, in *X-Ray Binaries*, ed. W. H. G. Lewin *et al.* (Cambridge University Press: Cambridge), 126.

VISHNIAC, E.T., & BRANDENBURG, A. 1996, *ApJ*, in press.

VISHNIAC, E.T., JIN, L., & DIAMOND, P.H. 1990, *ApJ*, **365**, 552.

VISHNIAC, E.T., & DIAMOND, P.H. 1992, *ApJ*, **398**, 561.

VISHNIAC, E.T., & DIAMOND, P.H. 1993, in *Accretion Disks in Compact Stellar Systems*, ed. J.C. Wheeler (World Scientific Press: Singapore), 41.

VISHNIAC, E. T. & WHEELER, J. C. 1996, *ApJ*, in press.

X-rays and gamma rays from active galactic nuclei

By ROLAND SVENSSON

Stockholm Observatory, S-133 36 Saltsjöbaden, Stockholm, Sweden

Various types of active galactic nuclei (AGN) are briefly discussed, with an emphasis on the theory of recent X-ray and γ-ray observations of the subclass, Seyfert 1 galaxies. The large radiation power from AGN is thought to originate from gravitational power released by matter accreting onto a supermassive black hole. The physical mechanisms responsible for the energy release and the geometry of the gaseous components are still uncertain in spite of three decades of observational and theoretical studies. The recent X-ray and γ-ray observations, however, start to provide useful constraints on the models. This kind of interpretation of the observations is possible due to theoretical developments during the last decade of radiative transfer of X-rays in both optically thick and thin media of various geometries. Particular attention is paid to various accretion disk-corona models. Recent work on exact radiative transfer in such geometries is reviewed.

1. Theoretical history

The underlying framework for almost all efforts to understand active galactic nuclei (AGN) is the accretion disk picture described in the classical papers by Novikov & Thorne (1973) and Shakura & Sunyaev (1973). This original picture was partly inspired by the extreme optical AGN luminosities. Here, effectively optically thick, rather cold matter rotates in a geometrically thin, differentially rotating Keplerian disk. The differential motion causes viscous dissipation of gravitational binding energy resulting in outward transportation of angular momentum and inward transportation of matter. The dissipated energy diffuses vertically and emerges as black body radiation, mostly in the optical-UV spectral range for the case of AGN.

Observations have also been the driving force in the discovery of two other solution branches.

a) Hard X-rays from the galactic black hole candidate, Cyg X-1, as well as the discovery of strong X-ray emission from most AGN led to the need for a hot accretion disk solution. Shapiro, Lightman & Eardley (1976) (SLE) found a hot, effectively optically thin, rather geometrically thin solution branch, where the ions and the electrons are in energy balance, with the ions being heated by dissipation and cooled through Coulomb exchange, leading to ion temperatures of order 10^{11} - 10^{12} K. The efficient cooling of electrons through a variety of mechanisms above 10^9 K, leads to electron temperatures being locked around 10^9 - 10^{10} K. The SLE-solution is thermally unstable.

b) Observations of Cyg X-1 and of radio galaxies led to two independent discoveries of the third solution branch. Ichimaru (1977) developed a model to explain the soft high and hard low state of Cyg X-1. The soft high state is due to the disk being in the optically thick cold state, and the hard low state occurs when the disk develops into a very hot, optically thin state. This solution branch is similar to the SLE-solution, except that now the ions are not in local energy balance. It was found that if the ions were sufficiently hot they would not cool on an inflow time scale, but rather the ions would heat up both by adiabatic compression and viscous dissipation and would carry most of that energy with them into the black hole. Only a small fraction would be transferred to the electrons, so the efficiency of the accretion is much less than the normal $\sim 10\%$. As the ion temperature was found to be close to virial, these disks are geometrically thick. Just as for the SLE-case above, the electrons decouple to be locked at 10^9 - 10^{10} K. The flows resemble the quasi-spherical dissipative flows studied by, e.g., Mészáros (1975) and Maraschi, Roasio & Treves (1982). Independently, Rees *et al.* (1982) and Phinney (1983) proposed that a similar inefficient accretion disk solution is responsible for the low nuclear luminosities in radio galaxies with large radio lobes and thus quite massive black holes. These geometrically thick disks were named *ion tori*. Rees *et al.* (1982) specified more clearly than Ichimaru (1977) the critical accretion rate above which the flow is dense enough for the ions to cool on an inflow time scale and the ion tori-branch would not exist. Further considerations of ion tori were made by Begelman, Sikora & Rees (1987). Ichimaru (1977) emphasized that his version of ion tori was thermally stable.

The ion-tori branch has been extensively studied and applied over the last two years (1995-1996) with more than 30 papers by Narayan and co-workers, Abramowicz and co-workers, as well as many others. These studies confirm and extend the original results, although some papers do not quote or recognize the original results. New terminology has been introduced based on an Eulerian viewpoint rather than a Lagrangian. Instead of the ions not cooling, it is said that the local volume is "advectively cooled" due to the ions carrying away their energy. The *ion tori* are therefore renamed as *advection-dominated disks*.

The three accretion disk branches were "unified" by Chen *et al.* (1996) and Björnsson *et al.* (1996), who determined where in parameter space the branches merge. Parallel to these developments there have been several other lines of research regarding the physics needed for realistic modelling of accretion flows. These areas include radiative processes in hot plasmas, radiation transfer in hot plasmas, radiation transfer of X-rays in cold plasma, MHD in differentially rotating gas, the origin of disk viscosity, magnetic flares, gas or MHD simulations of flows, and so on. Some of these research lines have already merged. Here we focus on the areas most important for interpreting the high energy spectra of AGN.

While the classical rates for bremsstrahlung, cyclo/synchrotron radiation, and Compton scattering in the nonrelativistic and relativistic limits were sufficient in the early disk models, the electron temperatures of 10^9 - 10^{10} K indicated by both observations and theory required the calculation of transrelativistic rates of the above processes as well as pair processes that become important at these temperatures. Rate calculations as well as studies of the properties of pair and energy balance in hot plasma clouds were done in the 1980s by Lightman, Svensson, Zdziarski and others. Methods to incorporate these sometimes complex hot plasma solutions into the simplest accretion disk models were developed by Björnsson & Svensson (1991, 1992). The Compton scattering kernel probably received its definite treatment in Nagirner & Poutanen (1994).

To obtain approximate spectra, the Kompaneets equation with relativistic corrections and with a simple escape probability replacing the radiative transfer is sufficient (Lightman & Zdziarski 1987). However, if one want to obtain constraints on the geometry from detailed observed spectra, then methods to obtain exact radiative transfer/Comptonization solutions in different accretion geometries must be developed. Here again, Igor Novikov had great influence as he and his group stimulated a particle physicist and expert on Monte Carlo simulations to develop a Monte Carlo code able to treat all the high energy radiative processes occurring in AGN. The first results were reported in Kardashev, Novikov & Stern (1986) and Novikov & Stern (1986), and an advanced version of the code was finally documented in Stern *et al.* (1995a). Faster methods were developed by Haardt (1993) (approximate treatment of the Compton scattering) and

Poutanen & Svensson (1996) (exact treatment) who solved the radiative transfer for each scattering order separately (the iterative scattering method). These codes are fast enough to be implemented in XSPEC, the standard X-ray spectral fitting package, and exactly computed model spectra can now be used when interpreting the observations. So far these methods have been used to study radiative transfer in two-phase media consisting of cold and hot gas with simple geometries, but they have yet to be integrated in the standard accretion disk models.

UV and X-ray observations indicate the coexistence if both cold and hot matter in AGN. Thus, one must be able to compute the reprocessing (both absorption, reflection, and transmission) of X-rays by the cold matter. The first approximate considerations of Compton reflection by Guilbert & Rees (1988) and White, Lightman, & Zdziarski (1988) have now led to quite accurate methods developed by Magdziarz & Zdziarski (1995) and Poutanen, Nagendra & Svensson (1996). Such Compton reflection forms an important ingredient in the radiative transfer calculations in two-phase media.

2. Observational history

The exploration of the spectra of AGN started some 30 years ago. It is mainly due to the developments in instrumental and satellite technology that we now in the 1990s are getting a full broad band picture of these spectra. In 1960s groundbased observations started exploring the optical and radio properties. The advent of X-ray astronomy allowed the 2-10 keV properties to be explored in the 1970s. The theoretical modelling then consisted of connecting the radio, optical, and the X-ray data points with one or two power laws and discussing the resulting fits within nonthermal models where relativistic power law electrons generate both power law synchrotron radiation and power law Compton-scattered radiation. The infrared spectral gap was covered in the 1980s. Only after the launch in April 1991 of the *Compton Gamma Ray Observatory* (*CGRO*) covering the gamma ray spectral range, did we get a full broad band view of AGN spectra and thus reliable estimates of their total luminosities and the spectral ranges where most of the luminosity was emitted. As described in Dermer & Gehrels (1995), AGN fell into two distinct classes, the γ-loud and the γ-weak AGN. The γ-loud AGN are the luminous blazars where we are looking down into a relativistic jet emerging from the AGN. Here, the power emerges in two broad band humps, one due to nonthermal synchrotron emission peaking in the IR-optical spectral range, the second due to Compton scattering peaking in the MeV-GeV spectral range. The γ-weak AGN, on the other hand, are the less luminous Seyfert and radio galaxies (being spiral and ellipticals, respectively), where the spectra extend from radio frequencies to the soft γ-rays (100s of keV). These spectra have approximately equal power per logarithmic frequency interval, but show many spectral features, in particular, three broad band spectral peaks at infrared, UV, and hard X-ray frequencies. The fast time variability of the UV to soft γ-ray spectra indicates that this emission is "nuclear" radiation originating in the accretion flow very close to the black hole. The slowly varying broad band infrared spectral peak originates at larger distances, most likely being due to nuclear radiation reprocessed by dust. The nuclear spectra have been studied for more than 20 years, yet little is known about the physical conditions close to the super massive black holes in AGNs.

Both Seyfert (normally being radio-weak) and radio galaxies come in two types based on optical classification. Seyfert 1 and the broad line radio galaxies show both Doppler-broadened broad (up to 10 000 km/s) and narrow (up to perhaps 1000 km/s) emission lines, while the Seyfert 2 galaxies and the narrow line radio galaxies show only the narrow emission lines. The broad lines originate in fast-moving photo-ionized 10^4 K gas clouds within 1 parsec of the central source, while the narrow lines originate in slower moving clouds at kiloparsec distances. Aspects of radio galaxies are discussed in the chapter by Malcolm Longair in this volume. Here, we will focus on the Seyfert galaxies only.

In the unified scheme for Seyfert galaxies (e.g. Antonucci 1993), we are directly viewing the central X-ray source and broad line region in Seyfert 1 galaxies, but our line-of-sight passes through an optically opaque molecular torus in Seyfert 2 galaxies obscuring both the central source and the broad line region but not the narrow line region. The photo-ionized narrow line gas lies in an ionization cone whose half-opening angle is determined by the geometry of the obscuring torus. In some Seyfert 2 galaxies we still see weak polarized broad lines being reflected by electrons in the ionization cone. In standard accretion disk scenarios for the central X-ray source, the unified models imply that our viewing angle (i.e. the angle between the disk normal and the line-of-sight) is less than the half-opening angle of the ionization cone for the case of Seyfert 1s. Half-opening angles inferred from observations are typically 30-40 degrees, which means that in Seyfert 1s we are viewing the accretion disk from directions that are closer to face-on than to edge-on.

3. X-ray and γ-ray spectra of Seyfert galaxies

The satellites (launch year and energy range in parenthesis) *ROSAT* (June 1990; 0.1-2 keV), *ASCA* (Feb 1993; 0.1-10 keV), *Ginga* (Feb 1987 - Nov 1991; 2-40 keV) , *SAX* (April 1996; 0.2-200 keV), *XTE* (Dec 1995; 2-200 keV), and *CGRO* (April 1991; 50 keV-30 GeV) have allowed broad band X/γ-ray studies of some 30 Seyfert galaxies in the 1990s. However, as AGNs vary in the X-rays on time-scales down to days or even hours simultaneous measurements are necessary. Such simultaneous observations unfortunately exist for only a handful objects, but the number is increasing. The launch of *INTEGRAL* (2 keV-10 MeV) in the year 2001 will finally provide a large number of truly simultaneous X/γ-spectra of AGN.

By combining observations in different energy bands, a picture has emerged where the overall shapes of the X-ray and γ-ray spectra of different Seyfert 1 galaxies are similar. The only Seyfert 1 galaxy for which the broad band X- and γ-ray spectrum is well determined is IC 4329A (Madejski *et al.* 1995) being the second brightest Seyfert galaxy in the 2-10 keV range. Simultaneous *ROSAT* and *CGRO* OSSE, as well as nonsimultaneous *Ginga* data are shown in Figure 1 (*left panel*). The observations were modelled by Madejski *et al.* (1995), Zdziarski *et al.* (1994), and Magdziarz & Zdziarski (1995). The spectrum consists of two components: 1) an intrinsic power law of slope $\alpha_{\rm int} \approx 0.9$ with an exponential cutoff energy of about 300 keV (dashed curve in Fig. 1), and 2) a reflection component caused by cold reflecting matter subtending a solid angle $\sim 1 - 2\pi$ as viewed from the X-ray source (dotted curve in Fig. 1). The reflection component consists of a fluorescent Fe line at 6.4 keV and a broad peak at about 30 keV causing the overall 2-18 keV spectral slope to be considerably flatter, $\alpha_{2-18} \approx 0.7$ and the apparent cutoff energy to be smaller, ≈ 50 keV. A small column depth of neutral material about $N_{\rm H} = 3 \times 10^{21}$ cm^{-2} along the line of sight causes absorption below 2 keV.

The average X/γ-spectrum for four Seyfert 1s using nonsimultaneous *Ginga* and OSSE data shows similar properties as the spectrum for IC 4329A (Zdziarski *et al.* 1995). The same holds for the nonsimultaneous average *EXOSAT*/OSSE spectrum for seven Seyfert 1s (Gondek *et al.* 1996).

Larger samples exist within narrower spectral ranges using a single experiment. These samples show spectra consistent with the smaller samples above. For example, 60 *Ginga* spectra of 27 Seyferts show $\alpha_{\rm int,2-18} \approx 0.95$ and the overall $\alpha_{2-18} \approx 0.73$, each having a dispersion of about 0.15 (Nandra & Pounds 1994).

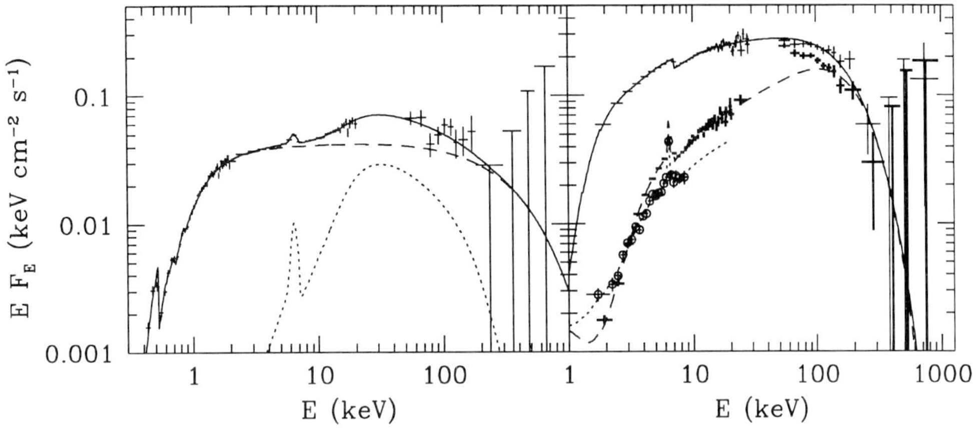

FIGURE 1. *Left* panel shows the observed broad band spectrum (*crosses*) of IC 4329A from *ROSAT*, *Ginga*, and *CGRO* OSSE (Madejski *et al.* 1995). The *dashed curve* shows the intrinsic model spectrum incident on both the cold matter and the observer, the *dotted curve* is the reflected component, and the *solid curve* is the observed sum. From Magdziarz & Zdziarski (1995). *Right* panel shows the observed broad band spectrum of NGC 4151 from *Ginga*, and *CGRO* OSSE (*crosses*) and from *EXOSAT* (*circled crosses*). The *EXOSAT* data are from April 1984, the highest and lowest *Ginga* data from May 1990 and May 1987, respectively, and the highest and lowest *CGRO* OSSE data from May and April 1993, respectively. *Solid*, *dashed*, and *dotted curves* show fits to the data using a simple thermal Comptonization model. From Zdziarski, Johnson & Magdziarz (1996).

The brightest Seyfert galaxy, the Seyfert 1.5 NGC 4151, once considered a prototype for Seyfert 1s, has now turned out to have unusual X-ray properties. Figure 1 (*right panel*) from Zdziarski, Johnson & Magdziarz (1996) shows the extreme spectral states of NGC 4151. In the 2-18 keV *Ginga*-range, the spectral index, α_{2-18}, varies in the range 0.3-0.8 instead of having the canonical Seyfert value of 0.7. The harder the spectrum is, the weaker is the 2-18 keV flux, and the spectrum seems to pivot around 100 keV. And indeed, OSSE observations during 5 viewing periods from 1991-1994 show that the OSSE flux is essentially constant. NGC 4151 shows no or a smaller reflection component than normal Seyfert 1s. This causes the deduced intrinsic power law component to have a smaller cut off energy than the standard Seyfert 1, although the total OSSE spectra are indistinguishable. NGC 4151 also has a larger column depth $N_H \sim 10^{23}$ cm^{-2} of absorbing gas along the line of sight compared to typical Seyfert 1s. This causes absorption below about 6 keV. Optical observations of an ionization cone indicate that we are not viewing NGC 4151 face-on but rather at a viewing angle of 65 degrees. NGC 4151 is clearly a freak object which requires special consideration to fit into the unified scheme of Seyfert galaxies. The standard interpretation is that we have a direct view of the central X-ray source, but it has also been proposed (Poutanen *et al.* 1996) that the central X-ray source is obscured and that the observed X-ray spectrum is a standard Seyfert 1 spectrum scattered by about 65 degrees into the line of sight. Compton recoil then decreases the spectral cut off energy.

The observed obscuring column depths in Seyfert 2 galaxies lie in the range $N_H \sim 2 \times 10^{22} - 5 \times 10^{23}$ cm^{-2} (Smith & Done 1996). An extreme case is NGC 4945 (Done, Madejski & Smith 1996) whose broad band X-ray spectrum is shown in Figure 2. Here, the absorbing column is very large, $N_H \sim 4 \times 10^{24}$ cm^{-2}, causing absorption of the direct spectrum below 20 keV. Below 10 keV the spectrum is dominated by a weak component

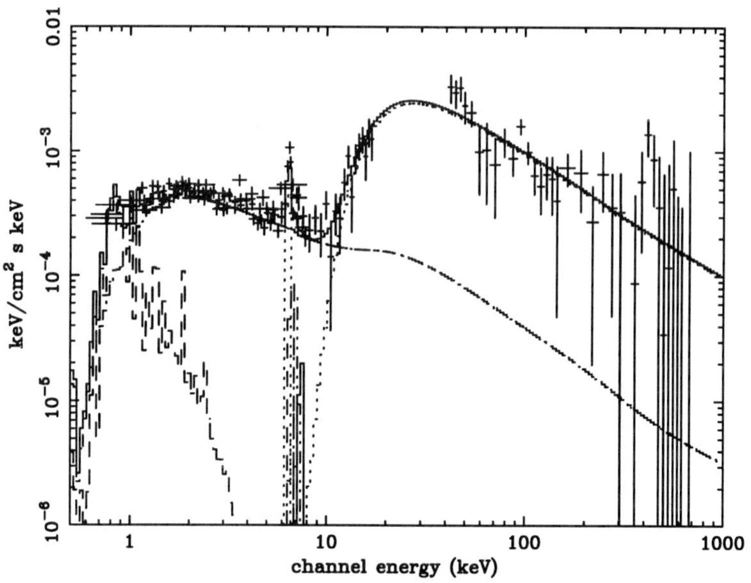

FIGURE 2. The broad band spectrum (*crosses*) of NGC 4945 from *ASCA*, *Ginga*, and *CGRO* OSSE. The *dotted curve* is the direct power law spectrum being absorbed below 20 keV by a column depth $N_{\rm H} \sim 4 \times 10^{24}$ cm^{-2}. The *dash-dotted curve* shows the component scattered into the line-of-sight by electrons in the ionization cone. The effects of a reflection component are included. From Done, Madejski & Smith (1996).

that has been scattered into the line-of-sight by electrons in the ionization cone. Although weak in the 2-10 keV range, NGC 4945 is the second brightest known Seyfert galaxy in the sky above 50 keV. In the Seyfert 2 galaxy, NGC 1068, the absorbing column is even larger and the scattered component dominates at least up to 20 keV (Koyama *et al.* 1989).

4. Outstanding questions regarding the X-ray emitting regions in Seyfert 1 galaxies

The obvious question is, of course, which radiation process generates the X-ray continuum. It is commonly believed that energetic electrons — either mildly relativistic thermal electrons or highly relativistic nonthermal electrons — Compton scatter soft UV-photons into the X-ray range.

Then there is the question of geometry, i.e. the spatial distribution of hot X-ray generating electrons and of cold ($T < 10^6$ K) reflecting matter. Possible geometries include a cold slab surrounded either by plane parallel coronal slabs of hot electrons, or by coronal patches (active regions) of unknown geometries.

Finally, there is the question whether it is thermal or nonthermal electrons or both that account for the X-ray emission. This question might be answered if the spectrum above a few hundred keV was known with certainty, but the best available spectrum of the brightest typical Seyfert 1, IC 4329A, does not have sufficient signal to noise at such energies. The spectrum of IC 4329A can be fit with both thermal and nonthermal models (Zdziarski *et al.* 1994).

The nonthermal models studied in the 1980s have several attractive features (for review, see Svensson 1994). In particular, it was predicted already in 1985 (Svensson 1986) that the pair cascades in the nonthermal models give rise to an X-ray spectrum with $\alpha_{\rm int,2-18} \approx 0.9 - 1.0$ in contradiction to the overall slope of $\alpha_{2-18} \approx 0.7$ but in agreement with the intrinsic spectral spectral slope later found by *Ginga* (e.g. Nandra & Pounds 1994).

Due to the non-detections by COMPTEL and the high energy cutoffs indicated by OSSE, attention is now focused on the thermal models.

5. Geometry of the X-ray source

In principle, the central X-ray source could have spherical symmetry with the source of UV-photons being small cold clouds uniformly distributed throughout the X-ray source. In this case, radiation models invoking spherical symmetry would be applicable (e.g., Ghisellini & Haardt 1994, Skibo *et al.* 1995; Pietrini & Krolik 1995). In some models, the central X-ray source is quasi-spherical but the UV-source is anisotropic. Here, the X-ray source can either be a hot two-temperature SLE-accretion disk or a two-temperature ion torus. The UV-source may then be the standard cold Shakura-Sunyaev disk located beyond some transition radius. Another proposed alternative is that the disk is hot at larger radii but that in the innermost region the cooling times become sufficiently short for the disk to collapse into a cold UV-emitting disk state. The resulting anisotropy of the UV-photons breaks the spherical symmetry in both of these scenarios. Then detailed radiative transfer calculations of how the UV-photons penetrate the hot disk are required in order to determine the emerging X-ray spectrum of Comptonized UV-photons at different viewing angles. This problem has not yet been solved.

The most commonly used scenario (mainly because of its simplicity) is the two-phase disk-corona model (e.g., Haardt & Maraschi 1991, 1993) in which a hot X-ray emitting corona is located above the cold UV-emitting disk of the canonical black hole model for AGNs. The power law X-ray spectrum is generated by thermal Comptonization of the soft UV-radiation. About half of the X-rays enter and are reprocessed by the cold disk, emerging mostly as black body disk radiation in the UV. Haardt & Maraschi (1991) emphasized the coupling between the two phases due to the reprocessing, as the soft disk photons influence the cooling of the corona. They showed that nearly all power must be dissipated in the corona in order to have $\alpha_{\rm int} \sim 0.9 - 1$. A consequence of this is that the soft disk luminosity, L_s, is of the same order as the hard X-ray luminosity, L_h. The disk-corona scenario is highly anisotropic as the UV-photons enter the corona from below only.

Observations show that L_s may be several times larger than L_h, in contradiction to the prediction of the uniform two-phase disk-corona model. This led Haardt, Maraschi, & Ghisellini (1994) to propose a patchy disk-corona model, where the corona consists of several localized active regions on the surface of the disk. Internal disk dissipation results in UV-radiation that leaves the disk without entering the active regions, thus enhancing the observed L_s relative to L_h. The patchy corona model is also highly anisotropic as the UV-radiation enters the active regions from below and through the sides. The active regions could also be elevated above the disk.

The full radiative transfer and Comptonization problem in these geometries has recently been solved (Haardt & Maraschi 1993, Stern *et al.* 1995b, Poutanen & Svensson 1996). We now turn to discuss the methods and the results.

6. Radiative transfer/Comptonization

Two different methods have been used to solve the full radiative transfer/Comptonization problem in mildly relativistic thermal plasmas accounting for energy and pair balance as well as reprocessing by the cold disk (including angular anisotropy and Klein-Nishina effects).

The first method is based on the Non-Linear Monte Carlo (NLMC) method developed by Stern (1985, 1988) and described in detail in Stern *et al.* (1995a). The Monte Carlo particles (particles and photons) make up the background so the Monte Carlo particles can interact with themselves (e.g. photon-photon pair production) causing nonlinearity. Any geometry can be treated including 2D and 3D geometries, but calculations so far (Stern et al 1995b) have been limited to coronal slabs (1D) or active regions (2D) in the shape of hemispheres or spheres at different elevations above the cold disc. Another advantage is the possibility to divide the region into several zones in order to study the inhomogeneous distributions of, e.g., temperature and pair density. The drawback is that each run takes a few hours on a Sun IPX.

The second method is based on the iterative scattering method (ISM), where the radiative transfer is exactly solved for each scattering order separately (e.g. Sunyaev & Titarchuk 1985, Haardt 1994). The code and its testing are briefly described in Poutanen and Svensson (1996, see also Poutanen 1994ab; Poutanen & Vilhu 1993). It was shown that the results of the NLMC and the ISM code are in excellent agreement. The ISM code is a 1D code but it can also treat quasi-2D radiative transfer in cylinders/pill boxes and hemispheres. The full Compton scattering matrix is used allowing to solve polarized radiative transfer in thermal relativistic plasmas. Fully angular dependent, polarized Compton reflection is implemented using a Green's matrix (Poutanen, Nagendra & Svensson 1996). The advantage of the ISM code is that it takes 10 minutes or less on a Sun IPX.

7. The Comptonization solution

The two codes have been used to study radiative transfer/Comptonization in pure pair coronae in energy and pair balance. For coronae of a given geometry and in energy balance, there exists a unique $T_e - \tau_T$ relation, where T_e is the volume-averaged coronal temperature and τ_T is a characteristic Thomson scattering optical depth of the coronal region. For slabs, τ_T is the vertical Thomson scattering optical depth. For hemispheres and for spheres, the radial τ_T is averaged over 2π solid angle for hemispheres, and 4π solid angle for spheres.

In Figure 3a, this relation is shown for different geometries. The results for slabs are shown by *rectangles*, for hemispheres located on the disk surface by *hemispheres*, for surface spheres located on the disk surface by *underlined spheres*, and for spheres located at heights, $0.5h$, $1h$, and $2h$, by *spheres*, *diamonds*, and *triangles*, respectively (h is the radius of the sphere). The results for active regions are connected by *dotted* curves. For comparison we also show the slab results from Stern *et al.* (1995b) using the ISM code (*dashed curve*). Each curve is characterized by an almost constant generalized Kompaneets parameter, $y \equiv \tau_T(1+\tau_T)(4\Theta + 16\Theta^2)$, where $\Theta \equiv kT_e/m_ec^2$, and where $y \sim 0.49$ for slabs, ~ 2.0 for hemispheres, and ~ 2.3, 3.7, 5.9, and 7.8 for the spheres. The larger the soft photon starvation (i.e. the fewer the number of reprocessed soft photons reentering the coronal region), the larger is y. There is very good agreement between the slab results from the NLMC and the ISM codes, which tests the accuracy of both codes and methods.

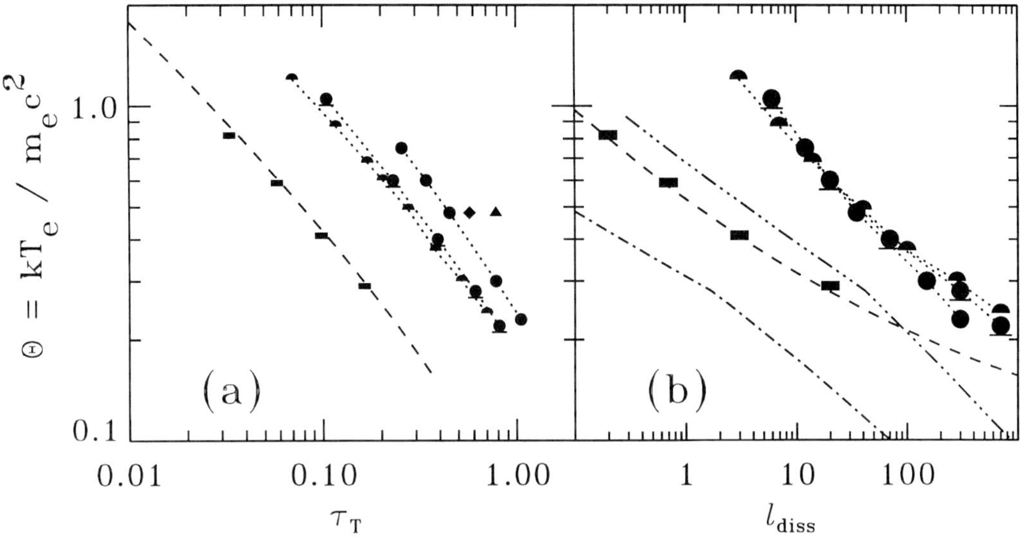

FIGURE 3. Dimensionless volume-averaged temperature, $\Theta \equiv kT_e/m_ec^2$, vs. Thomson scattering optical depth, τ_T, in panel (a), and vs. dissipation compactness, $\ell_{\rm diss} \equiv (L_{\rm diss}/h)\ (\sigma_T/m_ec^3)$ in panel (b), for a steady X-ray emitting plasma region in pair and energy balance on or above a cold disk surface of temperature, $kT_{\rm bb} = 5$ eV. The plasma Compton scatters reprocessed soft black body photons from the cold disk surface. *Solid rectangles* and *dashed curve* show results from NLMC code and ISM code, respectively, for the case of a plane-parallel slab corona. Results using the NLMC code for individual active pair regions are shown for *hemispheres* located on the disk surface; surface spheres also located on the surface (*underlined spheres*); spheres located at a height of $0.5h$ (*spheres*), $1h$ (*diamond*), and $2h$ (*triangle*), where h is the radius of the sphere. The results for each type of active region are connected by *dotted* curves. The *dash-dotted* and *dash-dot-dot-dotted* curves in panel (b) show the critical compactness as function of Θ above which thermalization by Moller and Bhaba scattering is not achieved for the cases of pair slabs and surface spheres, respectively. From Stern *et al.* (1997).

8. Pair balance and the compactness

Solving the pair balance for the combinations of (Θ, τ_T) obtained in § 7 gives a unique dissipation compactness, $\ell_{\rm diss}$ (see Ghisellini & Haardt 1994 for a discussion). Here, the local dissipation compactness, $\ell_{\rm diss} \equiv (L_{\rm diss}/h)(\sigma_T/m_ec^3)$, characterizes the dissipation with $L_{\rm diss}$ being the power providing uniform heating in a cubic volume of size h in the case of a slab of height h, or in the whole volume in the case of an active region of size h. The Figure 3b shows the Θ vs. $\ell_{\rm diss}$ relation obtained with the NLMC code for different geometries and for $kT_{\rm bb} = 5$ eV. The slab results with the ISM codes are also shown by the dashed curve. The parameter space to the right of respective curve is forbidden as pair balance cannot be achieved, and the parameter space to the left would contain solutions where the background coronal plasma dominates over the pairs, i.e. "pair free" solutions (e.g. Svensson 1984, Haardt & Maraschi 1993). Note that the pair temperatures are locked in the range $\Theta \sim 0.15 - 1$ (80 – 500 keV) for the considered compactnesses. The pairs thus act as a "thermostat" (e.g., Svensson 1984)

From Figure 3b we find that at a given Θ the active regions have a larger $\ell_{\rm diss}$ than the slabs. This is due to the longer escape times in slabs. In a local cubic volume in slabs, the four vertical surfaces act effectively as "reflecting" surfaces increasing the radiation energy density in the volume as compared to the active regions where the radiation escapes freely through the surface of the volume.

9. Thermalization

The question arises whether the electrons can thermalize or not for the conditions, Θ, τ_T, and ℓ_{diss}, in Figure 3. Energy exchange and thermalization through Moller ($e^{\pm}e^{\pm}$) and Bhaba (e^+e^-) scattering compete with various loss mechanisms, with Compton losses being the most important for our conditions. The thermalization is slowest and the Compton losses largest for the higher energy particles in the Maxwellian tail. Ghisellini, Haardt & Fabian (1993) compared average Maxwellian time scales to find the conditions when thermalization is not achieved. Instead we use the detailed simulations by Dermer & Liang (1989) (their Fig. 8) to find the critical compactness above which the deviation at the Maxwellian mean energy is more than a factor $e \approx 2.7$. The *dash-dotted* and *dash-dot-dot-dotted* curves in Figure 3b show this critical compactness for slabs and for surface spheres, respectively. In agreement with Ghisellini, Haardt & Fabian (1993), we find that Moller and Bhaba scattering cannot compete with Compton losses in our pair slabs and active regions. On the other hand, Ghisellini, Guilbert & Svensson (1988) and Ghisellini & Svensson (1990) found that cyclo/synchrotron self-absorption acts as a very efficient thermalizing mechanism as long as the magnetic energy density dominates the radiation energy density, a condition expected in the corona and in the magnetic flares on the surface of accretion disks.

10. Anisotropy effects

Figure 4 shows the the "face-on" emerging spectrum for a hemisphere with $\ell_{diss} = 20$ and $kT_{bb} = 5$ eV with the reflected component subtracted. The *solid curve* is the sum of the reprocessed soft black body spectrum and the Comptonized spectrum. The "face-on" spectrum is averaged over viewing angles $0.6 < \cos\theta < 1$. The numbered *dotted curves* show spectral profiles of the different scattering orders. As noted by Haardt (1993) and Haardt & Maraschi (1993), the first scattering order is significantly reduced in face-on directions due to anisotropic Compton scattering when Θ is mildly relativistic. This deficiency at low energies causes an *anisotropy break* close to the peak of the second scattering order. Below the anisotropy break, the spectrum is a power law with an *anisotropy slope*, α_{aniso}, that is harder than the standard *isotropy slope*, α_{iso}, above the break. It is the standard isotropy slope, α_{iso}, that has been fitted with several analytical expressions that are functions of Θ and τ_T (e.g., Zdziarski 1985, Titarchuk 1994). "Edge-on" spectra have a much weaker reflection component and no anisotropy break.

11. Spectra from active pair regions: hemispheres

Figure 5 shows emerging "face-one" and "edge-on" spectra from hemispheres at different ℓ_{diss} for $kT_{bb} = 5$ eV. We have $\alpha_{aniso} \approx 0.77$ while $\alpha_{iso} \approx 1.09$. The anisotropy break is therefore about 0.3. The anisotropy break moves through the 2-18 keV range for $\ell_{diss} \approx$ 10-100, and thus α_{2-18} is smaller than unity for ℓ_{diss} less than 100. At this low kT_{bb} the first Compton order hardly extends into the 2-18 keV range and there is little discreteness effect on the behavior of α_{2-18}.

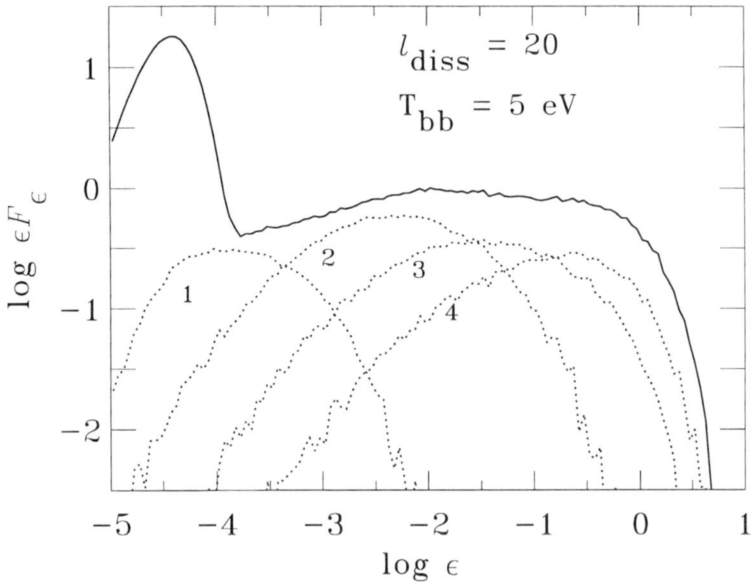

FIGURE 4. Emerging spectrum, ϵF_ϵ, where F_ϵ is the energy flux (arbitrary units) and $\epsilon \equiv h\nu/m_e c^2$ from hemisphere with compactness, $\ell_{\rm diss} = 20$, and cold disk temperature, $kT_{\rm bb} = 5$ eV. The spectrum is averaged over viewing angles $0.6 < \cos\theta < 1$. The *solid curve* shows the sum of the Comptonized spectrum from the hemisphere and the reprocessed black body spectrum (the reflection component is not included). The numbered *dotted curves* show the different scattering orders. Note that the first scattering order is significantly reduced due to anisotropic Compton scattering. This deficiency causes an *anisotropy break* close to the peak of the second scattering order. Below the anisotropy break, the spectrum is a power law with an *anisotropy slope* that is harder than the standard *isotropy slope* above the break. From Stern *et al.* (1997).

12. Diagnostics Using Compactness and *Ginga* Slopes

The least square overall spectral slope, α_{2-18}, for the 2-18 keV range was determined and is displayed in Figure 6 as a function of the dissipation compactness, $\ell_{\rm diss}$, for different geometries. The *right panel* of Figure 6 shows the observed distribution of α_{2-18} for *Ginga* spectra from 27 Seyfert galaxies (Nandra & Pounds 1994). The *crosses* represent 17 Seyfert galaxies that have both been observed by *Ginga* and have known estimates of their X-ray time variability (and thus lower limits of their compactnesses). The true crosses may lie to the right of the plotted ones.

One sees that the observations are more consistent with active surface regions (such as hemispheres or surface spheres) than with slabs. Active regions produce spectra covering the *observed ranges* of α_{2-18} ($\approx 0.4 - 0.9$) and cutoff energies ($\sim 2kT_e$) for the *observed range* of compactnesses.

13. Conclusions

Have we been able to throw some light upon the outstanding questions discussed in § 4?

Regarding the radiative transfer/Comptonization, the progress has been rapid and it is now possible to use full radiative transfer simulations when fitting X-ray observations, e.g. with XSPEC, at least for the simplest geometries (e.g., Poutanen *et al.* 1996 for the case of NGC 4151). As anisotropic effects are very important, i.e. as spectral shapes

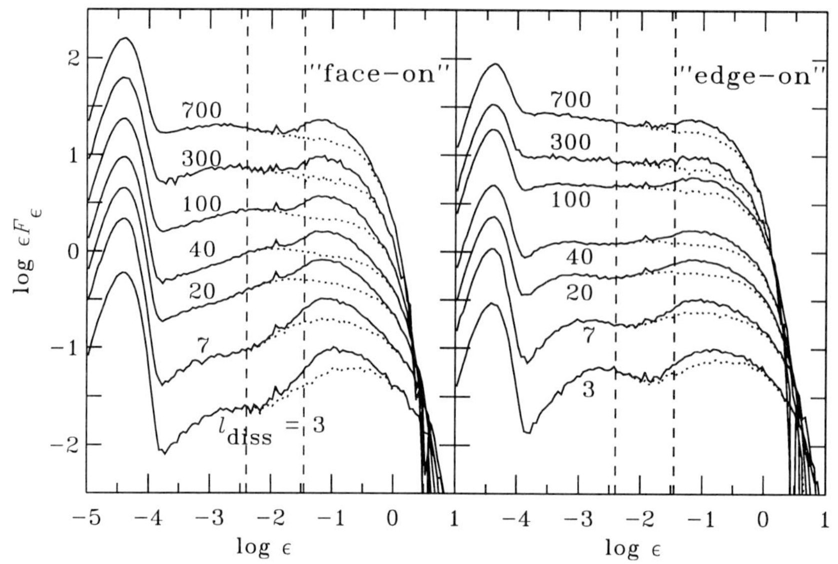

FIGURE 5. Emerging "face-on" and "edge-on" spectra, ϵF_ϵ, where F_ϵ is the energy flux (arbitrary units) and $\epsilon \equiv h\nu/m_e c^2$, from hemispheres of different compactnesses, $\ell_{\rm diss}$ and for cold disk temperature, $kT_{\rm bb} = 5$ eV. The "face-on" spectra are averaged over viewing angles $0.6 < \cos\theta < 1$, and the "edge-on" spectra over viewing angles $0 < \cos\theta < 0.5$. The *solid curves* show the total spectrum which is the sum of the Comptonized spectrum from the hemisphere itself (*dotted curves*), the reprocessed black body spectrum, and the reflection component. Vertical *dashed lines* show the 2-18 keV spectral range. The anisotropy break in the face-on spectra moves to lower photon energies as $\ell_{\rm diss}$ increases. From Stern et al. (1995b, 1997).

depend strongly on viewing angle, it will be possible to set constraints on the viewing angle. The ISM code also gives detailed predictions for the spectral X/γ-ray polarization, which may provide crucial constraints once observations become available.

The spectra also depend on the geometry of the coronal regions, so observed spectra can be used as diagnostics of the geometry. Presently, it seems that active regions are favoured over homogeneous slab coronae.

There is still no answer to the question whether the radiating electrons (and positrons) are thermal or nonthermal. High quality spectra above 200 keV are needed, and the question may not be settled before the launch of *INTEGRAL* in the year 2001.

The author acknowledges support from the Swedish Natural Science Research Council. He thanks Juri Poutanen, Marek Sikora, and Boris Stern for numerous valuable discussions. Boris Stern's research visits at the Stockholm Observatory were financed by the Swedish Institute, the Swedish Royal Academy of Sciences, and by a NORDITA Nordic project grant.

REFERENCES

ANTONUCCI, R. 1993 Unified models for active galactic nuclei and quasars. *Ann. Rev. Astr. Astroph.* **31**, 473–521.

BEGELMAN, M. C., SIKORA, M. & REES, M. J. 1987 Thermal and dynamical effects of pair production on two temperature accretion flows. *Astrophys. J.* **313**, 689–698.

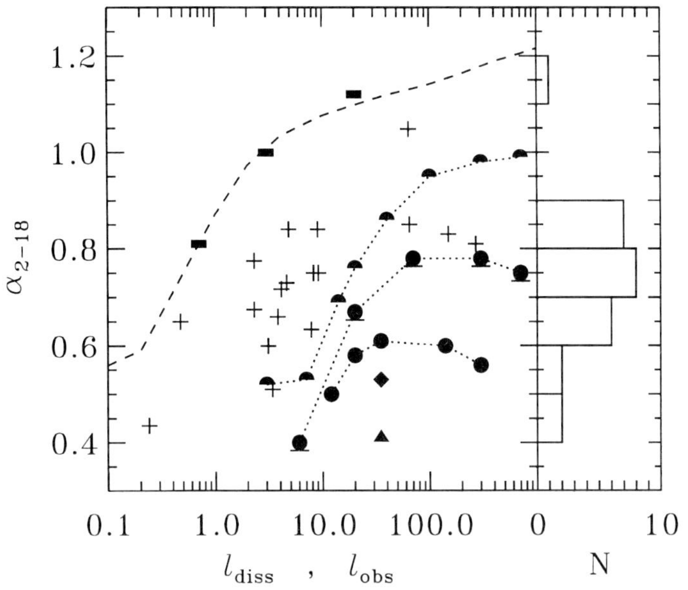

FIGURE 6. Overall spectral intensity index, α_{2-18}, least square fitted to the model spectra in the 2-18 keV range vs. the dissipation compactness, $\ell_{\rm diss}$. Same notation as in Fig. 3. The spectra from the NLMC code were averaged over viewing angles $0.6 < \cos\theta < 1.0$ before determining α_{2-18} with least square fitting. For the ISM code, face-on spectra (at $\cos\theta = 0.887$) were used. The right panel shows the observed distribution of α_{2-18} for *Ginga* spectra from 27 Seyfert galaxies (Nandra & Pounds 1994). The crosses represent 17 Seyfert galaxies that have both been observed by *Ginga* and have known estimates of their X-ray time variability, Δt, and thus of the observed compactness, $\ell_{\rm obs} \equiv (L_{\rm obs}/c\Delta t)(\sigma_{\rm T}/m_e c^3)$. From Stern *et al.* (1995b, 1997).

BJÖRNSSON, G. & SVENSSON, R. 1991 A recipe for making hot accretion disks. *Astrophys. J. Letters* **371**, L69–L72.

BJÖRNSSON, G. & SVENSSON, R. 1992 Hot pair dominated accretion disks. *Astrophys. J.* , **394**, 500–514.

BJÖRNSSON, G., CHEN, X., ABRAMOWICZ, M. A., & LASOTA, J. P. 1996 Hot accretion disks revisited. *Astroph. J.* , in press.

CHEN, X., ABRAMOWICZ, M. A., LASOTA, J. P., NARAYAN, R. & YI, I. 1996 Unified description of accretion flow around black holes. *Astroph. J. Letters* **443**, L61–L64.

DERMER, C. D. & GEHRELS, N. 1995 Two classes of gamma-ray emitting active galactic nuclei. *Astrophys. J.* **447**, 103–120.

DERMER, C. D. & LIANG, E. P. 1989 Electron thermalization and heating in relativistic plasmas. *Astrophys. J.* **339**, 512–528.

DONE, C., MADEJSKI, G. M. & SMITH, D. A. 1996 NGC 4945: The brightest Seyfert 2 galaxy at 100 keV. *Astrophys. J. Letters* **463**, L63–L66.

GHISELLINI, G. & HAARDT, F. 1994 On thermal Comptonization in $e^{\pm}$ pair plasmas. *Astroph. J. Letters* / **429**, L53–L56.

GHISELLINI, G. & SVENSSON, R. 1990 Synchrotron self-absorption as a thermalizing mechanism. in *Physical Processes in Hot Cosmic Plasmas* (eds. W. Brinkmann, A. C. Fabian & F. Giovannelli) pp. 395–400. Kluwer.

GHISELLINI, G., HAARDT, F. & FABIAN, A. C. 1993 On reacceleration, pairs and the high energy spectrum of AGN and Galactic black hole candidates. *Monthly Not. Roy. Astr. Soc.* **263**, L9–L12.

GHISELLINI, G., GUILBERT, P. W. & SVENSSON, R. 1988 The synchrotron boiler. *Astrophysical Journal Letters* **335**, L5–L8.

GONDEK, D., *et al.* 1996 The average X-ray/gamma-ray spectrum of radio-quiet Seyfert 1s. *Monthly Not. Roy. Astr. Soc.*, in press.

GUILBERT, P. W. & REES, M. J. 1988 'Cold' material in non-thermal sources. *Monthly Not. Roy. Astr. Soc.* **233**, 475–484.

HAARDT, F. 1993 Anisotropic Comptonization in thermal plasmas: spectral distribution in plane parallel geometry. *Astrophys. J.* **413**, 680–693.

HAARDT, F. 1994 High Energy Processes in Seyfert Galaxies. PhD dissertation, SISSA, Trieste.

HAARDT, F. & MARASCHI, L. 1991 A two-phase model for the X-ray emission from Seyfert galaxies. *Astrophys. J. Letters* **380**, L51–L54.

HAARDT, F. & MARASCHI, L. 1993 X-ray spectra from two-phase accretion disks. *Astrophys. J.* **413**, 507–517.

HAARDT, F., MARASCHI, L. & GHISELLINI, G. 1994 A model for the X-ray and UV emission from Seyfert galaxies and galactic black holes. *Astrophys. J. Letters* **432**, L95–L99.

ICHIMARU, S. 1977 Bimodal behavior of accretion disks: theory and application to Cyg X-1 transitions. *Astrophys. J.* **214**, 840–855.

KARDASHEV, N. S., NOVIKOV, I. D. & STERN, B. E. 1986 An electron-positron cauldron and the formation of the hard radiation of quasars and AGN. In *IAU Symp. 119: Quasars* (ed. G. Swarup & V. K. Kapahi) pp. 383–393. D. Reidel.

KOYAMA, K. *et al.* 1989 An intense iron line in NGC 1068. *Publ. Astron. Soc. Japan* **41**, 731–737.

LIGHTMAN, A. P. & ZDZIARSKI, A. A. 1987 Pair production and Compton scattering in compact sources and comparison to observations of active galactic nuclei. *Astrophys. J.* **319**, 643–661.

MADEJSKI, G. M. *et al.* 1995 Joint *ROSAT-COMPTON GRO* observations of the X-ray-bright Seyfert galaxy IC 4329A. *Astrophys. J.* **438**, 672–679.

MAGDZIARZ, P. & ZDZIARSKI, A. A. 1995 Angle-dependent Compton reflection of X-rays and gamma-rays. *Monthly Not. Roy. Astron. Soc.* **273**, 837–848.

MARASCHI, L., ROASIO, R. & TREVES, A. 1982 The effect of multiple Compton scattering on the temperature and emission spectra of accreting black holes. *Astrophys. J.* **253**, 312–317.

MÉSZÁROZ, P. 1975 Radiation from spherical accretion onto black holes. *Astr. Astroph.* **44**, 59–68.

NAGIRNER, D. J. & POUTANEN, J. 1993 Compton scattering of polarized light: scattering matrix for isotropic electron gas. *Astr. Astroph.* **275**, 325–336.

NAGIRNER, D. J. & POUTANEN, J. 1994 Single Compton scattering. *Astrophys. Space Phys.* **9**, 1–47.

NANDRA, K. & POUNDS, K. A. 1994 *Ginga* observations of the X-ray spectra of Seyfert galaxies. *Monthly Not. Roy. Astron. Soc.* **268**, 405–429.

NOVIKOV, I. D. & STERN, B. E. 1986 A possible mechanism of the formation of the hard spectrum of active galactic nuclei. In *Structure and Evolution of Active Galactic Nuclei* (ed. G. Gluricin *et al.*) pp. 149–171. D. Reidel.

NOVIKOV, I. D. & THORNE, K. S. 1973 Astrophysics of black holes. In *Black Holes, Les Houches* (ed. C. De Witt & B. DeWitt) pp. 343–450. Gordon & Breach.

PHINNEY, E. S. 1983 A theory of radio sources. PhD dissertation, University of Cambridge.

PIETRINI, P. & KROLIK, J. H. 1995 The inverse Compton thermostat in hot plasmas near accreting black holes. *Astrophys. J.* **447**, 526–544.

POUTANEN, J. 1994a Compton scattering matrix for relativistic Maxwellian electron distribution. *J. Quant. Spectrosc. Rad. Transfer* **51**, 813–822.

POUTANEN, J. 1994b Compton scattering of polarized light in active galactic nuclei and X-ray binaries. PhD dissertation, University of Helsinki.

POUTANEN, J. & SVENSSON, R. 1996 The two-phase pair corona model for active galactic nuclei and X-ray binaries: How to obtain exact solutions. *Astrophys. J.* in press.

POUTANEN, J. & VILHU, O. 1993 Compton scattering of polarized light in two-phase accretion discs. *Astr. Astroph.* **275**, 337–344.

POUTANEN, J., NAGENDRA, K. N. & SVENSSON, R. 1996 Green's matrix for Compton reflection of polarized radiation from cold matter. *Monthly Not. Roy. Astron. Soc.* , in press.

POUTANEN, J., SIKORA, M., BEGELMAN, M. C. & MAGDZIARZ, P. 1996 Compton mirror in NGC 4151. *Astrophys. J. Letters* **465**, L107–??.

REES, M. J., BEGELMAN, M. C., BLANDFORD, R. D. & PHINNEY, E. S. 1982 Ion-supported tori and the origin of radio jets. *Nature* **295**, 17–21.

SHAKURA, N. I. & SUNYAEV, R. A. 1973 Black holes in binary systems. Observational appearance. *Astr. Astroph.* **24**, 337–355.

SHAPIRO, S. L., LIGHTMAN, A. P. & EARDLEY D. N. 1976 A two-temperature accretion disk model for Cygnus X-1: structure and spectrum. *Astrophys. J.* **204**, 187–199.

SKIBO, J. G., DERMER, C. D., RAMATY, R. & MCKINLEY, J. M. 1995 Thermal Comptonization in mildly relativistic pair plasmas. *Astrophys. J.* **446**, 86–100.

SMITH, D. A. & DONE, C. 1996 Unified theories of AGN: A hard X-ray sample of Seyfert 2 galaxies. *Monthly Not. Roy. Astron. Soc.* , in press.

STERN, B. E. 1985 On the possibility of efficient production of electron-positron pairs near pulsars and accreting black holes. *Sov.Astr.* **29**, 306–313.

STERN, B. E. 1988 Nonthermal pair production in active galactic nuclei: A detailed radiation transfer model. *Nordita/88-51 A*, preprint.

STERN, B. E., BEGELMAN, M. C., SIKORA, M. & SVENSSON, R. 1995a A large-particle Monte Carlo code for simulating non-linear high-energy processes near compact objects. *Monthly Not. Roy. Astron. Soc.* **272**, 291–307.

STERN, B. E., POUTANEN, J., SVENSSON, R., SIKORA, M. & BEGELMAN M. C. 1995b On the geometry of the X-ray emitting region in Seyfert galaxies. *Astrophys. J. Letters* **449**, L13–L17.

STERN, B. E. *et al.* 1997, in preparation.

SUNYAEV, R. A. & TITARCHUK, L. G. 1985 Comptonization of low-frequency radiation in accretion disks: angular distribution and polarized hard radiation. *Astr. Astroph.* **143**, 374–388.

SVENSSON, R. 1984 Steady mildly relativistic thermal plasmas: processes and properties. *Monthly Not. Roy. Astron. Soc.* **209**, 175–208.

SVENSSON, R. 1986 Physical processes in active galactic nuclei. In *IAU Coll. 89: Radiation Hydrodynamics in Stars and Compact Objects* (ed. D. Mihalas & K-H. Winkler), pp. 325–345. Springer.

SVENSSON, R. 1994 The nonthermal pair model for the X-ray and gamma-ray spectra from active galactic nuclei. *Astrophys. J. Suppl.* **92**, 585–592.

TITARCHUK L. 1994 Generalized Comptonization models and applications to recent high-energy observations. *Astrophys. J.* **434**, 570–586.

WHITE, T. R., LIGHTMAN, A. P., ZDZIARSKI, A. A. 1988 Compton reflection of gamma rays by cold electrons. *Astrophys. J.* **331**, 939–948.

ZDZIARSKI, A. A. 1985 Power-law X-ray and gamma-ray emission from relativistic thermal plasmas. *Astrophys. J.* **289**, 514–525.

ZDZIARSKI, A. A. *et al.* 1994 Physical processes in the X-ray/gamma-ray source of IC 4329A. *Monthly Not. Roy. Astron. Soc.* **269**, L55-L60.

ZDZIARSKI, A. A., JOHNSON, W. N. & MAGDZIARZ, P. 1996 Broad-band gamma-ray and X-ray spectra of NGC 4151 and their implications for physical processes and geometry. *Monthly Not. Roy. Astron. Soc.* , in press.

ZDZIARSKI, A. A., JOHNSON, W. N., DONE, C., SMITH, D. & MCNARON-BROWN, K. 1995 The average X-ray/gamma-ray spectra of Seyfert galaxies from *Ginga* and OSSE and the origin of the comic X-ray background. *Astrophys. J. Letters* **438**, L63–L66.

Gamma-Ray Bursts: A Challenge to Relativistic Astrophysics

By MARTIN J. REES

Institute of Astronomy, Madingley Road, Cambridge, CB3 0HA, UK

Although they were discovered more than 25 years ago, gamma-ray bursts are still a mystery. All that we can be confident about is that they involve compact objects and relativistic plasma. Current ideas and prospects are briefly reviewed.

1. History

Astrophysics is a subject where the observers generally lead, and theorists follow behind. The topic of my talk is one where the lag is embarrassingly large. In some respects, therefore, the subject may be inappropriate at a meeting where we are honouring a distinguished theorist. However, gamma-ray bursts raise issues which are certainly fascinating to everyone involved in relativistic astrophysics.

Even though the history of gamma-ray bursts dates back more than 25 years, we still know neither where nor what they are. The story starts in the late 1960s, when American scientists at Los Alamos had developed a set of satellites aimed at detecting clandestine nuclear tests in space by the associated gamma-ray emission. Gamma-ray flashes, lasting a few seconds, were indeed detected. It took several years before these were realised to be natural, rather than sinister phenomena, and in 1973 a paper was published by Klebesadal, Strong & Olson entitled *Observations of Gamma-ray Bursts of Cosmic Origin.* In this classic paper they reported 16 short bursts of photons in the energy range between 0.2 and 1.5 MeV, which had been observed during a three-year period using widely separated spacecraft. The burst durations ranged from less than 0.1 second up to about 30 seconds, but significant fine time-structure was observed within the longer bursts. It was already clear that these came neither from the Earth nor from the Sun, but little else was clear at that time. It did not take long for the theorists to become enthusiastically engaged. At the Texas conference in December 1974, Ruderman (1975), gave a review of models and theories. He presented an exotic menu of alternatives that had already appeared in the literature, involving supernovae, neutron stars, flare stars, antimatter effects, relativistic dust, white holes, and some even more bizarre options. He noted also the tendency, still often apparent, for theorists to strive strenuously to fit new phenomena into their chosen specialities.

During the 1970s and 1980s, data accumulated on gamma-ray bursts, due to a number of satellites. Particular mention should be made of the contributions by Mazets and his colleagues in Leningrad. Also important were the extended observations made by the Pioneer Venus Orbiter (PVO).

During that period, three classes of models were pursued: those in which the bursts were respectively in the Galactic Disc (at distances of a few hundred parsecs), in the halo (at distances of tens of kiloparsecs), and at cosmological distances. The characteristic energies of each burst, according to these three hypotheses, are respectively 10^{37} ergs, 10^{41} ergs, and 10^{51} ergs. The most popular and widely-discussed option during the 1980s was that the bursts were relatively local, probably in our Galactic Disc, and due to magnetospheric phenomena or "glitches" on old neutron stars, defunct pulsars.

It was clear that there were two statistical clues which could in principle decide the location of gamma-ray bursts as soon as enough data had accumulated, and selection effects were understood. One was the number-versus-intensity of the events, which tells us whether they are uniformly distributed in Euclidean space, or whether we are in some sense seeing the edge of the distribution. The other is the degree of anisotropy.

There was already evidence that the counts of gamma-ray bursts were flatter than the classic Euclidean slope, since otherwise more faint bursts would have been detected by balloon experiments. This would not of course have been unexpected if the bursts were within the galaxy. However, the real surprise came with the launch of the Compton Gamma Ray Observatory (GRO) satellite, whose BATSE experiment provided the first systematic all-sky coverage. It is the data from BATSE which has transformed and extended the subject in recent years. The most remarkable and surprising result was that the bursts seemed to be isotropic over the sky. More than a thousand have now been recorded, and there is no statistical evidence for any anisotropy or repetition. The lack of any enhancement either towards the plane of the Galaxy, or towards the Galactic Centre, is a very severe constraint on the hypothesis that bursts come from the Galaxy. Note that the hypothesis that they are ultra-local objects within our galactic disc, which would naturally permit isotropy, is ruled out by the flatness of the number counts.

The experiments on GRO have produced evidence on the spectra and time structure of events. (For a recent review, see Fishman (1995) and references cited therein.) Despite the large variety, there is little doubt that gamma-ray bursts are a well-defined class of objects, distinguished spectrally from phenomena such as X-ray bursters, and also from the so-called 'soft gamma repeaters' which have substantially softer spectra.

The isotropy evidence has tilted the balance of opinion strongly towards a cosmological interpretation of the classical gamma-ray bursts. In this paper, I will discuss the challenge posed to theorists by that model. I will then discuss, more briefly, types of halo model that are compatible with the isotropy since these cannot be definitively excluded yet. In conclusion, I will list some observations which might in the near future settle the issue, or at least release the current level of perplexity.

2. Models for "Cosmological" Bursts

If the bursts are cosmological, then the sub-Euclidean counts imply that the typical burst has a redshift z of order 1. The precise redshift distribution depends on how much evolution there is in the population. The mean redshift would be less, for instance if the burst rate increased with cosmic time. However, we can confidently say that all but the very nearest of the observed bursts must have redshifts of at least 0.2. Otherwise evolution would need to be implausibly steep to explain the non-Euclidean counts, and nearby superclusters would show up in the distribution over the sky.

The event rate per unit volume is very low if we are sampling a population out to cosmological distances. It is of order 10^{-5} per year per galaxy, in other words a thousand times less than the supernova rate in galaxies. The energy release that is required amounts to 10^{51} ergs in a few seconds. (Both the estimates of the rates and of the energy per event would need to be modified in a straightforward way, of course, if the individual events were beamed in a small solid angle.)

The amount of energy released is not in itself a problem. After all, whenever a supernova goes off, the binding energy of a neutron star is released in a fraction of a second, and this amounts to 10^{53} ergs, a hundred times what is needed for the burst. But in a supernova most of this energy goes to waste as neutrinos; if there were any impulsive electromagnetic release it would not escape promptly, but would be degraded by adiabatic expansion of the envelope before, much later, it could leak out. So is it possible for some rare events to occur where the energy release can escape promptly, rather than being surrounded by an extensive opaque envelope? One possibility is coalescence of binary neutron stars (see, for example, Narayan, Paczynski and Piran 1992).

Systems such as the famous binary pulsar will eventually coalesce, when gravitational radiation drives them together. The final merger, leading probably to the production of a black hole, happens in a fraction of a second. The calculated event rates for such phenomena – and perhaps also for the coalescence of binaries consisting of a neutron star and a black hole, rather than two neutron stars – are uncertain but are probably high enough to supply the requisite rates of bursts.

How can the energy be transformed into gamma rays after such a coalescence event? There seem to be two options. The first is that some of the energy released as neutrinos is reconverted, when the neutrinos collide outside the dense core where they were produced, into electron positron pairs or photons. The second option is that strong magnetic fields are involved, and directly convert the rotational energy of the system into a directed outflow. This latter option requires that the magnetic fields be amplified to strengths of order 10^{15} Gauss.

The observed gamma rays seem to have a nonthermal spectrum. Moreover, they sometimes extend to energies above 1 MeV, the pair production threshold in the rest frame. These facts together imply that the emitting region must be relativistically expanding. We draw this conclusion firstly because, if the region were indeed only a light second across or less, as would be required to display the observed rapid variability in the absence of relativistic effects, the total mass of baryons in the region would need to be below about 10^{21} grams in order that the electrons associated with the baryons should not provide a large opacity: the rest mass energy of the baryons would need to be 10 orders of magnitude less than that of the radiation energy in the same volume. Not only is this a remarkably low figure, implying that only 10^{-12} of the material from the compact objects is mixed up in the emitting region, but it would in any case imply a relativistic expansion. Moreover, quite apart from the baryon constraint, a larger source volume is required in order to avoid opacity due to photon-photon collisions (via $\gamma + \gamma \rightarrow e^+ + e^-$), and this also indicates relativistic expansion.

If the emitting region is expanding relativistically, then for a given observed variation timescale the dimension R can be increased by γ^2. The opacity to electrons and pairs is then reduced by γ^4, and the threshold for pair production, in our frame, goes up by $\sim \gamma$ from its "rest" value of $\sim$ 1 MeV. A high γ will of course only be attained if the baryon loading is sufficiently low, such that the ratio of total energy to rest mass energy is larger than γ. A variety of models have been discussed, best-guess numbers are, for an energy of 10^{51} ergs, a Lorentz factor γ in the range 10^2 to 10^3, allowing the rapidly variable emission to occur at radii in the range 10^{14} to 10^{16} cms. The entrained baryonic mass would need to be below $10^{-6} M_\odot$ to allow these high relativistic expansion speeds.

Efficient conversion of kinetic energy into gamma rays then requires relativistic shocks. These could involve either impact on an external medium, which would randomise half of the initial energy merely by reducing the expansion Lorentz factor by a factor of 2, or else internal shocks within the outflow. If, for instance, the Lorentz factor in an outflowing wind varied by a factor more than 2, then the shocks that developed when fast material overtakes slower material would be internally relativistic (Rees & Mészáros 1994).

In the case of expansion into an external medium, the energy would be rethermalised after sweeping up external matter with rest mass $E/c^2\gamma^2$ (Rees & Mészáros 1992; Mészáros & Rees 1993). For $E = 10^{51}$ ergs and $\gamma = 10^3$, only $10^{-9} M_\odot$ of external matter need be swept up. In an unsteady wind, if γ were to vary by a factor of 2 on a timescale δt, internal shocks would develop at a distance $\gamma^2 c\delta t$, and randomise most of the energy. For instance, if γ ranged between 500 and 2000, on a timescale of 1 second, efficient dissipation would occur at 3×10^{16} cms.

Another important consequence of relativistic outflow is that only material moving within an angle γ^{-1} of the line of sight contributes to what we observe. Observations cannot therefore tell us if bursts are highly beamed. Transverse pressure gradients are only effective on angles below γ^{-1}, so material ejected in widely differing directions behaves quite independently. There are already a variety of models in the literature discussing the radiation from expanding fireballs and relativistic winds (see Piran 1996 for a recent review). The parameters are uncertain, and the relevant physics, involving for instance the coupling between electrons and ions in relativistic shocks, is not sufficiently well developed to allow accurate modelling of the radiation (see, for instance, Gallant *et al.* 1992). There are also various ideas about how the energy is channelled from the central object into the outflowing fireball or wind. Recent calculations by Ruffert *et al.*, 1995, suggest problems with releasing neutrino energy efficiently enough, and on a short enough timescale, to allow production of a fireball.

The options involving *magnetic* energy are rather less quantitative, but I still believe they are more promising. As discussed by Usov, (1994), and Thompson (1994), a millisecond pulsar with a $\sim 10^{15}$ Gauss field would be slowed down in 1 second, its spin energy being dumped in a pair-dominated relativistic wind. As these authors and others have discussed, internal processes in such a wind could explain gamma rays with the observed spectrum and variability characteristics.

My personal favourite model involves the debris from a disrupted neutron star orbiting around a black hole. If this debris contains a strong magnetic field, amplified perhaps by differential rotation, then an axial magnetically-dominated wind may be generated along the rotation axis, perpendicular to the plane of the disc. The advantage of this geometry is that it seems to offer the best chance of preventing baryon contamination, because the baryonic material would be precluded by angular momentum from getting near the axis, because it would first fall into the black hole. Such a configuration could arise from capture of a neutron star by a black of less than $5M_\odot$, this mass limit being required because otherwise the neutron star would be swallowed before disruption. Alternatively, it could be the outcome of the merger of two neutron stars, where most of the mass collapses to a black hole, leaving some fraction of the original material in orbit around it.

3. 'Extended Halo' Models

In the interests of balance, I would like to make a few remarks about the alternative idea that the bursts are not from cosmological distances, but instead come from within our own galaxy. Classical gamma-ray bursts could be isotropic enough to be consistent with the BATSE data if they came from neutron stars ejected from our Galactic Disc at more than 700 km s^{-1}, and remained active, bursting sporadically, for long enough to allow them to reach distances of at least 100 kiloparsecs. They may either escape from the galaxy, or be on very extended bound orbits. These high velocity objects could be a special subset of pulsars. If they form in our Galactic Disc at a rate of about 1 per thousand years, then each must produce 10^6 bursts, of typical energy 10^{41} ergs. Repetition would not necessarily be expected, since each neutron star could in principle continue bursting at a slow rate for more than a billion years. However, if the bursts came in groups, rather than being independent poissonian events, repetition would not be impossible.

Podsiadlowski has done detailed calculations of whether such a population can provide an isotropic distribution. He shows this is indeed possible for long-lived bursters whose orbits take them out beyond 100 kiloparsecs. An important feature of such orbits is that, because the galactic halo potential is not spherical, (and may indeed be rather irregular at such large distances) objects do not conserve their angular momentum and therefore, even if they started off near the centre of our Galaxy, they need not return so close to the centre in later orbits. This effect helps to ensure greater isotropy. Another possibility is that the typical objects have velocities above a thousand kilometres per second, and are escaping the Galaxy completely. In this case, the best fit is obtained if the bursts do not start until after a delay of around 10^7 years, by which time all neutron stars have reached distances of 30 kiloparsecs or more. I think it is fair to say that such models need to be carefully tuned in order to fit the existing isotropy data, but that, though perhaps unappealing, they cannot be ruled out. The constraints on orbital parameters would be eased in alternative schemes where the neutron stars *formed* far out in the halo (perhaps, as Woosley (1993) has discussed, as relics of an early population of halo stars) rather than being ejected from the disc.

If such a model is to be taken seriously, there must be an acceptable mechanism for producing the succession of 10^{41} erg bursts, spread over a very long timescale. Two options have been proposed (Podsiadlowski, Rees & Ruderman 1995).

The first possibility is that these neutron stars start off with a super-strong ($\gtrsim 10^{15}$ Gauss) magnetic field. This field, penetrating the core of the neutron star, would gradually rise towards the surface through buoyancy effects, thereby causing stress in the crust. The timescale for the buoyancy is estimated to be at least 10^6 years. Acceptable models require that it be $\gtrsim 10^9$ years. The total stored energy is $\sim 10^{47}(B/10^{15}G)$ ergs. The energy depends linearly on B, rather than quadratically, because the field in the core is concentrated into tubes where its strength has a standard value of $\sim 3 \times 10^{15}G$.

The crust gets stretched as the field drifts outwards. The units in which energy is released depend on how much stress can build up in the crust, and what fraction is released when the crust cracks. This is a complicated problem in asteroseismology. However, a release of 10^{41} ergs per event is plausible, in which case the total stored magnetic energy would be sufficient to supply the requisite 10^6 events.

The second very different option for triggering halo bursts involves asteroidal impacts on to a neutron star. Each event requires, on energetic grounds, the impact of 10^{21} grams. The main problem with this idea is that such asteroidal or cometary bodies would be tidally disrupted too far out to give a sudden enough event. A possible solution is that the debris from the disrupted body squashes down the magnetic field, which then rebounds, generating high electric fields and thereby a pair cascade. Alternatively, the debris may form a disc which accumulates before triggering a sudden electromagnetic release when it couples its rotation to that of the neutron star.

The total impacting mass, to get enough bursts per star, must be 10^{27} grams. It is not impossible (especially now we know that planetary systems can exist around pulsars) that a neutron star could carry with it $\gtrsim 10^{27}$ gm of asteroidal debris. However, a larger reservoir, plus at least one large planet, is needed in order for enough of these planetesimals to be perturbed on to near radial orbits. We know that at least one pulsar has a planetary system. This fact, plus the evidence that even typical pulsars may have velocities of 400 km s^{-1}, suggests that models of this kind should not be dismissed. Whatever the bursts turn out to be, the primary trigger, and the efficient conversion of its energy into gamma rays, involve physical conditions that are extreme and unfamiliar.

4. How can we settle the debate?

There is no convincing and fully worked out model for the bursts on either the halo or the cosmological hypothesis. Neither option, however, seems to violate any cherished beliefs in physics or relativistic astrophysics. The issue is one of plausibility, and how one weighs different lines of evidence. The isotropy would be a natural consequence of the cosmological hypothesis. But the level of isotropy so far revealed by BATSE, which restricts any dipole or quadrupole anisotropy below the few per cent level and shows no evidence for clumping on smaller scales, could be accommodated in a halo hypothesis if high speed neutron stars were implicated.

In April 1995 there was an interesting debate on the location of gamma-ray bursts, in which the two main protagonists were Don Lamb and Bohdan Paczynski (a written version of the argument appears in Lamb (1995) and Paczynski (1995)). I had the privilege of acting as the moderator in this debate, perhaps because I was one of the few people who had not already taken a firm stance on one side of the issue or the other. There was an agreement among all participants that the issue would be settled only by more data. Indeed, there was a broad consensus on some particular tests that could be crucial, or at least highly suggestive. Among these might be the following.

The most valuable information, of course, would be a firm identification of a burster with some other class of object. The stumbling black here is the poor positional accuracy of most gamma-ray detectors. BATSE itself has error circles of 1 or 2 degrees for the brightest bursts, and more than 5 degrees for the fainter ones. However, there are some bursts whose locations have been pinned down with a precision of minutes of arc or better by triangulation experiments involving deep space probes; this technique utilises the rapid time structure, which, when recorded and timed by detectors separated by 10 light minutes or more, allows accurate positioning. There is still no firm identification of any classical gamma-ray burst, though there are tantalising indications that some of the brighter bursts may be correlated with galaxies or clusters of galaxies, whose distances are not inconsistent with what is expected on the cosmological hypothesis.

Even though the gamma-ray positional information is poor, one might be able to pin down the position of the sources more accurately if they displayed concurrent emission in some other waveband. Various projects have been undertaken in the optical and radio band. Ground-based observers can be notified of a BATSE event within a few seconds, allowing a telescope that can be rapidly slewed to seek an optical counterpart within less than a minute. No such counterparts have been detected, nor have radio searches yet yielded positional or timing coincidences. The likely strength of gamma-ray bursts in the optical or radio band is uncertain and highly model-dependent. Indeed, any detection in these wavebands would have the bonus that it would help to narrow down the range of possible models and emission mechanisms. However, most theories predict that there should be substantial spectral extension from gamma-rays down towards the X-rays, so it would seem less of a gamble to seek X-ray counterparts. This has been done by a variety of X-ray satellites, including the recently-launched High-Energy Transient Explorer.

If genuine anisotropies were detected, this would tend to support a local rather than cosmological origin. A particularly crucial test would be feasible if bursts more than ten times fainter than those recorded by BATSE could be detected. It would then, according to the halo hypothesis, be feasible to detect bursts from the halo of Andromeda. There should be a definite excess of weak bursts from the direction of Andromeda, compared to other directions. The lack of such an effect would be a very severe problem for halo models. A specific proposal, by the name of Smallex, has been proposed to look at a 10 degree field around Andromeda with 20 times the sensitivity of BATSE.

One possibility that would confirm the cosmological interpretation would be detection of bursts lensed by intervening galaxies. If a suitable galaxy lay along the line of sight to a cosmologically distant burst, radiation would reach us by two or more different paths, whose light travel times would differ typically by weeks or months. We would therefore detect two bursts from the same direction. Even though the positions could not be pinned down accurately, the elaborate time structure of each burst is a distinctive "fingerprint", and if two bursts with identical "fingerprints" were detected from within the same error circle, this would be compelling evidence that they were actually separate gravitationally-lensed images of the same burst. (As a technical point, it should be noted that microlensing by stars or substellar objects would only introduce different time structure on timescales of a few microseconds and therefore would not lead to detectable differences between the profiles of the two images.)

Unfortunately, the probability that a galaxy lies along a random line of sight to a high redshift object is below one per cent, the exact value depending of course on the presumed redshift of the burst. Moreover, because BATSE can only observe a given direction in the sky for about 40 per cent of the time it is more likely than not that, if a lensed event occurred, the recurrence would be missed because it would occur during dead time. Taking these effects into account, it is rather marginal whether we would expect BATSE to detect a single instance of this lensing before it dies, even if the bursts indeed come from cosmological distances. However, if we were lucky, such a double burst could clinch the cosmological interpretation.

A further issue which has figured strongly in the debate on the location of gamma-ray bursts concerns the existence or otherwise of spectral features attributable to cyclotron lines. This is a technical controversy which I will not enter here. However, its relevance lies in the fact that halo models involve neutron stars, where the magnetic fields are expected to be in the range such that cyclotron lines should be in the hard X-ray band. On the other hand, the fields in the emitting regions of cosmological fireballs or relativistic winds would not, even when relativistic effects are taken into account, give rise to such features.

The controversies in the Shapley-Curtis debate were settled within a few years. Our knowledge of extragalactic astronomy thereby made a forward leap, and astronomers moved on to address more detailed issues. I'm enough of an optimist to believe that it will be only a few years before we know where and perhaps even what, the gamma-bursters are. Even if this optimism is misplaced, I am completely sure that these mysterious phenomena will serve as a challenge and stimulus to exciting theoretical work, by Igor Novikov and others, on challenging issues in relativistic astrophysics.

REFERENCES

FISHMAN, G.J. 1995, *PASP* **107**, 1145.

GALLANT, Y.A., HOSHINO, M., LANGDON, A.B., ARONS, J. AND MAX, C.E. 1992, *ApJ* **391**, 73.

KLEBESADAL, R.W., STRONG, I.B. AND OLSEN, R.A. 1973, *ApJ (Lett.)* **182**, L85.

LAMB, D. 1995, *PASP* **107**, 1152.

MÉSZÁROS, P. AND REES, M.J. 1993, *ApJ* **405**, 278.

NARAYAN, R., PACZYNSKI, B. AND PIRAN, T. 1992, *ApJ (Lett.)* **395**, L83.

PIRAN, T. 1996. In *Unsolved Problems in Astrophysics* (ed. J. Bahcall and J.P. Ostriker) Princeton U.P.

PACZYNSKI, B. 1995, *PASP* **107**, 1167.

PODSIADLOWSKI, P., REES, M.J. AND RUDERMAN, M. 1995, *MNRAS* **273**, 755.

REES, M.J. AND MÉSZÁROS, P. 1992, *MNRAS* **258**, 41P.

REES, M.J. AND MÉSZÁROS, P. 1994, *ApJ (Lett.)* **430**, L93.

RUDERMAN, M. 1975, *Ann. N.Y. Acad. Sci.* **262**, 164.

RUFFERT, M., JANKA, H.-T. AND SCHÄFER, G. 1995, *Astron. Astrophys.* (in press)

THOMPSON, C. 1994, *MNRAS* **270**, 480.

USOV, V.V. 1994, *MNRAS* **267**, 1035.

WOOSLEY, S.E. 1993, *ApJ* **405**, 273.

Probing black holes and other exotic objects with gravitational waves

By KIP S. THORNE

Theoretical Astrophysics, California Institute of Technology, Pasadena, CA 91125, USA

Gravitational-Wave observations, using high-frequency Earth-based instruments (LIGO, VIRGO, etc.) and the low-frequency, space-based LISA, have great promise for probing the spacetime geometry around black holes and other exotic objects, and for studying their dynamics.

1. Introduction

There is an enormous difference between gravitational waves, and the electromagnetic waves on which our present knowledge of the Universe is based:

• Electromagnetic waves are oscillations of the electromagnetic field that propagate through spacetime; gravitational waves are oscillations of the "fabric" of spacetime itself.

• Astronomical electromagnetic waves are almost always incoherent superpositions of emission from individual electrons, atoms, or molecules. Cosmic gravitational waves are produced by coherent, bulk motions of huge amounts of mass-energy—either material mass, or the energy of vibrating, nonlinear spacetime curvature.

• Since the wavelengths of electromagnetic waves are small compared to their sources (gas clouds, stellar atmospheres, accretion disks, ...), from the waves we can make pictures of the sources. The wavelengths of cosmic gravitational waves are comparable to or larger than their coherent, bulk-moving sources, so we cannot make pictures from them. Instead, the gravitational waves are like sound; they carry, in two independent waveforms, a stereophonic, symphony-like description of their sources.

• Electromagnetic waves are easily absorbed, scattered, and dispersed by matter. Gravitational waves travel nearly unscathed through all forms and amounts of intervening matter (Thorne 1987, Thorne 1983).

• Astronomical electromagnetic waves have frequencies that begin at $f \sim 10^7$ Hz and extend on *upward* by roughly 20 orders of magnitude. Astronomical gravitational waves should begin at $\sim 10^4$ Hz (1000-fold lower than the lowest-frequency astronomical electromagnetic waves), and should extend on *downward* from there by roughly 20 orders of magnitude.

These enormous differences make it likely that:

• The information brought to us by gravitational waves will be very different from (almost "orthogonal to") that carried by electromagnetic waves; gravitational waves will show us details of the bulk motion of dense concentrations of energy, whereas electromagnetic waves show us the thermodynamic state of optically thin concentrations of matter.

• Most (but not all) gravitational-wave sources that our instruments detect will not be seen electromagnetically, and conversely, most objects observed electromagnetically will never be seen gravitationally. Typical electromagnetic sources are stellar atmospheres, accretion disks, and clouds of interstellar gas—none of which emit significant gravitational waves, while typical gravitational-wave sources are black holes and other exotic objects which emit no electromagnetic wave at all, and the cores of supernovae which are hidden from electromagnetic view by dense layers of surrounding stellar gas.

• Gravitational waves may bring us great surprises. In the past, when a radically new window has been opened onto the Universe, the resulting surprises have had a profound, indeed revolutionary, impact. For example, the radio universe, as discovered in the 1940s, 50s and 60s, turned out to be far more violent than the optical universe; radio waves brought us quasars, pulsars, and the cosmic microwave radiation, and with them our first direct observational evidence for black holes, neutron stars, and the heat of the big bang (Sullivan 1984). It is reasonable to hope that gravitational waves will bring a similar "revolution".

Gravitational-wave detectors and detection techniques have been under vigorous development since about 1960, when Joseph Weber (1960) pioneered the field. These efforts have led to very promising sensitivities in four frequency bands:

• The Extremely Low Frequency Band (ELF), 10^{-15} to 10^{-18} Hz, in which the measured anisotropy of the cosmic microwave background radiation places strong limits on gravitational wave strengths — and may, in fact, have detected waves (Krauss & White 1992, Davis et al. 1992). The only waves expected in this band are relics of the big bang.

• The Very Low Frequency Band (VLF), 10^{-7} to 10^{-9} Hz, in which Joseph Taylor and others have achieved remarkable gravity-wave sensitivities by the timing of millisecond pulsars (Kaspi, Taylor& Ryba 1994). The only expected strong sources in this band are processes in the very early universe — the big bang, phase transitions of the vacuum states of quantum fields, and cosmic strings (vibrating, linear cracks in the geometry of space) (Zel'dovich 1980, Vilenkin 1981).

• The Low-Frequency Band (LF), 10^{-4} to 1 Hz, in which will operate the Laser Interferometer Space Antenna, LISA; see Sec. 3 below. This is the band of massive black holes ($M \sim 1000$ to $10^8 M_\odot$) in the distant universe, and of other hypothetical massive exotic objects (naked singularities, soliton stars), as well as of binary stars (ordinary, white dwarf, neutron star, and black hole) in our galaxy. Early universe processes should also have produced waves at these frequencies, as in the ELF and VLF bands.

• The High-Frequency Band (HF), 1 to 10^4Hz, in which operate earth-based gravitational-wave detectors such as LIGO; see Sec. 2 below. This is the band of stellar-mass black holes ($M \sim 1$ to $1000 M_\odot$ and of other conceivable stellar-mass exotic objects (naked singularities and boson stars) in the distant universe, as well as of supernovae, pulsars, and coalescing and colliding neutron stars. Early universe processes should also have produced waves at these frequencies, as in the ELF, VLF, and LF bands.

In this lecture I shall focus primarily on the HF and LF bands, because these are the ones in which we expect to study black holes and other modern-era (non-early-universe) exotic objects.

2. Ground-Based Laser Interferometers and Resonant-Mass Detectors

2.1. *Wave Polarizations, Waveforms, and How an Interferometer Works*

According to general relativity theory (which I shall assume to be correct in this paper), a gravitational wave has two linear polarizations, conventionally called + (plus) and × (cross). Associated with each polarization there is a gravitational-wave field, h_+ or $h_\times$, which oscillates in time and propagates with the speed of light. Each wave field produces tidal forces (stretching and squeezing forces) on any object or detector through which it passes. If the object is small compared to the waves' wavelength (as is the case for ground-based interferometers and resonant mass antennas), then relative to the object's center, the forces have the quadrupolar patterns shown in Figure 1. The names "plus"

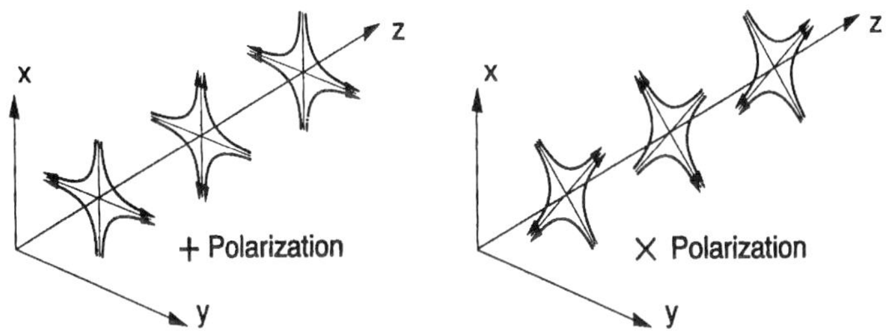

FIGURE 1. The lines of force associated with the two polarizations of a gravitational wave. (From Ref. Abramovici et al. 1992.)

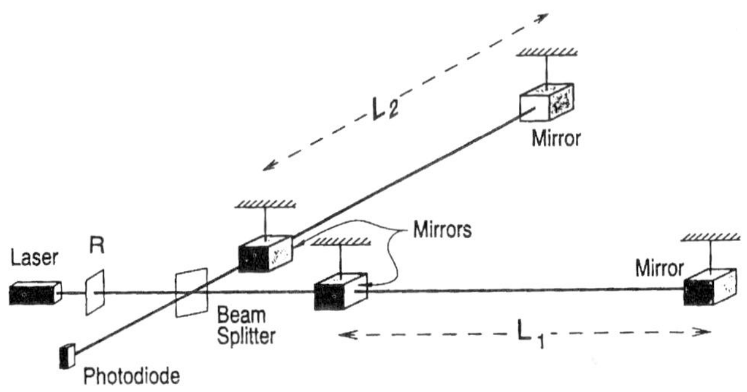

FIGURE 2. Schematic diagram of a laser interferometer gravitational wave detector. (From Ref. Abramovici et al. 1992.)

and "cross" are derived from the orientations of the axes that characterize the force patterns (Thorne 1987).

A laser interferometer gravitational wave detector ("interferometer" for short) consists of four masses that hang from vibration-isolated supports as shown in Figure 2, and the indicated optical system for monitoring the separations between the masses (Thorne 1987, Abramovici et al. 1992). Two masses are near each other, at the corner of an "L", and one mass is at the end of each of the L's long arms. The arm lengths are nearly equal, $L_1 \simeq L_2 = L$. When a gravitational wave, with frequencies high compared to the masses' $\sim$ 1 Hz pendulum frequency, passes through the detector, it pushes the masses back and forth relative to each other as though they were free from their suspension wires, thereby changing the arm-length difference, $\Delta L \equiv L_1 - L_2$. That change is monitored by laser interferometry in such a way that the variations in the output of the photodiode (the interferometer's output) are directly proportional to $\Delta L(t)$.

If the waves are coming from overhead or underfoot and the axes of the + polarization coincide with the arms' directions, then it is the waves' + polarization that drives the masses, and $\Delta L(t)/L = h_+(t)$. More generally, the interferometer's output is a linear combination of the two wave fields:

$$\frac{\Delta L(t)}{L} = F_+ h_+(t) + F_\times h_\times(t) \equiv h(t) \ . \tag{2.1}$$

The coefficients F_+ and $F_\times$ are of order unity and depend in a quadrupolar manner on the direction to the source and the orientation of the detector (Thorne 1987). The combination $h(t)$ of the two h's is called the *gravitational-wave strain* that acts on the detector; and the time evolutions of $h(t)$, $h_+(t)$, and $h_\times(t)$ are sometimes called *waveforms.*

2.2. *Wave Strengths and Interferometer Arm Lengths*

The strengths of the waves from a gravitational-wave source can be estimated using the "Newtonian/quadrupole" approximation to the Einstein field equations. This approximation says that $h \simeq (G/c^4)\ddot{Q}/r$, where $\ddot{Q}$ is the second time derivative of the source's quadrupole moment, r is the distance of the source from Earth (and G and c are Newton's gravitation constant and the speed of light). The strongest sources will be highly nonspherical and thus will have $Q \simeq ML^2$, where M is their mass and L their size, and correspondingly will have $\ddot{Q} \simeq 2Mv^2 \simeq 4E^{\rm ns}_{\rm kin}$, where v is their internal velocity and $E^{\rm ns}_{\rm kin}$ is the nonspherical part of their internal kinetic energy. This provides us with the estimate

$$h \sim \frac{1}{c^2}\frac{4G(E^{\rm ns}_{\rm kin}/c^2)}{r} \; ; \tag{2.2}$$

i.e., h is about 4 times the gravitational potential produced at Earth by the mass-equivalent of the source's nonspherical, internal kinetic energy—made dimensionless by dividing by c^2. Thus, in order to radiate strongly, the source must have a very large, nonspherical, internal kinetic energy.

The best known way to achieve a huge internal kinetic energy is via gravity; and by energy conservation (or the virial theorem), any gravitationally-induced kinetic energy must be of order the source's gravitational potential energy. A huge potential energy, in turn, requires that the source be very compact, not much larger than its own gravitational radius. Thus, the strongest gravity-wave sources must be highly compact, dynamical concentrations of large amounts of mass (e.g., colliding and coalescing black holes and neutron stars).

Such sources cannot remain highly dynamical for long; their motions will be stopped by energy loss to gravitational waves and/or the formation of an all-encompassing black hole. Thus, the strongest sources should be transient. Moreover, they should be very rare — so rare that to see a reasonable event rate will require reaching out through a substantial fraction of the Universe. Thus, just as the strongest radio waves arriving at Earth tend to be extragalactic, so also the strongest gravitational waves are likely to be extragalactic.

For highly compact, dynamical objects that radiate in the high-frequency band, e.g. colliding and coalescing neutron stars and stellar-mass black holes, the internal, nonspherical kinetic energy $E^{\rm ns}_{\rm kin}/c^2$ is of order the mass of the Sun; and, correspondingly, Eq. (2.2) gives $h \sim 10^{-22}$ for such sources at the Hubble distance (3000 Mpc, i.e., 10^{10} light years); $h \sim 10^{-21}$ at 200 Mpc (a best-guess distance for several neutron-star coalescences per year; see Section 4.2), $h \sim 10^{-20}$ at the Virgo cluster of galaxies (15 Mpc); and $h \sim 10^{-17}$ in the outer reaches of our own Milky Way galaxy (20 kpc). These numbers set the scale of sensitivities that ground-based interferometers seek to achieve: $h \sim 10^{-21}$ to 10^{-22}.

When one examines the technology of laser interferometry, one sees good prospects to achieve measurement accuracies $\Delta L \sim 10^{-16}$ cm (1/1000 the diameter of the nucleus of an atom). With such an accuracy, an interferometer must have an arm length $L = \Delta L/h \sim 1$ to 10 km, in order to achieve the desired wave sensitivities, 10^{-21} to 10^{-22}. This sets the scale of the interferometers that are now under construction.

FIGURE 3. Artist's conception of one of the LIGO interferometers. [Courtesy the LIGO Project.]

2.3. *LIGO, VIRGO, and the International Gravitational Wave Network*

Interferometers are plagued by non-Gaussian noise, e.g. due to sudden strain releases in the wires that suspend the masses. This noise prevents a single interferometer, by itself, from detecting with confidence short-duration gravitational-wave bursts (though it might be possible for a single interferometer to search for the periodic waves from known pulsars). The non-Gaussian noise can be removed by cross correlating two, or preferably three or more, interferometers that are networked together at widely separated sites.

The technology and techniques for such interferometers have been under development for nearly 25 years, and plans for km-scale interferometers have been developed over the past 14 years. An international network consisting of three km-scale interferometers, at three widely separated sites, is now in the early stages of construction. It includes two sites of the American LIGO Project ("Laser Interferometer Gravitational Wave Observatory") (Abramovici et al. 1992), and one site of the French/Italian VIRGO Project (named after the Virgo cluster of galaxies) (Bradaschia et al. 1990).

LIGO will consist of two vacuum facilities with 4-kilometer-long arms, one in Hanford, Washington (in the northwestern United States; Fig. 3) and the other in Livingston, Louisiana (in the southeastern United States). These facilities are designed to house many successive generations of interferometers without the necessity of any major facilities upgrade; and after a planned future expansion, they will be able to house several interferometers at once, each with a different optical configuration optimized for a different type of wave (e.g., broad-band burst, or narrow-band periodic wave, or stochastic wave).

The LIGO facilities and their first interferometers are being constructed by a team of about 80 physicists and engineers at Caltech and MIT, led by Barry Barish (the PI), Gary Sanders (the Project Manager), Albert Lazzarini, Rai Weiss, Stan Whitcomb, and Robbie Vogt (who directed the project during the pre-construction phase). Other research groups from many different universities are contributing to R&D for enhancements of the first interferometers, or are computing theoretical waveforms for use in data analysis, or are developing data analysis techniques. These groups are linked together by an organization called the *Ligo Research Community*. For further details, see the LIGO World Wide Web Site, http://www.ligo.caltech.edu/.

The VIRGO Project is building one vacuum facility in Pisa, Italy, with 3-kilometer-long arms. This facility and its first interferometers are a collaboration of more than a hundred physicists and engineers at the INFN (Frascati, Napoli, Perugia, Pisa), LAL (Orsay), LAPP (Annecy), LOA (Palaiseau), IPN (Lyon), ESPCI (Paris), and the University of Illinois (Urbana), under the leadership of Alain Brillet and Adalberto Giazotto.

Both LIGO and VIRGO are scheduled for completion in the late 1990s, and their first gravitational-wave searches are likely to be performed in 2001. The design sensitivity for LIGO's first interferometers is shown in Fig. 4, along with a benchmark sensitivity goal for more advanced interferometers that will operate in the LIGO vacuum system some years later (Abramovici et al. 1992). For each interferometer, the lower curve (bottom of stippled region) is the rms strain noise $h_{\rm rms} = \Delta L_{\rm rms}/L$ in a bandwidth equal to frequency, and the upper curve is the corresponding "sensitivity to bursts" that occur only once a year or so, $h_{\rm SB}(f)$.

This $h_{\rm SB}$ is to be compared with the "characteristic amplitude" $h_c(f) = h\sqrt{n}$ of the waves from a source; here h is the waves' amplitude when they have frequency f, and n is the number of cycles the waves spend in a bandwidth $\Delta f \simeq f$ near frequency f (Thorne 1987, Abramovici et al. 1992). Any source with $h_c > h_{\rm SB}$ should be detectable with high confidence, even if it arrives only once per year. As examples, Fig. 4 shows the characteristic amplitudes h_c of several binary systems made of $1.4 M_\odot$ neutron stars ("NS") or $10 M_\odot$ black holes ("BH"), which spiral together under the driving force of gravitational radiation reaction. As the bodies spiral inward, their waves sweep upward in frequency (rightward across the figure along the dashed lines). From the figure we see that LIGO's first interferometers should be able to detect waves from NS/NS binaries out to a distance of 30Mpc, and its advanced interferometers, NS/NS out to 1000Mpc and BH/BH out to 3000Mpc (a cosmological redshift of unity). We shall discuss the implications of these sensitivities in Sec. 4 below.

LIGO alone, with its two sites which have parallel arms, will be able to detect an incoming gravitational wave, measure one of its two waveforms, and (from the time delay between the two sites) locate its source to within a $\sim 1^\circ$ wide annulus on the sky. LIGO and VIRGO together, operating as a *coordinated international network*, will be able to locate the source (via time delays plus the interferometers' beam patterns) to within a 2-dimensional error box with size between several tens of arcminutes and several degrees, depending on the source direction and on the amount of high-frequency structure in the waveforms. They will also be able to monitor both waveforms $h_+(t)$ and $h_\times(t)$ (except for frequency components above about 1kHz and below about 10 Hz, where the interferometers' noise becomes severe).

The accuracies of the direction measurements and the ability to monitor more than one waveform will be severely compromised when the source lies anywhere near the plane formed by the three LIGO/VIRGO interferometer locations. To get good all-sky coverage will require a fourth interferometer at a site far out of that plane; Japan and Australia would be excellent locations, and research groups there are carrying out research and development on interferometric detectors, aimed at such a possibility. A 300-meter prototype interferometer called TAMA is under construction in Tokyo, and a 400-meter prototype called AIGO400 has been proposed for construction north of Perth.

Two other groups are major players in this field, one in Britain led by James Hough, the other in Germany, led by Karsten Danzmann. These groups each have two decades of experience with prototype interferometers (comparable experience to the LIGO team and far more than anyone else) and great expertise. Frustrated by inadequate financing for a kilometer-scale interferometer, they are constructing, instead, a 600 meter system called GEO600 near Hannover, Germany. Their goal is to develop, from the outset, an

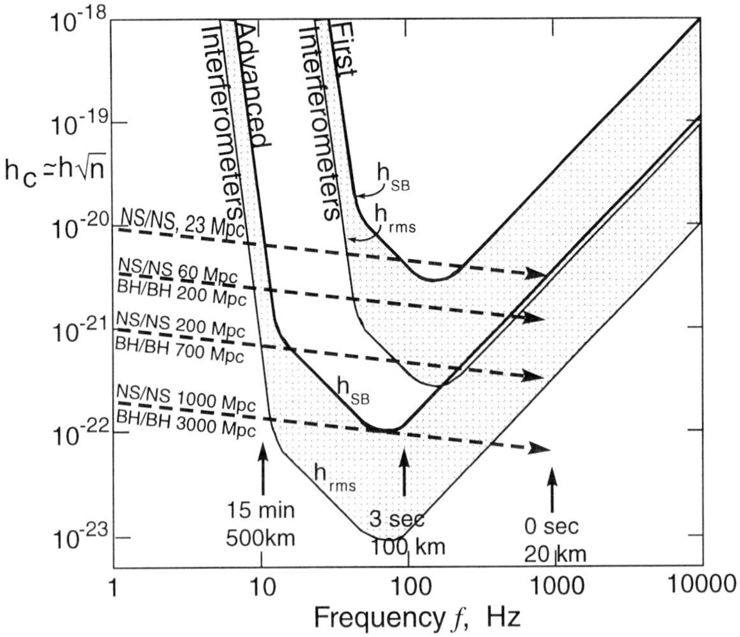

FIGURE 4. LIGO's projected broad-band noise $h_{\rm rms}$ and sensitivity to bursts $h_{\rm SB}$ (Ref. Abramovici et al. 1992) compared with the strengths of the waves from the last few minutes of inspiral of compact binaries. The signal to noise ratios are $\sqrt{2}$ higher than in Ref. (Abramovici et al. 1992) because of a factor 2 error in Eq. (29) of Ref. (Thorne 1987).

interferometer with the sort of advanced design that LIGO and VIRGO will attempt only as a "second-generation" instrument, and thereby achieve sufficient sensitivity to be full partners in the international network's first gravitational-wave searches; they then would offer a variant of their interferometer as a candidate for second-generation operation in the much longer arms of LIGO and/or VIRGO. It is a seemingly audacious plan, but with their extensive experience and expertise, the British/German collaboration might pull it off successfully.

2.4. *Narrow-Band Detectors*

At frequencies $f \gtrsim 1000$Hz, the interferometers' photon shot noise becomes a serious obstacle to wave detection. However, narrow-band detectors specially optimized for kHz frequencies show considerable promise. These include a special "dual recycled interferometer" invented by Brian Meers (1988), and huge spherical or icosahedral resonant-mass detectors (Johnson & Merkowitz 1993) that are modern variants of Joseph Weber's original "bar" detector (1960).

3. LISA: The Laser Interferometer Space Antenna

Turn, now, from the high-frequency band, 1—10^4 Hz, to the low-frequency band, 10^{-4}—1 Hz. At present, the most sensitive gravitational-wave searches at low frequencies are those carried out by researchers at NASA's Jet Propulsion Laboratory, using microwave-frequency Doppler tracking of interplanetary spacecraft. These searches are done at rather low cost, piggy-back on missions designed for other purposes. Although they have a possibility of success, the odds are against them. Their best past sensitivi-

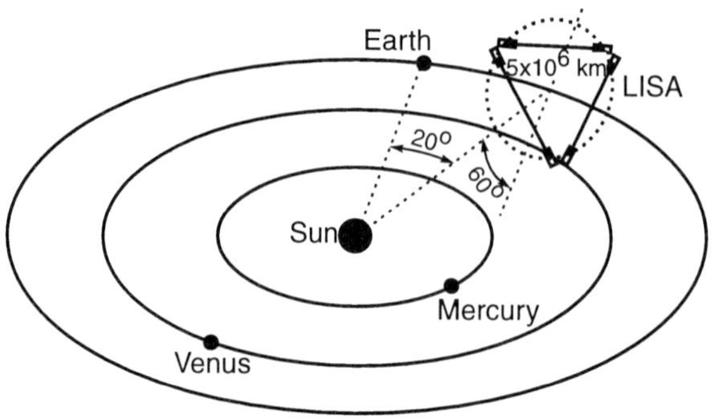

FIGURE 5. LISA's orbital configuration, with LISA magnified in arm length by a factor ~ 10 relative to the solar system.

ties to bursts, for example, have been $h_{\rm SB} \sim 10^{-14}$, and prospects are good for reaching $\sim 10^{-15}$—10^{-16} in the next 5 to 10 years. However, the strongest low-frequency bursts arriving several times per year might be no larger than $\sim 10^{-18}$; and the domain of an assured plethora of signals is $h_{\rm SB} \sim 10^{-19}$—10^{-20}.

To reach into this assured-detection domain will almost certainly require switching from microwave-frequency tracking of spacecraft (with its large noise due to fluctuating dispersion in the troposphere and interplanetary plasma) to optical tracking. Such a switch is planned for the 2014 time frame or sooner, when the European Space Agency (ESA) and/or NASA is likely to fly the *Laser Interferometer Space Antenna*, LISA.

3.1. *Mission Status*

LISA is largely an outgrowth of 15 years of studies by Peter Bender and colleagues at the University of Colorado. In 1990, NASA's Ad Hoc Committee on Gravitation Physics and Astronomy selected a LISA-type gravitational-wave detector as its highest priority in the large space mission category (Armstrong et al. 1990); and since then enthusiasm for LISA has continued to grow within the American gravitation community. Unfortunately, the prospects for NASA to fly such a mission did not look good in the early 1990s. By contrast, prospects in Europe looked much better, so a largely European consortium was put together in 1993, under the leadership of Karsten Danzmann (Hannover) and James Hough (Glasgow), to propose LISA to the European Space Agency. This proposal has met with considerable success; LISA might well achieve approval to fly as an ESA Cornerstone Mission around 2014 (Bender et al. 1994). Members of the American gravitation community and members of the LISA team hope that NASA will join together with ESA in this endeavor, and that working jointly, ESA and NASA will be able to fly LISA considerably sooner than 2014.

3.2. *Mission Configuration*

As presently conceived, LISA will consist of six compact, drag-free spacecraft (i.e. spacecraft that are shielded from buffeting by solar wind and radiation pressure, and that thus move very nearly on geodesics of spacetime). All six spacecraft would be launched simultaneously by a single Ariane rocket. They would be placed into the same heliocentric orbit as the Earth occupies, but would follow 20° behind the Earth; cf. Figure 5. The spacecraft would fly in pairs, with each pair at the vertex of an equilateral triangle that is inclined at an angle of 60° to the Earth's orbital plane. The triangle's arm

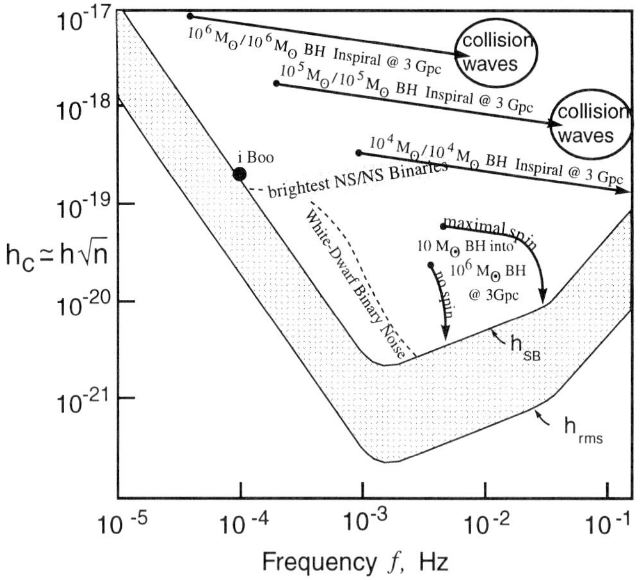

FIGURE 6. LISA's projected broad-band noise $h_{\rm rms}$ and sensitivity to bursts $h_{\rm SB}$, compared with the strengths of the waves from several low-frequency sources. [*Note:* When members of the LISA team plot curves analogous to this, they show the sensitivity curve (top of stippled region) in units of the amplitude of a periodic signal that can be detected with $S/N = 5$ in one year of integration; that sensitivity to periodic sources is related to the $h_{\rm SB}$ used here by $h_{\rm SP} = h_{\rm SB}/\sqrt{f \cdot 3 \times 10^7 \text{sec}}$.]

length would be 5 million km (10^6 times larger than LIGO's arms!). The six spacecraft would track each other optically, using one-Watt YAG laser beams. Because of diffraction losses over the 5×10^6km arm length, it is not feasible to reflect the beams back and forth between mirrors as is done with LIGO. Instead, each spacecraft will have its own laser; and the lasers will be phase locked to each other, thereby achieving the same kind of phase-coherent out-and-back light travel as LIGO achieves with mirrors. The six-laser, six-spacecraft configuration thereby functions as three, partially independent but partially redundant, gravitational-wave interferometers.

3.3. *Noise and Sensitivity*

Figure 6 depicts the expected noise and sensitivity of LISA in the same language as we have used for LIGO (Fig. 4). The curve at the bottom of the stippled region is $h_{\rm rms}$, the rms noise, in a bandwidth equal to frequency, for waves with optimum direction and polarization. The top of the stippled region is $h_{\rm SB} = 5\sqrt{5} h_{\rm rms}$, the sensitivity for high-confidence detection ($S/N = 5$) of a broad-band burst coming from a random direction, assuming Gaussian noise.

At frequencies $f \gtrsim 10^{-3}$Hz, LISA's noise is due to photon counting statistics (shot noise). The noise curve steepens at $f \sim 3 \times 10^{-2}$Hz because at larger f than that, the waves' period is shorter than the round-trip light travel time in one of LISA's arms. Below 10^{-3}Hz, the noise is due to buffeting-induced random motions of the spacecraft that are not being properly removed by the drag-compensation system. Notice that, in terms of dimensionless amplitude, LISA's sensitivity is roughly the same as that of LIGO's first interferometers (Fig. 4), but at 100,000 times lower frequency. Since the waves' energy flux scales as f^2h^2, this corresponds to 10^{10} better energy sensitivity than LIGO.

3.4. *Observational Strategy*

LISA can detect and study, simultaneously, a wide variety of different sources scattered over all directions on the sky. The key to distinguishing the different sources is the different time evolution of their waveforms. The key to determining each source's direction, and confirming that it is real and not just noise, is the manner in which its waves' amplitude and frequency are modulated by LISA's complicated orbital motion—a motion in which the interferometer triangle rotates around its center once per year, and the interferometer plane precesses around the normal to the Earth's orbit once per year. Most sources will be observed for a year or longer, thereby making full use of these modulations.

4. Binaries Made of Neutron Stars, Black Holes, and Other Exotic Objects: Inspiral, Collision, and Coalescence in the HF Band

The best understood of all gravitational-wave sources are coalescing, compact binaries composed of neutron stars (NS) and black holes (BH). These NS/NS, NS/BH, and BH/BH binaries may well become the "bread and butter" of the LIGO/VIRGO diet.

The Hulse-Taylor (Hulse & Taylor 1975, Taylor 1994) binary pulsar, PSR 1913+16, is an example of a NS/NS binary whose waves could be measured by LIGO/VIRGO, if we were to wait long enough. At present PSR 1913+16 has an orbital frequency of about 1/(8 hours) and emits its waves predominantly at twice this frequency, roughly 10^{-4} Hz, which is in LISA's low-frequency band (cf. Fig. 6). As a result of their loss of orbital energy to gravitational waves, the PSR 1913+16 NS's are gradually spiraling inward. If we wait roughly 10^8 years, this inspiral will bring the waves into the LIGO/VIRGO high-frequency band. As the NS's continue their inspiral, the waves will then sweep upward in frequency, over a time of about 15 minutes, from 10 Hz to $\sim 10^3$ Hz, at which point the NS's will collide and coalesce. It is this last 15 minutes of inspiral, with $\sim 16,000$ cycles of waveform oscillation, and the final coalescence, that LIGO/VIRGO seeks to monitor.

4.1. *Wave Strengths Compared to LIGO Sensitivities*

Figure 4 above compares the projected sensitivities of interferometers in LIGO (Abramovici et al. 1992) with the wave strengths from the last few minutes of inspiral of BH/BH, NS/BH, and NS/NS binaries at various distances from Earth.

As the binaries spiral inward, their waves sweep upward in frequency (left to right along the dashed lines if Fig. 4). The arrows along the bottom inspiral track indicate the time until final coalescence for an NS/NS binary and the separation between the NS centers of mass. Each NS is assumed to have a mass of 1.4 suns and a radius ~ 10 km; each BH, 10 suns and ~ 20 km.

Notice that the signal strengths in Figure 4 are in good accord with our rough estimates based on Eq. (2.2); at the endpoint (right end) of each inspiral, the number of cycles n spent near that frequency is of order unity, so the quantity plotted, $h_c \simeq h\sqrt{n}$, is about equal to h—and at distance 200 Mpc is roughly 10^{-21}, as we estimated in Section 2.2.

4.2. *Coalescence Rates*

Such final coalescences are few and far between in our own galaxy: about one every 100,000 years, according to 1991 estimates by Phinney (1991) and by Narayan, Piran, and

Shemi (1991), based on the statistics of binary pulsar searches in our galaxy which found three that will coalesce in less than 10^{10} years. Extrapolating out through the universe on the basis of the density of production of blue light (the color produced predominantly by massive stars), Phinney (1991) and Narayan et. al. (1991) infer that to see several NS/NS coalescences per year, LIGO/VIRGO will have to look out to a distance of about 200 Mpc (give or take a factor ~ 2); cf. the "NS/NS inspiral, 200 Mpc" line in Figure 4. Since these estimates were made, the binary pulsar searches have been extended through a significantly larger volume of the galaxy than before, and no new ones with coalescence times $\lesssim 10^{10}$ years have been found. This would drive the estimated event rate downward, except that revisions of other aspects of the Phinney/Narayan analyses have compensated, leaving the binary-pulsar-search-based best estimates unchanged, at several events per year out to 200 Mpc (van den Heuvel & Lorimer 1996).

A rate of one every million years in our galaxy is ~ 1000 times smaller than the birth rate of the NS/NS binaries' progenitors: massive, compact, main-sequence binaries (Phinney 1991, Narayan, Piran & Shemi 1991). Therefore, either 99.9 per cent of progenitors fail to make it to the NS/NS state (e.g., because of binary disruption during a supernova or forming TŻO's), or else they do make it, but they wind up as a class of NS/NS binaries that has not yet been discovered in any of the pulsar searches. Several experts on binary evolution have argued for the latter (Tutukov & Yungelson 1993, Yamaoka, Shigeyama & Nomoto 1993, Lipunov, Postnov & Prokhorov 1994): most NS/NS binaries, they suggest, may form with such short orbital periods that their lifetimes to coalescence are significantly shorter than normal pulsar lifetimes ($\sim 10^7$ years); and with such short lifetimes, they have been missed in pulsar searches. By modeling the evolution of the galaxy's binary star population, these binary experts arrive at best estimates as high as 3×10^{-4} coalescences per year in our galaxy, corresponding to several per year out to 60 Mpc distance (Tutukov & Yungelson 1993), though more recent and more conservative models (Zwart & Spreeuw 1996) give results more nearly in accord with the binary pulsar searches, several per year out to 200Mpc. Phinney (1991) describes other plausible populations of NS/NS binaries that could increase the event rate, and he argues for "ultraconservative" lower and upper limits of 23 Mpc and 1000Mpc for how far one must look to see several coalescence per year.

By comparing these rate estimates with the signal strengths in Figure 4, we see that (i) the first interferometers in LIGO/VIRGO have a possibility but not high probability of seeing NS/NS coalescences; (ii) advanced interferometers are almost certain of seeing them (the requirement that this be so was one factor that forced the LIGO/VIRGO arm lengths to be so long, several kilometers); and (iii) they are most likely to be discovered roughly half-way between the first and advanced interferometers—which means by an improved variant of the first interferometers several years after LIGO operations begin.

We have no good observational handle on the coalescence rate of NS/BH or BH/BH binaries. However, theory suggests that their progenitors might not disrupt during the stellar collapses that produce the NS's and BH's, so their coalescence rate could be about the same as the birth rate for their progenitors: $\sim 1/100,000$ years in our galaxy. This suggests that within 200 Mpc distance there might be several NS/BH or BH/BH coalescences per year. (Phinney 1991, Narayan, Piran & Shemi 1991, Tutukov & Yungelson 1993, Lipunov, Postnov & Prokhorov 1994). This estimate should be regarded as a plausible upper limit on the event rate and lower limit on the distance to look (Phinney 1991, Narayan, Piran & Shemi 1991).

If this estimate is correct, then NS/BH and BH/BH binaries will be seen before NS/NS, and might be seen by the first LIGO/VIRGO interferometers or soon thereafter; cf. Figure 4. However, this estimate is far less certain than the (rather uncertain) NS/NS estimates!

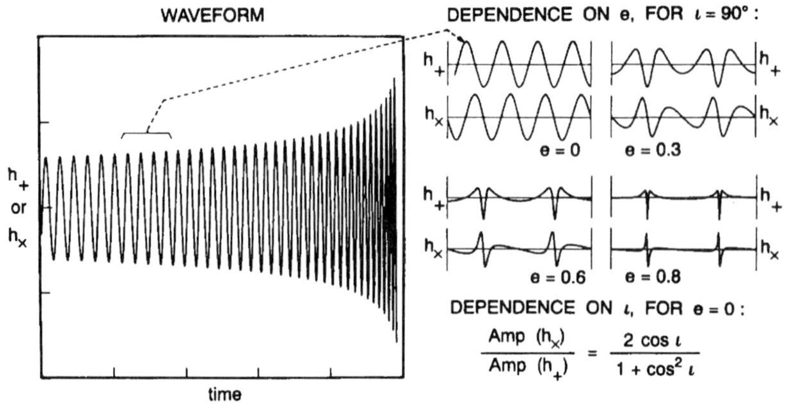

FIGURE 7. Waveforms from the inspiral of a compact binary, computed using Newtonian gravity for the orbital evolution and the quadrupole-moment approximation for the wave generation. (From Ref. Abramovici et al. 1992.)

Once coalescence waves have been discovered, each further improvement of sensitivity by a factor 2 will increase the event rate by $2^3 \simeq 10$. Assuming a rate of several NS/NS per year at 200 Mpc, the advanced interferometers of Figure 4 should see ~ 100 per year.

4.3. *Inspiral Waveforms and the Information They Can Bring*

Neutron stars and black holes have such intense self gravity that it is exceedingly difficult to deform them. Correspondingly, as they spiral inward in a compact binary, they do not gravitationally deform each other significantly until several orbits before their final coalescence (Kochanek 1992, Bildsten & Cutler 1992). This means that the inspiral waveforms are determined to high accuracy by only a few, clean parameters: the masses and spin angular momenta of the bodies, and the initial orbital elements (i.e. the elements when the waves enter the LIGO/VIRGO band).

Though tidal deformations are negligible during inspiral, relativistic effects can be very important. If, for the moment, we ignore the relativistic effects—i.e., if we approximate gravity as Newtonian and the wave generation as due to the binary's oscillating quadrupole moment (Thorne 1987), then the shapes of the inspiral waveforms $h_+(t)$ and $h_\times(t)$ are as shown in Figure 7.

The left-hand graph in Figure 7 shows the waveform increasing in amplitude and sweeping upward in frequency (i.e., undergoing a "chirp") as the binary's bodies spiral closer and closer together. The ratio of the amplitudes of the two polarizations is determined by the inclination ι of the orbit to our line of sight (lower right in Fig. 7). The shapes of the individual waves, i.e. the waves' harmonic content, are determined by the orbital eccentricity (upper right). (Binaries produced by normal stellar evolution should be highly circular due to past radiation reaction forces, but compact binaries that form by capture events, in dense star clusters that might reside in galactic nuclei (Quinlan & Shapiro 1987), could be quite eccentric.) If, for simplicity, the orbit is circular, then the rate at which the frequency sweeps or "chirps", df/dt [or equivalently the number of cycles spent near a given frequency, $n = f^2(df/dt)^{-1}$] is determined solely, in the Newtonian/quadrupole approximation, by the binary's so-called *chirp mass*, $M_c \equiv (M_1 M_2)^{3/5}/(M_1 + M_2)^{1/5}$ (where M_1 and M_2 are the two bodies' masses). The amplitudes of the two waveforms are determined by the chirp mass, the distance to the source, and the orbital inclination. Thus (in the Newtonian/quadrupole approximation),

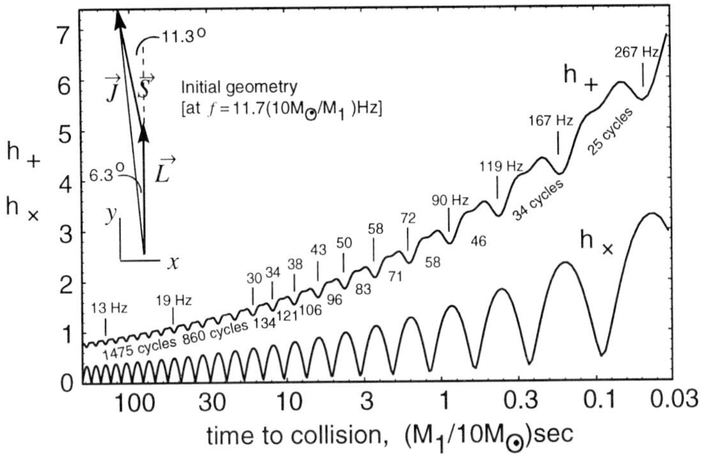

FIGURE 8. Modulational envelope for the waveform from a $1M_\odot$ nonspinning NS spiraling into a $10M_\odot$, rapidly spinning Kerr black hole (spin parameter $a = 1$). The orbital angular momentum $\mathbf{L}$ is inclined by $\alpha = 11.3$ degrees to the hole's spin angular momentum $\mathbf{S}$, and the two precess around $\mathbf{J} = \mathbf{L} + \mathbf{S}$, whose direction remains fixed in space as $L = |\mathbf{L}|$ shrinks and $S = |\mathbf{S}| = M_{\rm BH}a$ remains constant. The precession modulates the waves by an amount that depends on (i) the direction to Earth (here along the initial $\mathbf{L} \times \mathbf{S}$, i.e. out of the paper) and (ii) the orientation of the detector's arms (here parallel to the figure's initial $\mathbf{L}$ and to $\mathbf{L}\times$(direction to Earth) for h_+, and rotated 45 degrees for $h_\times$). The figure shows the waveforms' modulational envelopes (in arbitrary units, the same for h_+ and $h_\times$), parametrized by the wave frequency f and the number of cycles of *oscillation* between the indicated f's. The total number of *precessions* from f to coalescence is $N_{\rm prec} \simeq (5/64\pi)(Ma/\mu)(\pi M f)^{-2/3} \simeq 20(f/10{\rm Hz})^{-2/3}$. (From Cutler et al. 1993, Apostolatos et al. 1994.)

by measuring the two amplitudes, the frequency sweep, and the harmonic content of the inspiral waves, one can determine as direct, resulting observables, the source's distance, chirp mass, inclination, and eccentricity (Schutz 1986, Schutz 1989).

As in binary pulsar observations (Taylor 1994), so also here, relativistic effects add further information: they influence the rate of frequency sweep and produce waveform modulations in ways that depend on the binary's dimensionless ratio $\eta = \mu/M$ of reduced mass $\mu = M_1M_2/(M_1 + M_2)$ to total mass $M = M_1 + M_2$ (Lincoln & Will 1990) and on the spins of the binary's two bodies (Kidder, Will & Wiseman 1993). These relativistic effects are reviewed and discussed at length in Refs. (Cutler et al. 1993, Will 1994a). Two deserve special mention: (i) As the waves emerge from the binary, some of them get backscattered one or more times off the binary's spacetime curvature, producing wave *tails*. These tails act back on the binary, modifying its inspiral rate in a measurable way. (ii) If the orbital plane is inclined to one or both of the binary's spins, then the spins drag inertial frames in the binary's vicinity (the "Lense-Thirring effect"), this frame dragging causes the orbit to precess, and the precession modulates the waveforms (Cutler et al. 1993, Apostolatos et al. 1994, Kidder 1995). Figure 8 shows the resulting modulation for a $1M_\odot$ NS spiraling into a rapidly spinning, $10M_\odot$ BH.

Remarkably, the relativistic corrections to the frequency sweep will be measurable with very high accuracy, even though they are typically $\lesssim 10$ per cent of the Newtonian contribution, and even though the typical signal to noise ratio will be only ~ 9 even after optimal signal processing. The reason is as follows (Cutler & Flanagan 1994, Finn & Chernoff 1993, Cutler et al. 1993):

The frequency sweep will be monitored by the method of "matched filters"; in other words, the incoming, noisy signal will be cross correlated with theoretical templates. If the signal and the templates gradually get out of phase with each other by more than $\sim 1/10$ cycle as the waves sweep through the LIGO/VIRGO band, their cross correlation will be significantly reduced. Since the total number of cycles spent in the LIGO/VIRGO band will be $\sim 16,000$ for a NS/NS binary, ~ 3500 for NS/BH, and ~ 600 for BH/BH, this means that LIGO/VIRGO should be able to measure the frequency sweep to a fractional precision $\lesssim 10^{-4}$, compared to which the relativistic effects are very large. (This is essentially the same method as Joseph Taylor (1994) and colleagues use for high-accuracy radio-wave measurements of relativistic effects in binary pulsars.)

Preliminary analyses, using the theory of optimal signal processing, predict the following typical accuracies for LIGO/VIRGO measurements based solely on the frequency sweep (i.e., ignoring modulational information) (Poisson & Will 1995, Cutler & Flanagan 1994, Finn & Chernoff 1993, Jaranowski & Krolak 1994, Cutler et al. 1993): (i) The chirp mass M_c will typically be measured, from the Newtonian part of the frequency sweep, to $\sim 0.04\%$ for a NS/NS binary and $\sim 0.3\%$ for a system containing at least one BH. (ii) *If* we are confident (e.g., on a statistical basis from measurements of many previous binaries) that the spins are a few percent or less of the maximum physically allowed, then the reduced mass μ will be measured to $\sim 1\%$ for NS/NS and NS/BH binaries, and $\sim 3\%$ for BH/BH binaries. (Here and below NS means a $\sim 1.4M_\odot$ neutron star and BH means a $\sim 10M_\odot$ black hole.) (iii) Because the frequency dependences of the (relativistic) μ effects and spin effects are not sufficiently different to give a clean separation between μ and the spins, if we have no prior knowledge of the spins, then the spin/μ correlation will worsen the typical accuracy of μ by a large factor, to $\sim 30\%$ for NS/NS, $\sim 50\%$ for NS/BH, and a factor ~ 2 for BH/BH (Poisson & Will 1995, Cutler & Flanagan 1994). These worsened accuracies might be improved somewhat by waveform modulations caused by the spin-induced precession of the orbit (Apostolatos et al. 1994, Kidder 1995), and even without modulational information, a certain combination of μ and the spins will be determined to a few per cent. Much additional theoretical work is needed to firm up the measurement accuracies.

To take full advantage of all the information in the inspiral waveforms will require theoretical templates that are accurate, for given masses and spins, to a fraction of a cycle during the entire sweep through the LIGO/VIRGO band. Such templates are being computed by an international consortium of relativity theorists (Blanchet and Damour in France, Iyer in India, Will and Wiseman in the U.S., and others) (Will 1994a, Blanchet et al. 1995), using post-Newtonian expansions of the Einstein field equations. This enterprise is rather like computing the Lamb shift to high order in powers of the fine structure constant, for comparison with experiment. The terms of leading order in the mass ratio $\eta = \mu/M$ are being checked by a Japanese-American consortium (Nakamura, Sasaki, Tagoshi, Tanaka, Poisson) using the Teukolsky formalism for weak perturbations of black holes (Poisson & Sasaki 1995, Shibata et al. 1996). These small-η calculations have been carried to very high post-Newtonian order for circular orbits and no spins (Tagoshi & Nakamura 1994, Tagoshi & Sasaki 1994), and from those results Cutler and Flanagan (Cutler & Flanagan 1997) have estimated the order to which the full, finite-η computations must be carried in order that systematic errors in the theoretical templates will not significantly impact the information extracted from the LIGO/VIRGO observational data. The answer appears daunting: radiation-reaction effects must be computed to three full post-Newtonian orders [six orders in v/c =(orbital velocity)/(speed of light)] beyond the leading-order radiation reaction, which itself is 5 orders in v/c beyond the Newtonian theory of gravity.

It is only about ten years since controversies over the leading-order radiation reaction

(Ashtekar, ed. 1983) were resolved by a combination of theoretical techniques and binary pulsar observations. Nobody dreamed then that LIGO/VIRGO observations will require pushing post-Newtonian computations onward from $O[(v/c)^5]$ to $O[(v/c)^{11}]$. This requirement epitomizes a major change in the field of relativity research: At last, 80 years after Einstein formulated general relativity, experiment has become a major driver for theoretical analyses.

Remarkably, the goal of $O[(v/c)^{11}]$ is achievable. The most difficult part of the computation, the radiation reaction, has been evaluated to $O[(v/c)^9]$ beyond Newton by the French/Indian/American consortium (Blanchet et al. 1995) and as of this writing, rumors have it that $O[(v/c)^{10}]$ is coming under control.

These high-accuracy waveforms are needed only for extracting information from the inspiral waves, after the waves have been discovered; they are not needed for the discovery itself. The discovery is best achieved using a different family of theoretical waveform templates, one that covers the space of potential waveforms in a manner that minimizes computation time instead of a manner that ties quantitatively into general relativity theory (Cutler et al. 1993). Such templates are in the early stage of development (Sathyaprakash 1994, Królak, Kokkotas & Schäfer 1995, Apostolatos 1995).

4.4. *Testing GR, Measuring the Cosmological Universe, Mapping Black Holes, and Searching for Exotic Objects*

LIGO/VIRGO observations of compact binary inspiral have the potential to bring us far more information than just binary masses and spins:

- They can be used for high-precision tests of general relativity. In scalar-tensor theories (some of which are attractive alternatives to general relativity (Damour & Nordvedt 1993), radiation reaction due to emission of scalar waves places a unique signature on the gravitational waves that LIGO/VIRGO would detect—a signature that can be searched for with high precision (Will 1994c).
- They can be used to measure the Universe's Hubble constant, deceleration parameter, and cosmological constant (Schutz 1986, Schutz 1989, Markovic 1993, Chernoff & Finn 1993). The keys to such measurements are that: (i) Advanced interferometers in LIGO/VIRGO will be able to see NS/NS out to cosmological redshifts $z \sim 0.3$, and NS/BH out to $z \sim 2$. (ii) The direct observables that can be extracted from the observed waves include the source's luminosity distance r_L (measured to accuracy ~ 10 per cent in a large fraction of cases), and its direction on the sky (to accuracy ~ 1 square degree)—accuracies good enough that only one or a few electromagnetically-observed clusters of galaxies should fall within the 3-dimensional gravitational error boxes, thereby giving promise to joint gravitational/electromagnetic statistical studies. (iii) Another direct gravitational observable is $(1+z)M$ where z is redshift and M is any mass in the system (measured to the accuracies quoted above). Since the masses of NS's in binaries seem to cluster around $1.4M_\odot$, measurements of $(1+z)M$ can provide a handle on the redshift, even in the absence of electromagnetic aid.
- For a NS or small BH spiraling into a massive ~ 50 to $500M_\odot$ compact central body, the inspiral waves will carry a "map" of the massive body's spacetime geometry. This map can be used to determine whether the massive body is a black hole (in which case the geometry must be that of Kerr, with all its features uniquely determined by its mass and angular momentum) or is some other kind of exotic compact object, e.g. a soliton star or naked singularity (Ryan, Finn & Thorne 1997). As we shall see in Section 6.2 below, this type of black-hole study and search for exotic objects can be carried out with much higher precision by LISA at large masses and low frequencies, than by LIGO/VIRGO at low masses and high frequencies.

4.5. *Coalescence Waveforms and their Information*

The waves from the binary's final coalescence can bring us new types of information.

BH/BH Coalescence

In the case of a BH/BH binary, the coalescence will excite large-amplitude, highly nonlinear vibrations of spacetime curvature near the coalescing black-hole horizons—a phenomenon of which we have very little theoretical understanding today. Especially fascinating will be the case of two spinning black holes whose spins are not aligned with each other or with the orbital angular momentum. Each of the three angular momentum vectors (two spins, one orbital) will drag space in its vicinity into a tornado-like swirling motion—the general relativistic "dragging of inertial frames," so the binary is rather like two tornados with orientations skewed to each other, embedded inside a third, larger tornado with a third orientation. The dynamical evolution of such a complex configuration of coalescing spacetime warpage (as revealed by its emitted waves) might bring us surprising new insights into relativistic gravity (Abramovici et al. 1992). Moreover, if the sum of the BH masses is fairly large, ~ 40 to $200 M_{\odot}$, then the waves should come off in a frequency range $f \sim 40$ to 200 Hz where the LIGO/VIRGO broad-band interferometers have their best sensitivity and can best extract the information the waves carry.

To get full value out of such wave observations will require (Flanagan & Hughes 1997) having theoretical computations with which to compare them. There is no hope to perform such computations analytically; they can only be done as supercomputer simulations. The development of such simulations is being pursued by several research groups, including an eight-university American consortium of numerical relativists and computer scientists called the Two-Black-Hole Grand Challenge Alliance (NRGCA 1995) (Co-PIs: Richard Matzner and Jim Browne, U. Texas Austin; Larry Smarr, Ed Seidel, Paul Saylor, Faisal Saied, U. Illinois Urbana; Geoffrey Fox, Syracuse U.; Stu Shapiro and Saul Teukolsky, Cornell U.; Jim York and Charles Evans, U. North Carolina; Sam Finn, Northwestern U.; Pablo Laguna, Pennsylvania State U.; and Jeff Winicour, U. Pittsburgh). I have a bet with Matzner, the lead PI of this alliance, that LIGO/VIRGO will discover waves from such coalescences with misaligned spins before the Alliance is able to compute them.

NS/NS Coalescence

The final coalescence of NS/NS binaries should produce waves that are sensitive to the equation of state of nuclear matter, so such coalescences have the potential to teach us about the nuclear equation of state (Abramovici et al. 1992, Cutler et al. 1993). In essence, LIGO/VIRGO will be studying nuclear physics via the collisions of atomic nuclei that have nucleon numbers $A \sim 10^{57}$—somewhat larger than physicists are normally accustomed to. The accelerator used to drive these nuclei up to the speed of light is the binary's self gravity, and the radiation by which the details of the collisions are probed is gravitational.

Unfortunately, the final NS/NS coalescence will emit its gravitational waves in the kHz frequency band ($800\text{Hz} \lesssim f \lesssim 2500\text{Hz}$) where photon shot noise will prevent them from being studied by the standard, "workhorse," broad-band interferometers of Figure 4. However, a specially configured ("dual-recycled") interferometer invented by Brian Meers (Meers 1988), which could have enhanced sensitivity in the kHz region at the price of reduced sensitivity elsewhere, may be able to measure the waves and extract their equation of state information, as might massive, spherical, resonant-mass detectors

(Cutler et al. 1993, Kennefick, Laurence & Thorne 1997). Such measurements will be very difficult and are likely only when the LIGO/VIRGO network has reached a mature stage.

A number of research groups (Shibata, Nakamura & Oohara 1993, Rasio & Shapiro 1992, Nakamura 1994, Zhuge & Centrella 1994, Davies et al. 1994) are engaged in numerical astrophysics simulations of NS/NS coalescence, with the goal not only to predict the emitted gravitational waveforms and their dependence on equation of state, but also (more immediately) to learn whether such coalescences might power the γ-ray bursts that have been a major astronomical puzzle since their discovery in the early 1970s.

NS/NS coalescence is currently a popular explanation for the γ-ray bursts because (i) the bursts are isotropically distributed on the sky, (ii) they have a distribution of number versus intensity that suggests they might lie at near-cosmological distances, and (iii) their event rate is roughly the same as that predicted for NS/NS coalescence ($\sim$ 1000 per year out to cosmological distances, if they are cosmological). If LIGO/VIRGO were now in operation and observing NS/NS inspiral, it could report definitively whether or not the γ-bursts are produced by NS/NS binaries; and if the answer were yes, then the combination of γ-burst data and gravitational-wave data could bring valuable information that neither could bring by itself. For example, it would reveal when, to within a few msec, the γ-burst is emitted relative to the moment the NS's first begin to touch; and by comparing the γ and gravitational times of arrival, we might test whether gravitational waves propagate with the speed of light to a fractional precision of $\sim 0.01\text{sec}/3 \times 10^9\,\text{lyr} = 10^{-19}$.

NS/BH Coalescence

A NS spiraling into a BH of mass $M \gtrsim 10 M_\odot$ should be swallowed more or less whole. However, if the BH is less massive than roughly $10 M_\odot$, and especially if it is rapidly rotating, then the NS will tidally disrupt before being swallowed. Little is known about the disruption and accompanying waveforms. To model them with any reliability will likely require full numerical relativity, since the circumferences of the BH and NS will be comparable and their physical separation at the moment of disruption will be of order their separation. As with NS/NS, the coalescence waves should carry equation of state information and will come out in the kHz band, where their detection will require advanced, specialty detectors.

Christodoulou Memory

As the coalescence waves depart from their source, their energy creates (via the nonlinearity of Einstein's field equations) a secondary wave called the "Christodoulou memory" (Christodoulou 1991, Thorne 1992, Wiseman & Will 1991). Whereas the primary waves may have frequencies in the kHz band, the memory builds up on the timescale of the primary energy emission profile, which is likely to be of order 0.01 sec, corresponding to a memory frequency in the optimal band for the LIGO/VIRGO workhorse interferometers, $\sim$ 100Hz. Unfortunately, the memory is so weak that only very advanced interferometers have much chance of detecting and studying it—and then, perhaps only for BH/BH coalescences and not for NS/NS or NS/BH (Kennefick 1994).

5. Other High-Frequency Sources

5.1. *Stellar Core Collapse and Supernovae*

When the core of a massive star has exhausted its supply of nuclear fuel, it collapses to form a neutron star or black hole. In some cases, the collapse triggers and powers

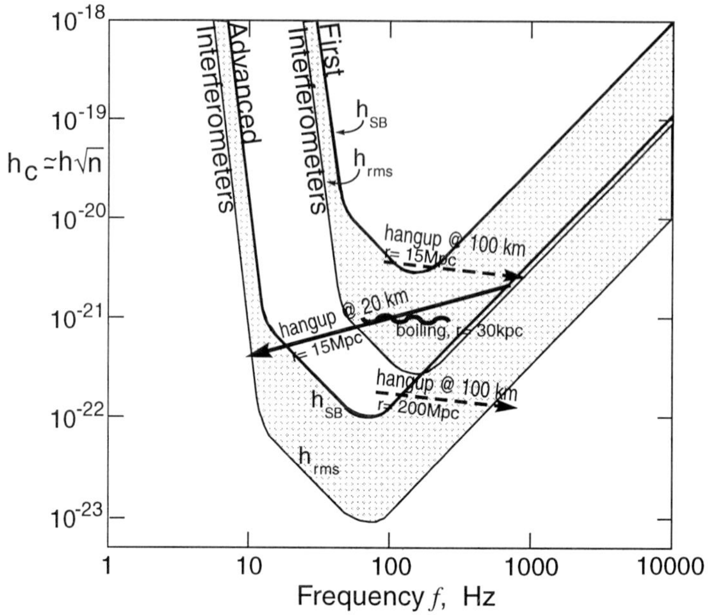

FIGURE 9. Characteristic amplitudes of the gravitational waves from various processes accompanying stellar core collapse and supernovae, compared with projected sensitivities of LIGO's interferometers.

a subsequent explosion of the star's mantle—a supernova explosion. Despite extensive theoretical efforts for more than 30 years, and despite wonderful observational data from Supernova 1987A, theorists are still far from a definitive understanding of the details of the collapse and explosion. The details are highly complex and may differ greatly from one core collapse to another (Petschek, ed. 1990).

Several features of the collapse and the core's subsequent evolution can produce significant gravitational radiation in the high-frequency band. We shall consider these features in turn, the most weakly radiating first.

Boiling of the Newborn Neutron Star

Even if the collapse is spherical, so it cannot radiate any gravitational waves at all, it should produce a convectively unstable neutron star that "boils" vigorously (and non-spherically) for the first $\sim$ 0.1 second of its life (Bethe 1990). The boiling dredges up high-temperature nuclear matter ($T \sim 10^{12}$K) from the neutron star's central regions, bringing it to the surface (to the "neutrino-sphere"), where it cools by neutrino emission before being swept back downward and reheated. Burrows estimates (Burrows, Hayes & Fryxell 1995, Burrows 1995) that the boiling should generate $n \sim$ 10 cycles of gravitational waves with frequency $f \sim$ 100Hz and amplitude $h \sim 3 \times 10^{-22}(30\text{kpc}/r)$ (where r is the distance to the source), corresponding to a characteristic amplitude $h_c \simeq h\sqrt{n} \sim 10^{-21}(30\text{kpc}/r)$; cf. Figure 9. LIGO/VIRGO will be able to detect such waves only in the local group of galaxies, where the supernova rate is probably no larger than $\sim$ 1 each 10 years. However, neutrino detectors have a similar range, and there could be a high scientific payoff from correlated observations of the gravitational waves emitted by the boiling's mass motions and neutrinos emitted from the boiling neutrino-sphere.

Axisymmetric Collapse, Bounce, and Oscillations

Rotation will centrifugally flatten the collapsing core, enabling it to radiate as it implodes. If the core's angular momentum is small enough that centrifugal forces do not halt or strongly slow the collapse before it reaches nuclear densities, then the core's collapse, bounce, and subsequent oscillations are likely to be axially symmetric. Numerical simulations (Finn 1991, Mönchmeyer, Schäfer & Müller 1991) show that in this case the waves from collapse, bounce, and oscillation will be quite weak: the total energy radiated as gravitational waves is not likely to exceed $\sim 10^{-7}$ solar masses (about 1 part in a million of the collapse energy) and might often be much less than this; and correspondingly, the waves' characteristic amplitude will be $h_c \lesssim 3\times10^{-21}(30\text{kpc}/r)$. These collapse-and-bounce waves will come off at frequencies $\sim$ 200 Hz to $\sim$ 1000 Hz, and will precede the boiling waves by a fraction of a second. Like the boiling waves, they probably cannot be seen by LIGO/VIRGO beyond the local group of galaxies and thus will be a very rare occurrence.

Rotation-Induced Bars and Break-Up

If the core's rotation is large enough to strongly flatten the core before or as it reaches nuclear density, then a dynamical and/or secular instability is likely to break the core's axisymmetry. The core will be transformed into a bar-like configuration that spins end-over-end like an American football, and that might even break up into two or more massive pieces. In this case, the radiation from the spinning bar or orbiting pieces *could* be almost as strong as that from a coalescing neutron-star binary, and thus could be seen by the LIGO/VIRGO first interferometers out to the distance of the Virgo cluster (where the supernova rate is several per year) and by advanced interferometers out to several hundred Mpc (supernova rate $\sim 10^4$ per year); cf. Figure 9. It is far from clear what fraction of collapsing cores will have enough angular momentum to break their axisymmetry, and what fraction of those will actually radiate at this high rate; but even if only $\sim$ 1/1000 or $1/10^4$ do so, this could ultimately be a very interesting source for LIGO/VIRGO.

Several specific scenarios for such non-axisymmetry have been identified:

Centrifugal hangup at $\sim$ 100km radius: If the pre-collapse core is rapidly spinning (e.g., if it is a white dwarf that has been spun up by accretion from a companion), then the collapse may produce a highly flattened, centrifugally supported disk with most of its mass at radii $R \sim$ 100km, which then (via instability) may transform itself into a bar or may bifurcate. The bar or bifurcated lumps will radiate gravitational waves at twice their rotation frequency, $f \sim$ 100Hz — the optimal frequency for LIGO/VIRGO interferometers. To shrink on down to $\sim$ 10km size, this configuration must shed most of its angular momentum. *If* a substantial fraction of the angular momentum goes into gravitational waves, then independently of the strength of the bar, the waves will be nearly as strong as those from a coalescing binary. The reason is this: The waves' amplitude h is proportional to the bar's ellipticity e, the number of cycles n of wave emission is proportional to $1/e^2$, and the characteristic amplitude $h_c = h\sqrt{n}$ is thus independent of the ellipticity and is about the same whether the configuration is a bar or is two lumps (Schutz 1989). The resulting waves will thus have h_c roughly half as large, at $f \sim$ 100Hz, as the h_c from a NS/NS binary (half as large because each lump might be half as massive as a NS), and the waves will chirp upward in frequency in a manner similar to those from a binary.

It is rather likely, however, that most of excess angular momentum does *not* go into gravitational waves, but instead goes largely into hydrodynamic waves as the bar or lumps, acting like a propeller, stir up the surrounding stellar mantle. In this case, the radiation will be correspondingly weaker.

Centrifugal hangup at $\sim$ 20km radius: Lai and Shapiro (1995) have explored the case of centrifugal hangup at radii not much larger than the final neutron star, say $R \sim 20$km. Using compressible ellipsoidal models, they have deduced that, after a brief period of dynamical bar-mode instability with wave emission at $f \sim 1000$Hz [explored by Houser, Centrella and Smith (1994)], the star switches to a secular instability in which the bar's angular velocity gradually slows while the material of which it is made retains its high rotation speed and circulates through the slowing bar. The slowing bar emits waves that sweep *downward* in frequency through the LIGO/VIRGO optimal band $f \sim 100$Hz, toward $\sim$ 10Hz. The characteristic amplitude (Fig. 9) is only modestly smaller than for the upward-sweeping waves from hangup at $R \sim 100$km, and thus such waves should be detectable near the Virgo Cluster by the first LIGO/VIRGO interferometers, and at distances of a few 100Mpc by advanced interferometers.

Successive fragmentations of an accreting, newborn neutron star: Bonnell and Pringle (1995) have focused on the evolution of the rapidly spinning, newborn neutron star as it quickly accretes more and more mass from the pre-supernova star's inner mantle. If the accreting material carries high angular momentum, it may trigger a renewed bar formation, lump formation, wave emission, and coalescence, followed by more accretion, bar and lump formation, wave emission, and coalescence. Bonnell and Pringle speculate that hydrodynamics, not wave emission, will drive this evolution, but that the total energy going into gravitational waves might be as large as $\sim 10^{-3} M_\odot$. This corresponds to $h_c \sim 10^{-21}(10\text{Mpc}/r)$.

5.2. *Spinning Neutron Stars; Pulsars*

As the neutron star settles down into its final state, its crust begins to solidify (crystalize). The solid crust will assume nearly the oblate axisymmetric shape that centrifugal forces are trying to maintain, with poloidal ellipticity $\epsilon_p \propto$(angular velocity of rotation)^2. However, the principal axis of the star's moment of inertia tensor may deviate from its spin axis by some small "wobble angle" θ_w, and the star may deviate slightly from axisymmetry about its principal axis; i.e., it may have a slight ellipticity $\epsilon_e \ll \epsilon_p$ in its equatorial plane.

As this slightly imperfect crust spins, it will radiate gravitational waves (Zimmermann & Szedenits 1979): ϵ_e radiates at twice the rotation frequency, $f = 2f_{\text{rot}}$ with $h \propto \epsilon_e$, and the wobble angle couples to ϵ_p to produce waves at $f = f_{\text{rot}} + f_{\text{prec}}$ (the precessional sideband of the rotation frequency) with amplitude $h \propto \theta_w \epsilon_p$. For typical neutron-star masses and moments of inertia, the wave amplitudes are

$$h \sim 6 \times 10^{-25} \left(\frac{f_{\text{rot}}}{500\text{Hz}}\right)^2 \left(\frac{1\text{kpc}}{r}\right) \left(\frac{\epsilon_e \text{ or } \theta_w \epsilon_p}{10^{-6}}\right) . \tag{5.3}$$

The neutron star gradually spins down, due in part to gravitational-wave emission but perhaps more strongly due to electromagnetic torques associated with its spinning magnetic field and pulsar emission. This spin-down reduces the strength of centrifugal forces, and thereby causes the star's poloidal ellipticity ϵ_p to decrease, with an accompanying breakage and resolidification of its crust's crystal structure (a "starquake") (Shapiro & Teukolsky 1983). In each starquake, θ_w, ϵ_e, and ϵ_p will all change suddenly, thereby changing the amplitudes and frequencies of the star's two gravitational "spectral lines" $f = 2f_{\text{rot}}$ and $f = f_{\text{rot}} + f_{\text{prec}}$. After each quake, there should be a healing period in which the star's fluid core and solid crust, now rotating at different speeds, gradually regain synchronism. By monitoring the amplitudes, frequencies, and phases of the two gravitational-wave spectral lines, and by comparing with timing of the electromagnetic pulsar emission, one might learn much about the physics of the neutron-star interior.

How large will the quantities ϵ_e and $\theta_w \epsilon_p$ be? Rough estimates of the crustal shear moduli and breaking strengths suggest an upper limit in the range $\epsilon_{\max} \sim 10^{-4}$ to 10^{-6}, and it might be that typical values are far below this. We are extremely ignorant, and correspondingly there is much to be learned from searches for gravitational waves from spinning neutron stars.

One can estimate the sensitivity of LIGO/VIRGO (or any other broad-band detector) to the periodic waves from such a source by multiplying the waves' amplitude h by the square root of the number of cycles over which one might integrate to find the signal, $n = f\hat{\tau}$ where $\hat{\tau}$ is the integration time. The resulting effective signal strength, $h\sqrt{n}$, is larger than h by

$$\sqrt{n} = \sqrt{f\hat{\tau}} = 10^5 \left(\frac{f}{1000\text{Hz}}\right)^{1/2} \left(\frac{\hat{\tau}}{4\text{months}}\right)^{1/2} . \tag{5.4}$$

This $h\sqrt{n}$ should be compared (i) to the detector's rms broad-band noise level for sources in a random direction, $\sqrt{5}h_{\text{rms}}$, to deduce a signal-to-noise ratio, or (ii) to h_{SB} to deduce a sensitivity for high-confidence detection when one does not know the waves' frequency in advance (Thorne 1987). Such a comparison suggests that the first interferometers in LIGO/VIRGO might possibly see waves from nearby spinning neutron stars, but the odds of success are very unclear.

The deepest searches for these nearly periodic waves will be performed by narrow-band detectors, whose sensitivities are enhanced near some chosen frequency at the price of sensitivity loss elsewhere—e.g., dual-recycled interferometers (Meers 1988) or resonant-mass antennas (Section 2.4). With "advanced-detector technology," dual-recycled interferometers might be able to detect with confidence all spinning neutron stars that have (Thorne 1987)

$$(\epsilon_e \text{ or } \theta_w \epsilon_p) \gtrsim 3 \times 10^{-10} \left(\frac{500\text{Hz}}{f_{\text{rot}}}\right)^2 \left(\frac{r}{1000\text{pc}}\right)^2 . \tag{5.5}$$

There may well be a large number of such neutron stars in our galaxy; but it is also conceivable that there are none. We are extremely ignorant.

Some cause for optimism arises from several physical mechanisms that might generate radiating ellipticities large compared to 3×10^{-10}:

• It may be that, inside the superconducting cores of many neutron stars, there are trapped magnetic fields with mean strength $B_{\text{core}} \sim 10^{13}$G or even 10^{15}G. Because such a field is actually concentrated in flux tubes with $B = B_{\text{crit}} \sim 6 \times 10^{14}$G surrounded by field-free superconductor, its mean pressure is $p_B = B_{\text{core}} B_{\text{crit}}/8\pi$. This pressure could produce a radiating ellipticity $\epsilon_e \sim \theta_w \epsilon_p \sim p_B/p \sim 10^{-8} B_{\text{core}}/10^{13}$G (where p is the core's material pressure).

• Accretion onto a spinning neutron star can drive precession (keeping θ_w substantially nonzero), and thereby might produce measurably strong waves (Schutz 1995).

• If a neutron star is born rotating very rapidly, then it may experience a gravitational-radiation-reaction-driven instability. In this "CFS" [Chandrasekhar (1970), Friedman & Schutz (1978)] instability, density waves propagate around the star in the opposite direction to its rotation, but are dragged forward by the rotation. These density waves produce gravitational waves that carry positive energy as seen by observers far from the star, but negative energy from the star's viewpoint; and because the star thinks it is losing negative energy, its density waves get amplified. This intriguing mechanism is similar to that by which spiral density waves are produced in galaxies. Although the CFS instability was once thought ubiquitous for spinning stars (Friedman & Schutz 1978, Wagoner 1984), we now know that neutron-star viscosity will kill it, stabilizing the star and turning off the waves, when the star's temperature is above some limit $\sim 10^{10}$K (Lindblom 1990) and below some limit $\sim 10^{9}$K (Lindblom & Mendell 1990); and correspondingly, the instability should operate only during the first few years of a neutron star's life, when $10^9\text{K} \lesssim T \lesssim 10^{10}\text{K}$.

6. Low-Frequency Gravitational-Wave Sources

6.1. *Waves from the Coalescence of Massive Black Holes in Distant Galaxies*

LISA would be a powerful instrument for studying massive black holes in distant galaxies. Figure 6 shows, as examples, the waves from several massive black hole binaries at 3Gpc distance from Earth (a cosmological redshift of unity). The waves sweep upward in frequency (rightward in the diagram) as the holes spiral together. The black dots show the waves' frequency one year before the holes' final collision and coalescence, and the arrowed lines show the sweep of frequency and characteristic amplitude $h_c = h\sqrt{n}$ during that last year. For simplicity, the figure is restricted to binaries with equal-mass black holes: $10^4 M_\odot/10^4 M_\odot$, $10^5 M_\odot/10^5 M_\odot$, and $10^6 M_\odot/10^6 M_\odot$.

By extrapolation from these three examples, we see that LISA can study much of the last year of inspiral, and the waves from the final collision and coalescence, whenever the holes' masses are in the range $3 \times 10^4 M_\odot \lesssim M \lesssim 10^8 M_\odot$. Moreover, LISA can study the final coalescences with remarkable signal to noise ratios: $S/N \gtrsim 1000$. Since these are much larger S/N's than LIGO/VIRGO is likely to achieve, we can expect LISA to refine the experimental understanding of black-hole physics, and of highly nonlinear vibrations of warped spacetime, which LIGO/VIRGO initiates—*provided* the rate of massive black-hole coalescences is of order one per year in the Universe or higher. The rate might well be that high, but it also might be much lower.

By extrapolating Figure 6 to lower BH/BH masses, we see that LISA can observe the last few years of inspiral, but not the final collisions, of binary black holes in the range $100 M_\odot \lesssim M \lesssim 10^4 M_\odot$, out to cosmological distances.

Extrapolating the BH/BH curves to lower frequencies using the formula (time to final coalescence) $\propto f^{-8/3}$, we see that equal-mass BH/BH binaries enter LISA's frequency band roughly 1000 years before their final coalescences, more or less independently of their masses, for the range $100 M_\odot \lesssim M \lesssim 10^6 M_\odot$. Thus, if the coalescence rate were to turn out to be one per year, LISA would see roughly 1000 additional massive binaries that are slowly spiraling inward, with inspiral rates df/dt readily measurable. From the inspiral rates, the amplitudes of the two polarizations, and the waves' harmonic content, LISA can determine each such binary's luminosity distance, redshifted chirp mass $(1+z)M_c$, orbital inclination, and eccentricity; and from the waves' modulation by LISA's orbital motion, LISA can learn the direction to the binary with an accuracy of order one degree.

6.2. *Waves from Compact Bodies Spiraling into Massive Black Holes or Exotic Objects in Distant Galaxies*

When a compact body with mass μ spirals into a much more massive black hole with mass M, the body's orbital energy E at fixed frequency f (and correspondingly at fixed orbital radius a) scales as $E \propto \mu$, the gravitational-wave luminosity $\dot{E}$ scales as $\dot{E} \propto \mu^2$, and the time to final coalescence thus scales as $t \sim E/\dot{E} \propto 1/\mu$. This means that the smaller is μ/M, the more orbits are spent in the hole's strong-gravity region, $a \lesssim 10GM/c^2$, and thus the more detailed and accurate will be the map of the hole's spacetime geometry, which is encoded in the emitted waves.

For holes observed by LIGO/VIRGO, the most extreme mass ratio that we can hope for is $\mu/M \sim 1M_\odot/300M_\odot$, since for $M > 300M_\odot$ the inspiral waves are pushed to frequencies below the LIGO/VIRGO band. This limit on μ/M seriously constrains the accuracy with which LIGO/VIRGO can hope to map out the spacetime geometries of black holes and test the black-hole no-hair theorem (Ryan, Finn & Thorne 1997) (end of Section 4.3). By contrast, LISA can observe the final inspiral waves from objects of any mass $M \gtrsim 0.5M_\odot$ spiraling into holes of mass $3 \times 10^5 M_\odot \lesssim M \lesssim 3 \times 10^7 M_\odot$.

Figure 6 shows the example of a $10M_\odot$ black hole spiraling into a $10^6 M_\odot$ hole at 3Gpc distance. The inspiral orbit and waves are strongly influenced by the hole's spin. Two cases are shown (Finn & Thorne 1997): an inspiraling circular orbit around a non-spinning hole, and a prograde, circular, equatorial orbit around a maximally spinning hole. In each case the dot at the upper left end of the arrowed curve is the frequency and characteristic amplitude one year before the final coalescence. In the nonspinning case, the small hole spends its last year spiraling inward from $r \simeq 7.4GM/c^2$ (3.7 Schwarzschild radii) to its last stable circular orbit at $r = 6GM/c^2$ (3 Schwarzschild radii). In the maximal spin case, the last year is spent traveling from $r = 6GM/c^2$ (3 Schwarzschild radii) to the last stable orbit at $r = GM/c^2$ (half a Schwarzschild radius). The $\sim 10^5$ cycles of waves during this last year should carry, encoded in themselves, rather accurate values for the massive hole's lowest few multipole moments (Ryan 1995). If the measured moments satisfy the "no-hair" theorem (i.e., if they are all determined uniquely by the measured mass and spin in the manner of the Kerr metric), then we can be sure the central body is a black hole. If they violate the no-hair theorem, then (assuming general relativity is correct), either the central body was an exotic object (e.g. soliton star or naked singularity) rather than a black hole, or else an accretion disk or other material was perturbing its orbit (Molteni, Gerardi & Chakrabarti 1994). From the evolution of the waves one can hope to determine which is the case, and to explore the properties of the central body and its environment (Ryan, Finn & Thorne 1997).

Models of galactic nuclei, where massive holes reside, suggest that inspiraling stars and small holes typically will be in rather eccentric orbits (Hils & Bender 1995). This is because they get injected into such orbits via gravitational deflections off other stars, and by the time gravitational radiation reaction becomes the dominant orbital driving force, there is not enough inspiral left to fully circularize their orbits. Such orbital eccentricity will complicate the waveforms and complicate the extraction of information from them. Efforts to understand the emitted waveforms are just now getting underway.

The event rates for inspiral into massive black holes are not at all well understood. However, since a significant fraction of all galactic nuclei are thought to contain massive holes, and since white dwarfs and neutron stars, as well as small black holes, can withstand tidal disruption as they plunge toward the massive hole's horizon, and since LISA can see inspiraling bodies as small as $\sim 0.5M_\odot$ out to 3Gpc distance, the event rate is likely to be interestingly large.

7. Conclusion

It is now 35 years since Joseph Weber initiated his pioneering development of gravitational-wave detectors (Weber 1960) and 25 years since Forward (Moss, Miller & Forward 1971) and Weiss (1972) initiated work on interferometric detectors. Since then, hundreds of talented experimental physicists have struggled to improve the sensitivities of these instruments. At last, success is in sight. If the source estimates described in this review article are approximately correct, then the planned interferometers should detect the first waves in 2001 or several years thereafter, thereby opening up this rich new window onto the Universe.

8. Acknowledgments

My group's research on gravitational waves and their relevance to LIGO/VIRGO and LISA is supported in part by NSF grants AST-9417371 and PHY-9424337 and by NASA grant NAGW-4268. This manuscript was largely adapted from my longer review article (Thorne 1995).

REFERENCES

ABRAMOVICI, A. et al. 1992 *Science* **256**, 325.

APOSTOLATOS, T. A., CUTLER, C., SUSSMAN, G. J. & THORNE, K. S. 1994 *Phys. Rev. D* **49**, 6274.

APOSTOLATOS, T. A. 1995 *Phys. Rev. D* **52**, 605.

ARMSTRONG, J., BENDER, P., AND EARDLEY, D., EVERITT, C. W. F., HELLINGS, R., NORDTVEDT, K., PAIK H. J., REASENBERT, R., SHAPIRO, I. (CHAIR), TAYLOR, J., THORNE, K., WEISS, R. & WILL, C. 1990 *Report of the Ad Hoc Committee on Gravitation Physics and Astronomy.* Astrophysics Division, NASA Headquarters, Washington DC.

ASHTEKAR, A (ed.) 1983 In *Gravitational Radiation* (ed. N. Deruelle & T. Piran), pp. 421–433. North Holland.

BAILES, M. 1995 In *Proceedings of I. A. U. Symposium 165, Compact Stars in Binaries* (ed. J. van Paradijs, E. van den Heuvel & E. Kuulkers), p. 213. Kluwer Academic Publishers, Dordrecht.

BASSAN, M. 1994 *Class. Quant. Grav.* **11S**, A39.

BENDER, P., BRILLET, A., CIUFOLINI, I., DANZMANN, K., HELLINGS, R., HOUGH, J., LOBO, A., SANDFORD, M., SCHUTZ, B. & TOUBOUL, P. 1994 *LISA, Laser Interferometer Space Antenna for Gravitational Wave Measurements: ESA Assessment Study Report.* R. Reinhard, ESTEC.

BETHE, H. A. 1990 *Rev. Mod. Phys.* **62**, 801.

BILDSTEN, L. & CUTLER, C. 1992 *Astrophys. J.* **400**, 175.

BLANCHET, L., DAMOUR, T., IYER, B. R., WILL, C. M. & WISEMAN, A. G. 1995 *Phys. Rev. Lett.* **74**, 3515.

BONNELL, I. A. & PRINGLE, J. E. 1995 *Mon. Not. Roy. Astron. Soc.* **273**, L12.

BRADASCHIA, C. et al. 1990 *Nucl. Instrum. & Methods A* **289**, 518

BRAGINSKY, V. B., MITROFANOV, V. P. & OKHRIMENKO, O. A. 1993 *Phys. Lett. A* **175**, 82.

BURROWS, A., HAYES, J. & FRYXELL, B. A. 1995 *Astrophys. J.* **450**, 830.

BURROWS, A. 1995 Private communication.

CHANDRASEKHAR, S. 1970 *Phys. Rev. Lett.* **24**, 611.

CHERNOFF, D. F. & FINN, L. S. 1993 *Astrophys. J. Lett* **411**, L5.

CHRISTODOULOU, D. 1991 *Phys. Rev. Lett.* **67**, 1486.

CUTLER, C., APOSTOLATOS, T. A., BILDSTEN, L., FINN, L. S., FLANAGAN, E. E., KENNEFICK, D., MARKOVIC, D. M., ORI, A., POISSON, E., SUSSMAN, G. J. & THORNE, K. S. 1993 *Phys. Rev. Lett.* **70**, 1984.

CUTLER, C. & FLANAGAN, E. E. 1994 *Phys. Rev. D* **49**, 2658.

CUTLER, C. & FLANAGAN, E. E. 1997 Paper in preparation.

DAMOUR, T. & TAYLOR, J. 1992 *Phys. Rev. D* **45**, 1840.

DAMOUR, T. & NORDTVEDT, K 1993 *Phys. Rev. D* **48**, 3436.

DAVIES, M. N., BENZ, W., PIRAN, T. & THIELEMANN, F. K. 1994 *Astrophys. J.* **431**, 742.

DAVIS, R. L., HODGES, H. M., SMOOT, G. F., STEINHARDT, P. J. & TURNER, M. S. 1992 *Phys. Rev. Lett.* **69**, 1856.

DREVER, R. W. P 1983 Interferomagnetic detectors of gravitational radiation. In *Gravitational Radiation* (ed. N. Deruelle & T. Piran), pp. 321–338, and references therein. North Holland.

FINN, L. S. 1991 *Ann. N. Y. Acad. Sci.* **631**, 1.

FINN, L. S. & CHERNOFF, D. F. 1993 *Phys. Rev. D* **47**, 2198.

FINN, L. S. & THORNE, K. S. 1997 Paper in preparation.

FLANAGAN, E. E. 1993 *Phys. Rev. D* **48**, 2389.

FLANAGAN, E. E. & HUGHES, S. A. 1997 Paper in preparation

FRIEDMAN, J. L. & SCHUTZ, B. F. 1978 *Astrophys. J.* **222**, 781.

GILLESPIE, A. & RAAB, F. 1994 *Phys. Lett. A* **190**, 213.

GRISHCHUK, L. P. 1974 *Soviet Physics–JETP* **40**, 409.

GRISHCHUK, L. P. 1993 *Class. Quant. Grav.* **10**, 2449.

HEUVEL, E. P. J. VAN DEN & LORIMER, D. R. 1996 *Mon. Not. Roy. Astron. Soc.* in press.

HILS, D., BENDER, P., & WEBBINK, R F. 1990 *Astrophys. J.* **360**, 75.

HILS, D. & BENDER, P. 1995 *Astrophys. J. Lett.* **445**, L7.

HOUGH, J. & DANZMANN, K. ET AL. 1994 *GEO600, Proposal for a 600 m Laser-Interferometric Gravitational Wave Antenna.* (unpublished report)

HOUSER, J. L., CENTRELLA, J. M. & SMITH, S. C. 1994 *Phys. Rev. Lett.* **72**, 1314.

HULSE, R. A. & TAYLOR, J. H. 1975 *Astrophys. J.* **324**, 355.

JARANOWSKI, P. & KROLAK, A. 1994 *Phys. Rev. D* **49**, 1723.

JOHNSON, W. W. & MERKOWITZ, S. M. 1993 *Phys. Rev. Lett.* **70**, 2367.

KASPI,V. M., TAYLOR, J. H. & RYBA, M. F. 1994 *Astrophys. J.* **428**, 713.

KENNEFICK, D. 1994 *Phys. Rev. D* **50**, 3587.

KENNEFICK, D., LAURENCE, D. & THORNE, K. S. 1997 Paper in preparation.

KIDDER, L. E., WILL, C. M. & WISEMAN, A. G. 1993 *Phys. Rev. D* **47**, 3281.

KIDDER, L. E. 1995 *Phys. Rev. D* **52**, 821.

KOCHANEK, C 1992 *Astrophys. J.* **398**, 234.

KOSOWSKY, A. & TURNER, M. S. 1994 *Phys. Rev. D* **49**, 2837.

KRAUSS, L. M. & WHITE, M. 1992 *Phys. Rev. Lett.* **69**, 969.

KRÓLAK, A., KOKKOTAS, K. D. & SCHÄFER G. 1995 In *Proceedings of the 17th Texas Symposium on Relativistic Astrophysics, Ann. N. Y. Acad. Sci.* **759**, 493.

LAI, D. & SHAPIRO, S. L. 1995 *Astrophys. J.* **442**, 259.

LINCOLN, C. W. & WILL, C. M. 1990 *Phys. Rev. D* **42**, 1123.

LINDBLOM, L. 1995 *Astrophys. J.* **438**, 265.

LINDBLOM , L. & MENDELL, G. 1995 *Astrophys. J.* **444**, 804.

LIPUNOV, V. M., POSTNOV, K. A. & PROKHOROV, M. E. 1994 *Astrophys. J.* **423**, L121.

MARKOVIC, D. 1993 *Phys. Rev. D* **48**, 4738.

MEERS, B. J. 1988 *Phys. Rev. D* **38**, 2317.

MOLTENI, D., GERARDI, G. & CHAKRABARTI, S. K. 1994 *Astrophys. J.* **436**, 249.

MOSS, G. E., MILLER, L. R. & FORWARD, R. L. 1971 *Applied Optics* **10**, 2495.

MÖNCHMEYER, R., SCHÄFER, G. & MÜLLER, E. 1991 *Astron. Astrophys.* **256**, 417.

NAKAMURA, T 1994 In *Relativistic Cosmology* (ed. M. Sasaki), p. 155. Universal Academy Press.

NUMERICAL RELATIVITY GRAND CHALLENGE ALLIANCE 1995 References and information on the World Wide Web, http://jean-luc.ncsa.uiuc.edu/GC

NARAYAN, R., PIRAN, T., & SHEMI, A. 1991 *Astrophys. J.* **379**, L17.

PAIS, A. 1982 *'Subtle is the Lord...' The Science and the Life of Albert Einstein.* Clarendon Press, Oxford.

PETSCHEK, A. G. (ed.) 1990 *Supernovae.* Springer Verlag.

PHINNEY, E. S. 1991 *Astrophys. J.* **380**, L17.

PIZELLA, G. 1992 *Nucl. Phys. B, Suppl.* **28A**, 48.

POISSON, E. & SASAKI, M. 1995 *Phys. Rev. D* **51**, 5753.

POISSON, E. & WILL, C. M. 1995 *Phys. Rev. D* **52**, 848.

QUINLAN, G. & SHAPIRO, S. L. 1987 *Astrophys. J.* **321**, 199.

RASIO, F. A. & SHAPIRO, S. L. 1992 *Astrophys. J.* **401**, 226.

RYAN, F. D. 1995 *Phys. Rev. D* **53**, 3064.

RYAN, F. D., FINN, L. S. & THORNE, K. S. Paper in preparation.

SATHYAPRAKASH, B. S. 1994 *Phys. Rev. D* **50**, R7111.

SAULSON, P 1990 *Phys. Rev. D* **42**, 2437.

SCHUTZ, B. F. 1986 *Nature* **323**, 310.

SCHUTZ, B. F. 1989 *Class. Quant. Grav.* **6**, 1761.

SCHUTZ, B. F. 1995 Private communication.

SHAPIRO, S. L. & TEUKOLSKY, S. A. 1983 *Black Holes, White Dwarfs and Neutron Stars.* Wiley, Interscience.

SHIBATA, M., NAKAMURA, T. & OOHARA, K. 1993 *Prog. Theor. Phys.* **89**, 809.

SHIBATA, M., SASAKI, M., TAGOSHI, H. & TANAKA, T. 1996 *Physical Review D* **54**, 1439.

SULLIVAN, W. J. 1984 *The Early Years of Radio Astronomy.* Cambridge University Press.

TAGOSHI, H. & NAKAMURA, T. 1994 *Phys. Rev. D* **49**, 4016.

TAGOSHI, H. & SASAKI, M. 1994 *Prog. Theor. Phys.* **92**, 745.

TAYLOR, J. H. 1994 *Rev. Mod. Phys.* **66**, 711.

THORNE, K. S. 1983 The theory of gravitational radiation: an introductory overview. In *Gravitational Radiation* (ed. N. Deruelle & T. Piran), pp. 1–58. North Holland.

THORNE, K. S. 1987 In *Three Hundred Years of Gravitation* (ed. S. W. Hawking & W. Israel), pp. 330–458. Cambridge University Press.

THORNE, K. S. 1992 *Phys. Rev. D* **45**, 520.

THORNE, K. S. 1994 In *Relativistic Cosmology* (ed. M. Sasaki), p. 67. Universal Academy Press.

THORNE, K. S. 1995 In *Proceedings of the Snowmass 95 Summer Study on Particle and Nuclear Astrophysics and Cosmology* (ed. E. W. Kolb & R. Peccei), p. 398. World Scientific.

TUTUKOV, A. V. & YUNGELSON, L. R. 1993 *Mon. Not. Roy. Astron. Soc.* **260**, 675.

VACHASPATI, T. & VILENKIN, A. 1985 *Phys. Rev.* **31**, 3052.

VILENKIN, A. 1981 *Phys. Rev. D* **24**, 2082.

VILENKIN, A. 1987 In *Three Hundred Years of Gravitation* (ed. S. W. Hawking & W. Israel), pp. 499–523. Cambridge University Press.

WAGONER, R. V. 1984 *Astrophys. J.* **278**, 345.

WEBER, J. 1960 *Phys. Rev.* **117**, 306.

WEISS, R. 1972 *Quarterly Progress Report of RLE, MIT* **105**, 54.

WHITCOMB, S. E. AND THE LIGO R&D SCIENCE TEAM 1994 *LIGO Report* No. 94-7.

WILL, C. M. 1994a In *Relativistic Cosmology* (ed. M. Sasaki), p. 83. Universal Academy Press.

WILL, C. M. 1994b *Theory and Experiment in Gravitational Physics.* Cambridge University Press.

WILL, C. M. 1994c *Phys. Rev. D* **50**, 6058.

WISEMAN, A. G. & WILL, C. M. 1991 *Phys. Rev. D* **44**, R2945.

WITTEN, E. 1984 *Phys. Rev. D* **30**, 272.

YAMAOKA, H., SHIGEYAMA, T. & NOMOTO, K. 1993 *Astron. Astrophys.* **267**, 433.

ZHUGE, X. & CENTRELLA, J. M. 1994 *Phys. Rev. D* **50**, 6247.

ZEL'DOVICH, YA. B. 1980 *Mon. Not. Roy. Astron. Soc.* **192**, 663.

ZEL'DOVICH, YA. B. & NOVIKOV, I. D. 1983 *Relativistic Astrophysics, Volume 2: The Structure and Evolution of the Universe.* pp. 160–163. U. Chicago Press.

ZIMMERMANN, M. & SZEDENITS, E. 1979 *Phys. Rev. D* **20**, 351

ZWART, S. F. P. & SPREEUW, H. N. 1996 *Astron. Astrophys.* **312**, 670.

The Past and Future of Relativistic Astrophysics

IGOR D. NOVIKOV[1,2,3,4]

[1]Theoretical Astrophysics Center, Juliane Maries Vej 30, DK-2100 Copenhagen O, Denmark
[2]University Observatory, DK-2100 Copenhagen O, Denmark
[3]NORDITA, Blegdamsvej 17, DK-2100 Copenhagen O, Denmark
[4]Lebedev Physical Institute, Profsoyusnaya 84/32, 117810 Moscow, Russia

1. Introduction

This meeting is not an ordinary one for me and so this talk is not my usual talk: I have chosen a subject which I think, is appropriate for this event. Historically, it so happened that I played a part in various stages of the history of research on Relativistic Astrophysics and Black Holes. In this essay I share recollections about my participation in projects and about my meetings and work with the creators of relativistic astrophysics. I will not touch upon my research on cosmology. This is basically a historical and subjective sketch, and not an exposition of the subject itself,and thus I try to keep technicalities to a minimum. In the last part of the essay I shall outline some future perspectives. For the preparation of this paper I have used some material from my book "River of time" Novikov (1996).

2. The Past

I began my scientific career in the 1960's, more than 30 years ago. What was Relativistic Astrophysics 30-35 years ago? There was no Relativistic Astrophysics at all! Neutron stars, black holes and quasars had not yet been discovered. There was no X-ray and γ- ray astronomy, no theory of black holes, not even the name "black hole". Theorists knew only the Schwarzschild solution which describes (as we know today) the gravitational field of the simplest not-rotating black hole. But even in this simplest of cases there was no understanding of the nature of the boundary of a black hole, the so called Schwarzschild singularity at the value of the radius r equal to the gravitational radius $r_g = 2GM/c^2$, where M is the mass of the black hole. Stephen Hawking began his career approximately at the same time as I did. He wrote about those early days [Hawking (1993)]: "There has been a great change in the status of general relativity and cosmology in the last thirty years. When I began research in the Department of Applied Mathematics and Theoretical Physics (DAMTP) at Cambridge in 1962, general relativity was regarded as a beautiful but impossibly complicated theory that had practically no contact with the real world. Cosmology was thought of as a pseudo-science where wild speculation was unconstrained by any possible observations".

I studied general relativity while still a student under the guidance of Dr. A.Zelmanov. During this period a question formed in my mind: what if the size of the massive body is less than its gravitational radius? Of course, I went to the teacher, who said that such bodies were thought to be physically impossible, but that he had never come across an explanation for this assumption. I discovered later that neither Zelmanov, not anyone else in the world had bothered to look into this matter. The problem was simply left lying off the main road that science was pursuing. The traditional belief was that the "Schwarzschild singularity" was somehow unphysical. Typical conclusions were as follows: "The "Schwarzschild singularity" does not appear for the reason that matter cannot be concentrated arbitrarily" Einstein (1939). "In nature, mass is never sufficiently concentrated to permit a Schwarzschild singularity to occur in empty space. Einstein considered a system of many mass points. ... The investigation showed that even before the critical concentration of particles is reached, some of the particles (the ones on the outside) begin to move with the velocity of light..." Bergman (1942). "Every gravitating particle has a ring-fence around it, which no other body can penetrate" Whittaker (1949).

Astronomers did not know of such dense objects in the Universe. Any arguments in this field were regarded as useless and unfounded; moreover, very few people had mastered general relativity at that time. Astronomers assumed that this science was of no use of them: it reigned in the realm of super strong gravitational fields,while no such fields were observed in the Universe. Nevertheless, I did not forget the impression it made on me, and having become a postgraduate, I chose to tackle the problem in earnest.

When my term of postgraduate studies was almost up, I wrote a paper [Novikov (1961)] about the region of the space-time inside the Schwarzschild sphere and this is still one of my pet papers. The gist of this discovery of mine was that once we switch from the external space to the space inside a black hole, the time coordinate in the formulas is simply replaced by the spatial radial coordinate, and vice versa. In other words, time transforms into the spatial radial distance, and this distance is time itself!

When I had written the corresponding equations (the resulting one was only a line long), As was the custom, I immediately went to show them to Zelmanov. One brief glance at my text was enough; Zelmanov understood everything and a couple of seconds later ordered: "To be sent to a journal, right away". This was an overwhelming compliment from a man who usually insisted on months of rethinking and recalculating and who was honing his own publications for years. This was how I started my research in black hole physics.

At the end of 1950's, the problem of the inner space of black holes attracted young physicists in other countries: D. Finkelstein and M. Kruskal, then others (see the essay by Israel (1987)). However, I was not aware of this during the early stages of my work. I was going forward, and succeeded in the finding how a free body would move within the event horizon. I gave this region in which nothing can be at rest and where everything is forced to fall towards the center, the name T *-region.* This name emphasized the mandatory dependence on time, T. Such regions are often referred to in this way even now.

When astrophysicists developed a serious interest in black holes in the 1960's, theoretical physicists faced new and very complex problems. Robert Oppenheimer and Hartland Snyder (1939) described the birth of a black hole in the compression of a perfectly spherical mass. However, nature never creates an absolutely ideal spherical body. What if the collapsing body is not spherical?

The question attracted me when I completed my postgraduate term and came to work in professor (Academician) Yakov Zeldovich's group. Yakov Borisovich was a genius in physics, and we - his young colleagues and students - were lucky to work with him. But it was not easy! Kip Thorne described Zeldovich's style in the following way K.Thorne (1994): "Zeldovich whipped his team into shape with a firm hand, a constant barrage of his own ideas, and joint exploitation of his team's ideas." "Zeldovich - still infused with the spirit of the race for the superbomb - sought the fastest pace possible, whatever the expense."

I would like to add one more trait which characterizes Zeldovich's style. Yakov Borisovich liked to write letters to his younger colleagues. The contents of these letters were science, calculations, plans, but mainly - criticism. As an example I'll give a few quotations from one of these letters. It was written in 1972 when our group was at the height of its success, and probably was the best in the world in relativistic astrophysics. The letter was addressed to Novikov, Doroshkevich, Lukash, Polnarev and Shandarin (all of whom are participants in this meeting). Zeldovich (1972) wrote (in my translation from Russian into English): "Your group is in a crisis situation...Why have you not caught on to confronting theory and observations for ^3He, D, ^7Li? (as was done by Reeves, but not by you)... Why did you not grind out estimates of gravitational waves using fluctuations of relic radiation and peculiar motions [of galaxies] (as was done by Dautcourt and Rees, but not by you)?... If there is no turning-point you will in big trouble!... Well, you are 5 men! Compare - Silk + Rees + Hawking=3 ONLY".

Zeldovich was a maximalist, he always demanded more, and more, and still more. But it was great happiness to work with him.

Let us come back to the 60's. Our team leader, a friend of mine of the same age as myself, Andrei Doroshkevich and I, started to work on the problem of the gravitational collapse of a non-spherical mass.† Calculations demonstrated that the compression of a (non-rotating) non-symmetrical body produces a black hole which very rapidly becomes perfectly symmetrical. Any deviation from sphericity in the gravitational field must be radiated away in the formation of the black hole and fly away with gravitational waves. The emerging boundary of the black hole - the event horizon - is spherical and only spherical! On the other hand the collapse of a rotating mass leads to a specific rotating black hole [Doroshkevich, Zeldovich & Novikov (1965)].

I reported this result at an International Conference in the summer of 1965. This was my first journey abroad, the first serious discussion of these problems with Western specialists, the first presentation of results to international scientific community by the recently formed group of Academician Zeldovich. The debut was successful. It became clear to me that owing to Zeldovich's tremendous physical intuition, his persistence and overwhelming industriousness which fascinated and conquered his students and which was based on his literally childish love of nature's secrets, our small group was leading in a new field of science - relativistic astrophysics.

After the talk I was surrounded by a crowd of colleagues who wanted to know details of calculations. Among these enthusiasts I immediately spotted a tall, lean, reddish-haired youngster, a typical American, as I imagined. After my presentation Kip Thorne (the name of the youth) helped me to make myself understood to people who wanted to discuss my talk (my English at that time was far below the grade of "far from perfect").

We then went on with our conversation. We quickly established common areas of scientific interest and, what counted at least as much, discovered important similarities in our outlook on the world (what a Russian would call "related souls"). Soon we grew to be real friends. This was the beginning of the finest friendship of my life.

I believe that this paper in a book dedicated to my jubilee is an appropriated opportunity to outline briefly events in our friendship which took place many years after our first meeting and which saved my life and allowed me to do physics now and, I hope, in the future.

In 1990 I had serious problems with my heart and an operation could be performed made only in the West. I had no currency at all for the operation. Kip immediately arranged for everything, including depositing his own money, and writing letters to other physicists about financial support. As a result dozens and dozens of colleagues sent

† In our publications we always used the alphabetical order for the names of the authors (according to the Russian alphabet if it is published in Russia). I believed in the past and I believe now that only this order is reasonable and not subjective or even discriminatory.

contributions to cover the cost of the operation. I came to California and before the operation lived in Kip's house. Kip dropped everything he was doing and together with Carolee, his wife, never left my side.

The cardiologist, the surgeon and the leading anesthetist of the Huntington Memorial Hospital in Pasadena decided to forego the treatment fees. When, after the successful operation, consciousness began to return to me, the first thing I understood was Kip's voice repeating (in Russian!) "Igor, everything is fine, just fine!" Even though my daughter, with whom I was always very close, was sitting by my bed, I perceived as something self-evident that it should be the voice of Kip Thorne which brought me back to life.

Kip, his wife, my colleagues around the world, and of course, American doctors granted me a second life. My thanks to the world for that.

Let us come back to the 60's. My work during that period appealed to Professor Andrei Dmitrievich Sakharov, who was getting more and more interested in problems of gravitation and cosmology. We were discussing the problems of black holes, the topology of the space-time, and cosmology between ourselves and with Yakov Zeldovich. As a result we published our papers in a report of the Institute of Applied Mathematics "Relativistic collapse and topological structure of the Universe" Novikov and Sakharov (1970). Soon after this publication, Sakharov acted as an official referee of my D.Sc dissertation. Our discussions with Sakharov were very important for me at the beginning of my scientific career.

In the late 60's Sakharov often paid visits to our laboratory. His purpose was to talk physics with Yakov Zeldovich, but Doroshkevich and I, still very young collaborators of Zeldovich, were also present (but mostly kept our mouths shut). Most of what Sakharov discussed was hardly known or understandable to us. To me (and very likely to others) the subject seemed extremely abstract and far removed from any conceivable reality. Today the world knows how uncompromising Sakharov was in the crucial matters of social life, in fact, a similar independence of any outside pressure could be felt in the evolution of his scientific ideas. It seemed to me that in the inner development of his ideas he never accepted any "compromise" that might "smooth out" their unconventionality. His approach to solving scientific problems was more than just unconventional.

For example, Sakharov suggested theoretically calculating the coefficient of universal gravitation - one of the few "fundamental" constants of nature - from the assumption that it describes the "elasticity of the vacuum". Odd, isn't it? However from the standpoint of today's physics (decades after Sakharov's work) this idea does not appear to be all that outlandish.

I do not know whether this unusual approach to scientific problems is fruitful or not, but there can be no doubt that no one else could work in the way. Perhaps Sakharov resembled Einstein in this respect but this is something on which I do not risk offering an opinion.

When Sakharov progressed from elementary particle physics and field theory, in which I was not very knowledgeable, to the physics of black holes, where I felt quite strong, the possibility of understanding increased considerably. However, even then the logic of Sakharov's inferences often baffled me. In fact, not only me. The following story comes to mind.

Andrei Sakharov was giving a talk at our seminar about the possible properties of a "gorge" in space (which we now call a "wormhole"). By that time there were about ten young colleagues in our group. We listened in silence. Yakov Zeldovich always treated this topic, to put it mildly, with disapproval. Furthermore, he was annoyed by Sakharov's quaint approach to treating the problem. To some extent, the "strangeness" of Sakharov approach was multiplied by the "strangeness" of the topic itself. Zeldovich felt that this produced a "perversity" in physics. Some time after Sakharov began his talk, Zeldovich started to show signs of impatience. Nevertheless, he did not interrupt the speaker even though questions and debates during presentation were a matter of course at our seminars. Quite often, heated debates flared up in the middle of a talk, and lasted longer than the talk itself. The unusual thing this time was rather the silence of the audience, and, above all, of Zeldovich. At last he stood up, went behind his chair and leaned on it. Sakharov looked at him inquiringly but Zeldovich merely waved him on: "don't pay attention to me, go ahead with the talk". Sakharov completed the presentation. "Is that all?" asked Zeldovich. "Yes, that is all I planned to describe" replied Sakharov softly and blinked. Zeldovich immediately asked for a newspaper, carefully spread it on the floor and, under our unbelieving stares, knelt on it in front of Sakharov. Placing his palms in a position of worship, he exclaimed: "Andrei Dmitrievich, please, stop doing this nonsense!" Fortunately, Sakharov did not stop and obtained very unusual results in the field.

Among other results of this early period of my research I should mention three more. The first was the paper of Zeldovich and Novikov (1964) entitled "The mass of quasi-stellar objects". In this paper we conjectured that the main engine of a quasar is a supermassive black hole with gas accretion onto it. On the basis of data on 3C273 we gave estimates of the central mass. For that we hypothesized that the radiation pressure balances gravity, so the central mass is approximately $10^8 M_\odot$. We discussed also physical processes in the vicinity of the central mass and many of those ideas remain popular today.

The second was the prediction of a possible way to detect black holes with stellar masses in the Universe. In 1966 Zeldovich and I pointed out that black holes (and neutron stars) can act as extremely powerful sources of X-ray radiation. This situation occurs if there is an ordinary (normal) star quite close to a black hole. In the paper Novikov & Zeldovich (1966) we wrote: "In a double system the accretion on the collapsed star is felt by the stellar wind from the normal star. It will lead to strong X-ray and γ-ray emission". According to the picture generally accepted now, the gravitational pull of the black hole will then force the gas from the atmosphere of the normal companion to flow in a spiral towards the black hole forming a compact gas disk. As a result of friction between gas layers, the gas in the disk heats up to a temperature of tens of millions of degrees and, before it sinks into the black hole, emits X-rays. This X-ray emission makes the black hole visible.

X-ray sources in binary stellar systems were first discovered in 1972. Some of these are, in the opinion of the majority of experts, black holes.

The third result concerns primordial black holes. These black holes could be formed in the very early Universe from initial perturbations. Gravitational forces could then slow down the expansion in the corresponding volumes and some time later give rise to compression, producing the black holes. Zeldovich and I (Zeldovich and Novikov (1966)) in 1966 and Hawking (1971) in 1971 pointed to this possibility. In the scientific literature the possibility of forming such primordial black holes, especially black holes with mass much less than the mass of the Sun and other stars, is often ascribed to Stephen Hawking. I believe that the reason for it is the following. Even though Yakov Zeldovich and I clearly understood and often mentioned it years before Hawking's paper was published we never emphasized with sufficient clarity and decisiveness that black holes with masses much smaller than stellar masses could be formed in such a way. I believe that clarity is mandatory if authors wish to be understood correctly, otherwise their pronouncements have a chance to remain "things unto themselves", comprehensible only to the authors. Or as Zeldovich used to say, they remain a V-sign given behind your back.

Probably I should stop here in my recollection of the beginning of relativistic astrophysics. Now a few words about future.

3. The Future

What can we expect in the coming century? What problems of relativistic astrophysics and the physics of black holes seem now to be especially important and interesting? I will give a few lists of such problems which were worked out recently by some participants of our conference. The first list is the most important problems of Astronomy according to Ginzburg (1995).

1. Experimental verification of the general theory of relativity.
2. Gravitational waves.
3. The cosmological problem; the connection between cosmology and high energy physics.
4. Neutron stars and pulsars.
5. Black holes.
6. Quasars and galactic nuclei; formation of galaxies; hidden mass (dark matter) and its detection.
7. The origin of cosmic rays and cosmic X-rays and gamma-ray emission; gamma-ray astronomy at super-high energies.
8. Neutrino astronomy.

The second list was published by Kardashev (1992)

1. Higher forms of intelligence in the Universe.
2. The anthropic principle and the multiplicity of Universes.
3. A cosmological model and the evolution of the present Universe.
4. Hidden mass.
5. A cosmological model and the evolution of the early Universe.
6. Galactic nuclei.
7. Planetary systems and solid bodies in the Universe.
8. Gravitational wave astrophysics and relic gravitational waves.
9. Neutrino astrophysics and relic neutrinos.
10. The origin of cosmic rays.

The author of the third list is Kip Thorne K.Thorne (1994). I made this list by slightly abridging the description of the most important problems of the physics in his book "Black holes and Time Warps" which are related with the space-time, Universe and black holes.

1. Gravitational - wave astronomy. Black holes will become objects of detailed experimental scrutiny.
2. Quantum gravity.
3. How our Universe's spacetime came into being.
4. How spacetime gets destroyed in the singularity of a black hole or in the Big Crunch.
5. Do the laws of quantum gravity permit time machines?
6. Grand unification of all physical forces.

In conclusion, the list I made myself.

1. The physical nature of the dark matter.

2. The origin of large scale structure.
3. A realistic theory of physical processes in the vicinity of black holes.
   a) in binaries,
   b) in galactic nuclei,

and confrontation with coming observations.

4. The internal structure of black holes.
5. Are there primordial black holes?
6. The nature of γ-ray bursters.
7. Gravitational-wave astronomy.

Of course, the lists are different in style and partly in subjects, reflecting the tastes and personal interests of the authors. Nevertheless, one can observe common points and tendencies in all four lists. Thus the future looks very intriguing and fascinating.

Of course, the development of science depends not only on scientists but on social factors. For example, the conditions for scientific work in my native country, the former USSR, are far from perfect now. The country, the former enormous empire, broke into pieces and is trying with untold hardship for its peoples, to claw its way out of the frightening historical abyss into which it was plunged. Still we hope for the best in the future.

In any case, scientists will work under any conditions just because they are scientists. As an outstanding physicist Peter Kapitza wrote in his letter to I.Stalin in the very difficult period of his life (I quote after Sagdeev (1994): "One is not a scientist who simply carries on scientific research, but a scientist is one who cannot **not** carry on scientific research."

REFERENCES

BERGMAN, P.G. 1942 An Introduction to the Theory of Relativity. *Prentice-Hall: New York* p.p. 203-204.

DOROSHKEVICH A.G., ZELDOVICH YA.B. & NOVIKOV I.D. 1965 *Zhurnal Eksperimentalnoi i Teoreticheskoi Fiziki,* **49**, 170. *English translation in Soviet Physics - JETP,* **22**, 122, 1966

EINSTEIN, A. 1939 *Ann. Math. (Princeton),* **40**, 922.

GINSBURG V.L. 1992 on Physics and Astrophysics (in Russian) *Buro quantum, Moscow,* page 5

HAWKING, S. 1993 Hawking on the Big Bang and Black Holes. *World Scientific Publishing Co. Pte. Ltd.,* page 1.

HAWKING, S.W. 1971 *Monthly Notices Roy. Astron. Soc.,* **152**, 75.

ISRAEL W. 1987 300 Years of Gravitation, ed. by S.Hawking and W.Israel. *Cambridge University Press* p. 199.

KARDASHEV N.S. 1992 Astrophysics on the Threshold of the 21st Century. *ed. by N.S.Kardashev, Gordon and Breach,* page 357.

NOVIKOV I.D. 1961 *Soviet Astronomy - A.J.,* **5**, 423.

NOVIKOV I.D. 1996 River of Time. *Cambridge University Press,* (in Press).

NOVIKOV I.D. & A. SAKHAROV 1970 Relativistic collapse and topological structure of the Universe *Report of the Institute of Applied Mathematics N7, Moscow.*

NOVIKOV I.D. & YA.B.ZELDOVICH 1966 *Nuovo Cimento. Suppl.,* **4**, 810.

OPPENHEIMER J.R. & H.SNYDER 1939 *Phys. Rev.,* **56**, 455.

SAGDEEV, R.Z. 1994 The Making of a Soviet Scientist, Wiley.

THORNE K. 1994 Black Holes and Time Wraps, Einstein Outrageous Legacy, W.V. Norton and Company, Inc., NY.

WHITTAKER E.T. 1949 From Euclid to Eddington *Cambridge University Press: Cambridge,* p.124.

ZELDOVICH YA.B. 1972 a letter to Novikov, Doroshkevich, Polnarev *unpublished*

ZELDOVICH YA.B. & NOVIKOV I.D. 1964 *Sov. Phys. Dokl.*, **9**, 834.

ZELDOVICH YA.B. & NOVIKOV I.D. 1966 *Astron. Zh.*, **43**, 758.

I. D. Novikov's scientific papers and books

Papers

- "VZ Cygni", *Peremennie zvezdi (Variable stars)*, **8**, 75, (1952).
- "BD+622332", *Peremennie zvezdi (Variable stars)*, **12**, 101, (1959). (With B.V. Kukarkin).
- "On the stress-energy pseudo-tensor for the gravitational field", *Vestnik Moskovskogo Universiteta (Herald of Moscow University)*, s.III, N2, 59, (1960).
- "Space-time metric inside the Schwarzschild singular sphere", *Astron. Zh.*, **38**, 564, (1961).
- "Properties of the comoving space curvature of cosmological models in the post-Newtonian approximation", *Astron. Zh.*, **38**, 961, (1961).
- "Relativistic effects in interstellar flights" in Proceeding of the 3rd meeting of the Astronomical-Geodesic Society (USSR), Moscow, Fizmatgiz, p.147, (1961).
- "The transformation of the space-time metric when crossing the singular sphere", *Publications of the Shternberg Astronomical Institute (USSR)*, N120, 42, (1962).
- "On some properties of the solution of the Einstein equations for spherically symmetric gravitational fields", *Vestnik Moskovskogo Universiteta (Herald of Moscow University)*, s.III, N5, 90, (1962).
- "On the motion of a spherical symmetric mass in general relativity", *Vestnik Moskovskogo Universiteta (Herald of Moscow University)*, s.III, N6, 66, (1962).
- "The relativistic theory of interstellar flights", Znanie (Knowledge), Moscow (1962).
- "The propagation of light outside and inside the Schwarzschild sphere", *Doklady Akad. Nauk (Report of Acad. of Science of USSR)*, **150**, 1019, (1963). (With L.M. Ozernoy).
- "The evolution of the semiclosed Universe", *Astron. Zh.*, **40**, 772, (1963).
- "R and T regions of space-time with spherical space", *Publications of the Shternberg Institute (USSR)*, N132, 3, (1964).
- "Some properties of the spherical comoving space", *Publications of the Shternberg Institute (USSR)*, N132, 43, (1964).
- "The number of collapsed stars in the Galaxy", *Publications of the Lebedev Physical Institute*, A-17, (1964). (With L.M. Ozernoy).
- "The origin of large-scale inhomogeneities in the expanding Universe", *JETF*, **46**, 686, (1964).
- "Average density of radiation in the Metagalaxy and the problems of relativistic cosmology", *Doklady Akad. Nauk (Report of Acad. of Science of USSR)*, **154**, 809, (1964). (With A.G. Doroshkevich).
- "Radiation of gravitational waves by bodies moving in the field of a collapsed star", *Doklady Acad. Nauk (Report of Acad. of Science of USSR)*, **155**, 1033, (1964). (With Ya.B. Zeldovich).
- "The mechanism of energy release during the collapse of a superstar", *Astron. Zh.*, N290, (1964).
- "The delay of the explosion of part of the Friedman Universe and superstars", *Astron. Zh.*, **41**, 1075, (1964).
- "Estimate of the mass of a superstar", *Doklady Akad. Nauk (Report of Acad. of Science of USSR)*, **158**, 811 (1964). (With Ya.B. Zeldovich).

- "Relativistic astrophysics", Part I, *Uspekhi Fiz. Nauk*, **84**, 877, (1964). With Ya.B. Zeldovich).
- Relativistic astrophysics", Part II, *Uspekhi Fiz. Nauk*, **86**, 447, (1965). (With Ya.B. Zeldovich).
- "Gravitational collapse of a non-symmetric and rotating masses", *JETF*, **49**, 171, (1965). (With A.G. Doroshkevich, Ya.B. Zeldovich).
- "Physics of relativistic collapse", *Supplement al Nuovo Cimento*, I, **4**, 810, (1966). (With Ya.B. Zeldovich).
- "General relativity and astrophysics", *Einsteinovskiy sbornik (USSR)*, Nauka, Moscow, **1**, 18, (1966). (With Ya.B. Zeldovich)..
- "Change of relativistic collapse into anti-collapse and kinematics of the charge sphere", *Pisma (Letters) JETF*, **3**, 223, (1966).
- "The hypothesis of lagging cores and the hot Universe", *Astron. Zh.*, **43**, 758, (1966). (With Ya.B. Zeldovich).
- "Change of relativistic collapse into expansion and the physical singularity", *Astron. Zh.*, **43**, 911, (1966).
- "The methods of the investigation and the cosmological significance of helium in the intergalactic medium", *Astron. Zh.*, N3, 1, (1966). (With R.A. Sunyaev, Ya.B. Zeldovich).
- "Relativistic stages of stellar evolution" in Proceedings of the II gravitational meeting (Tbilisi), 336, (1967).
- "The origin of galaxies in the expanding Universe", *Astron. Zh.*, **44**, 295, (1967). (With A.G. Doroshkevich, Ya.B. Zeldovich).
- "An explanation of the He^3 anomaly in the star C Co A", *Astron. Zh.*, **44**, 336, (1967). (With R.A. Sunyaev).
- "Free particles in anisotropic cosmological models", *Pisma (Letters) JETF*, **5**, 119, (1967). (With A.G. Doroshkevich, Ya.B. Zeldovich).
- "Astronomical test of strong anisotropy during the early stages of expansion of the Universe", *Astron. Zh.*, N442, (1967). (With A.G. Doroshkevich, Ya.B. Zeldovich).
- "Weakly interacting particles in anisotropic cosmology", *JETF*, **53**, 644, (1967). (With A.G. Doroshkevich, Ya.B. Zeldovich).
- "Physical restrictions on the possible topology of the Universe", *Pisma (Letters) JETF*, **6**, 772, (1967). (With Ya.B. Zeldovich).
- "The evolution of supermassive stars stabilized by large scale motions", *Astron. Zh.*, **44**, 525, (1967). (With G.S. Bisnovaty-Kogan, Ya.B. Zeldovich).
- "Cosmology", *Annual Rev. Astron. and Astrophys.*, **5**, 627, (1967). (With Ya.B. Zeldovich).
- "The origin of galaxies", *Publications of the Institute of Applied Mathematics*, (1967).
- "On the interpretation of the cosmic microwave radiation", *Astron. Zh.*, **44**, 663, (1967). (With Ya.B. Zeldovich).
- "Anisotropy of cosmological radiation in the homogeneous anisotropic model", *Astron. Zh.*, **45**, 538, (1968).
- "Anisotropy of the early stages of cosmological expansion and the relic radiation", *JETF*, **55**, 6(12), (1968). (With L.P. Grischuk, Ya.B. Zeldovich).
- "Neutrinos in anisotropic cosmological solutions", *Pisma (Letters) JETF*, **8**, 95, (1968). (With A.G. Doroshkevich, Ya.B. Zeldovich).
- "Hot Universe, origin of galaxies, non-uniformities and superdense bodies", Proceeding of IAU Symp. No.29, Erevan, 1966, (1968). (With Ya.B. Zeldovich).

- "Relic Neutrinos", *Neutrino physics, International symposium*, v.2, p.208, (1968). (With A.G. Doroshkevich, Ya.B. Zeldovich).
- "Kinetic theory of neutrinos", *Problems of Physics*, Moscow, p.15, (1969). (With A.G. Doroshkevich, Ya.B. Zeldovich).
- "The disturbances of the metric when a collapsing sphere passes below the Schwarzschild phase", *JETF*, **57**, 949, (1969).
- "Hypothesis on the initial metric perturbations in the Friedman model", *Astron. Zh.*, **46**, 960, (1969). (With Ya.B. Zeldovich).
- "Kinetic theory of neutrinos", *Astrofizika (USSR)*, **5**, 539, (1969). (With A.G. Doroshkevich, Ya.B. Zeldovich).
- "Evolution of the electrical charge sphere after crossing the Schwarzschild sphere", *JETF*, **58**, 262, (1970).
- "Vortex perturbations in Friedman's model", *Astrofizika (USSR)*, **6**, 379, (1970). (With Ya.B. Zeldovich).
- "Mixmaster model and the cosmological problem", *Astron. Zh.*, **47**, 948, (1970). (With A.G. Doroshkevich).
- "Stellar clusters", *Publications of the Institute of Applied Mathematics*, N23, (1970). (With Ya.B. Zeldovich).
- "The origin of the matter motion in uniform models and the chemical composition of primordial matter", *Astron. Zh.*, 47, 1203, (1970).
- "Origin of Helium in different cosmological models", *Highlight of Astronomy*, **2**, 318, (1970). (With A.G. Doroshkevich, Ya.B. Zeldovich).
- "Perturbations in a uniform anisotropic Universe", *JETF*, **60**, 1201, (1971). (With A.G. Doroshkevich, Ya.B. Zeldovich).
- "The impossibility of mixture in Bianchi Type IX cosmological models", *JETF*, **60**, 3, (1971). (With A.G. Doroshkevich, V.N. Lukash).
- "Gravitational collapse, white holes and particle creation", *GRG*, **3**, 118, (1971). (With Ya.B. Zeldovich).
- "Gravitational collapse (black holes) and the search for it in the Universe", *Vestnik Akademii Nauk (Herald of Acad. of Science of USSR)*, N4 (1972). (With Ya.B. Zeldovich).
- "The gravitational field of collapsed bodies", *Uspekhi Fiz. Nauk*, **107**, 521, (1972).
- "Modern theory of black holes" in the Proceedings of the Soviet grav. meeting, Erevan, (1972).
- "Synchrotron gravitational radiation", *JETF*, **63**, 2538, (1972). (With A.G. Doroshkevich, A.G. Polnarev).
- "The isotropization of uniform cosmological models", *JETF*, **64**, 1457, (1973). (With A.G. Doroshkevich, V.N. Lukash).
- "Gravitational collapse", *Soviet encyclopedia*, (in Russian), **14**, 422, (1973).
- "Black hole astrophysics " in "Black Holes (Les Astres Occlus)", eds. C. DeWitt and B. DeWitt, (Gordon and Breach), New York, (1973). (With K.S. Thorne).
- "A possible explanation of the 36-day period of Her X-1", *Astron. Zh.*, **50**, 59, (1973).
- "An electromagnetic detector for gravitational waves", *JETF*, **65**, 1729, (1973). (With V.B. Braginsky, A.G. Doroshkevich, Ya.B. Zeldovich, M.V. Sazhin).
- "Physical processes in the vicinity of the cosmological singularity", *Annual Rev. of Astr. and Astrophys.*, **11**, 387, (1973). (With Ya.B. Zeldovich).
- "Quantum explosion of a white hole", *Publications of the Institute of Applied Mathematics*, N98, (1973).
- "Isotropization of uniform cosmological models" in "Confrontation of cosmological

theories with observational data", ed. M. Longair, IAU Symposium No.63., D. Reidel Publ. Comp., p.349, (1974).

- "The problem of modern cosmology", *Voprosy filosofii (USSR) (Questions of philosophy),* N4, (1974). (With Ya.B. Zeldovich).
- "Search for observational evidence for black holes" in "Astrophysics and Gravitation", Proceedings of the 16th Solvay Conf. on Physics, 1973, edition de l'Universite de Bruxelles, p.317, (1974).
- "Quantum effects in white holes", *JETF*, **66**, 1897, (1974). (With Ya.B. Zeldovich, A.A. Starobinsky).
- "Relic radiation in the uniform anisotropic Universe", *Astron. Zh.*, **66**, 1897, (1974). (With A.G. Doroshkevich, V.N. Lukash).
- "On the physical nature of galactic coronaes", *Pisma (Letters) Astron. Zh.*, **1**, N3, (1975). (With B.V. Komberg).
- "Component masses and inclination of a binary system containing a pulsar: determination from relativistic effects", *Pisma (Letters) Astron. Zh.*, **1**, N1, 2, (1975). (With V.A. Brumberg, Ya.B. Zeldovich, N.I. Shakura).
- "Gravitational radiation from a star collapsing into a disk", *Astron. Zh.*, **52**, 657, (1975).
- "Particle creation in the vortex cosmological models", *JETF*, **69**, 1584, (1975). (With A.A. Starobinsky, V.N. Lukash).
- "Modern development of cosmology", *Voprosy filosofii (USSR) (Questions of philosophy),* N6, (1975). (With Ya.B. Zeldovich).
- "Particle creation in cosmology", III meeting of the Europ. Phys. Soc., Bucharest, (1975). (With Ya.B. Zeldovich, A.A. Starobinsky).
- "Pancake formation from the dust medium without rotation", *Astron. Zh.,*, **52**, 1038, (1975).
- "Quantum effects and the evolution of cosmological model", *Nuovo Cimento*, 35B, 293, (1975). (With Ya.B. Zeldovich, A.A. Starobinsky).
- "Microwave background radiation anisotropy in anisotropic cosmological models and galaxy formation", Proceedings of the XV IAU General Assembly, Grenoble, p.42, (1976). (With Ya.B. Zeldovich).
- "Hawking process on the boundary of and inside a black hole", *JETF*, **71**, 393, (1976).
- "Fluctuation of MBR, gravitational waves and anisotropy of the Universe", Proceedings of IAU Symposium No.74, Cambridge, (1976). (With Ya.B. Zeldovich).
- "Black holes", *Uspekhi Fiz. Nauk*, **120**, 514, (1976).
- "Parallaxes of objects at cosmological distances", *Astron. Zh.*, **54**, 722, (1977).
- "Temperature fluctuations of the relic radiation and gravitational waves", *Astron. Zh.*, **54**, 932, (1977). (With A.G. Doroshkevich, A.G. Polnarev).
- "Hydrodynamics of primordial black hole formation", Proceedings of the 8th International Conf. of the GRG Soc., p.32, (1977). (With D.K. Nadezhin, A.G. Polnarev).
- "The temperature fluctuations in the microwave background due to the gravitational waves" in *"Experimental gravity"*, Proceedings of the International Conf., Pavia, p.91, (1977). (With A.G. Doroshkevich, A.G. Polnarev).
- "Gravity", *Soviet encyclopedia*, (in Russian), **26**, 419, (1977).
- "Space-time and physical fields inside a black hole", *JETF*, **74**, 9, (1978). (With A.G. Doroshkevich).
- "Hydrodynamics of primordial black hole formation", *Astron. Zh.*, **55**, 231, (1978). (With D.K. Nadezhin, A.G. Polnarev).
- "Numerical simulation of the nonlinear stage of one-dimensional gravitational col-

lapse", *Publications of the Institute of Applied Mathematics*, N82, (1978). (With A.G. Doroshkevich, E.V. Kotok, U.S. Sigov, S.F. Shandarin).

- "The problem of relativistic cosmology", *The results of the astronomical researches in the USSR, Astrosovet AN USSR*, p.113, (1978).
- "Two dimensional model of the formation of the large-scale structure of the Universe", *Publications of the Institute of Applied Mathematics*, N83, (1978). (With A.G. Doroshkevich, E.V. Kotok, A.N. Polyudov, U.S. Sigov, S.F. Shandarin).
- "On the parallax distance to objects in a locally inhomogeneous Universe", *Astron. Zh.*, **55**, 1137, (1978). (With A.D. Popova).
- "Evolution of scalar perturbations near the Cauchy horizon of a charged black hole", *Phys. Rev. D*, **19**, 413, (1979). (With Y. Gursel, V. Sandberg, A. Starobinsky).
- "The final state of evolution of the inner part of a charged black hole", *Phys. Rev. D*, **20**, 1260, (1979). (With Y. Gursel, V. Sandberg, A. Starobinsky).
- "Primordial black holes", *Astron. Astrophys.*, **80**, 104, (1979). (With Ya.B. Zeldovich, A.G. Polnarev, A.A. Starobinsky).
- "On electromagnetic detection of gravitational waves", *GRG*, **11**, N6, (1979). (With V. Braginsky, L. Grischuk, A. Mensky, M. Sazhin, Ya.B. Zeldovich, A. Doroshkevich).
- "Physics and cosmology", *Astronomija, metodologia, mirovozrenie (Astronomy, methodology, world outlook)*, p.121, Nauka, Moscow, (1979). (With Ya.B. Zeldovich).
- "Photometric properties of the object SS433", *Astron. Zh.*, N1086, (1979). (With S.A. Gladishev, N.E. Kurochkin, A.M. Cherepashuk).
- "Hydrodynamics of primordial black hole formation and dependence of the process on the equation of state" in *"Source of Gravitational Radiation"*, Cambridge, p.173, (1979). (With A.G. Polnarev).
- "Quantum electrodynamic effects inside the charged black hole and the Cauchy horizon problem", *JETF*, **78**, 3, (1980). (With A.A. Starobinsky).
- "Hydrodynamics of primordial black hole formation and dependence of the process on the equation of state", *Astron. Zh.*, **57**, 237, (1980). (With A.G. Polnarev).
- "Numerical simulation of gravitational systems", *MNRAS*, **192**, 321, (1980). (With A.G. Doroshkevich, E.V. Kotok, A.N. Polyudov, U.S. Sigov, S.F. Shandarin).
- "Kinetic simulation of gravitating systems in the expanding Universe" in *"Numerical simulation of collective processes in plasma"*, ed. M.V. Maslennikov, Institute of Applied Mathematics, Moscow, p.224, (1980). (With A.G. Doroshkevich, E.V. Kotok, A.N. Polyudov, S.F. Shandarin).
- "Cosmology with neutrinos that have a nonzero rest mass", *Astron. Zh.*, **57**, 899, (1980). (With G.S. Bisnovaty-Kogan).
- "The origin of the structure of the Universe" in *"Variability in stars and galaxies"*, Proceedings of the 5th IAU Meeting, Liege, (1980). (With G.S. Bisnovaty-Kogan, V.N. Lukash).
- "The structure of space-time and physical processes inside black holes", Proceedings of the 9th International Meeting on GRG, p.262, (1980). (With A.A. Starobinsky).
- "Gamma burst coordinate transformation to a stellar catalogue", *Astrophys. and Space Science*, **75**, 219, (1981). (With G.S. Bisnovaty-Kogan, I.V. Estulin, A. Havinson, V.G. Kurt, G. Mersov).
- "A model of the peculiar object SS433", *Astron. Zh.*, **58**, 3, (1981). (With G.S. Bisnovaty-Kogan, B.V. Komberg, A.F. Illarionov).
- Electric charge in the Kruskal metric and Jeans' hypothesis", *GRG*, **14**, 1116, (1982). (With M. Demiansky).
- "The origin of the large scale structure of the Universe", Proceedings of the International School and Workshop on Plasma Astrophysics, Varenna, Como, (1981).

- "Einstein's gravitational theory and black holes" in *"Einstein i ego vklad v nauku" (Einstein and his input in the science)*, "Nauka i iskusstvo", Sofia, Bulgaria, p.76, (1981).
- "Numerical values of the fundamental constants and the anthropic principle", *Izvestia Academii Nauk Est. S.S.R.*, seria Fisika i Matematika, 31, 284, (1982). (With A.G. Polnarev, I.L. Rosental).
- "Creation of primordial inhomogeneities of the Universe", *Astron. Zh.*, **59**, 424, (1982). (With D.A. Kompaneets, V.N. Lukash).
- "Gamma-gun in the center of the Galaxy", *Astron. Zh.*, **60**, 209, (1983). (With N.S. Kardashev, A.G. Polnarev, B.E. Stern).
- "Primordial cosmological perturbation formation in the expanding Universe" in Proceedings of the 10th International Meeting on GRG, (Italy), v.2, p.844, (1983). (With V.N. Lukash).
- "Creation of primordial perturbations in the early Universe" in *"The very early Universe"*, Proceedings of the workshop, Cambridge, (1982), Cambridge University Press, p.311, (1983).
- "Properties of the electric field vector flux in a multiply connected space and radiation of a uniformly moving charge", *Publications of the Space Research Institute*, N827, (1983).
- "Gamma-gun in the center of the Galaxy" in *"Positron-electron pairs in astrophysics"*, eds. M. Burns, K. Harding, R. Ramaty, Goddard Space Flight Center, (1983). (With N.S. Kardashev, A.G. Polnarev, B.E. Stern).
- "Perturbations in the Universe with massive " in *"Early Evolution of the Universe and its Present Structure"*, IAU Symp. No.104, Crete, 1982, eds. G. Abell, C. Chincarini, Reidel Publ. Company, p.327, (1983). (With G.S. Bisnovaty-Kogan, V.N. Lukash).
- "The inflationary Universe, primordial sound waves and galaxy formation" in *"Early Evolution of the Universe and its Present Structure"*, eds. G. Abell, C. Chincarini, Reidel Publ. Company, (IAU Symp. No.104, Crete, 1982), p.457, (1983). (With V.N. Lukash).
- "Phase transitions of the cosmological vacuum and primordial black holes" in *"Early Evolution of the Universe and its Present Structure"*, eds. G. Abell, C. Chincarini, Reidel Publ. Company, (IAU Symp. No.104, Crete, 1982), p.463, (1983). (With N.S. Kardashev).
- "Exploding neutron stars in close binaries", *Pisma (Letters) Astron. Zh.*, **10**, 422, (1984). (With S.I. Blinnikov, T.V. Perevodchikova, A.G. Polnarev).
- "$e^{\pm}$ - photon interactions in the magnetospheres of neutron stars", *Astron. Zh.*, **60**, 1113, (1984). (With N.S. Kardashev, I.G. Mitrofanov).
- "Thermal regime of neutron stars heated by nucleon decay", *Astron. Zh.*, **60**, 935, (1984). (With T.G. Perevodchikova).
- "Cosmological parameters and cosmological models" in *"General Relativity and Gravitation"*, ed. B. Bertotti et al., p.337, (1984).
- "The capture of particles in Einstein-Straus Universe. A Newtonian approach", *Letter al Nuovo Cimento*, ser.2, **39**, 185, (1984). (With S. Fillipi, R. Ruffini).
- "Models of the annihilation $\gamma-$ line and the continuous gamma-ray spectrum in the center of the Galaxy" in *"X-ray and UV Emission from Active Galactic Nuclei"*, eds. W. Brinkmann, J. Trumper, p.211, (1984).
- "Spottiness in the large scale structure of the microwave background", *Nature*, **316**, 46, (1985). (With V.N. Lukash).
- "Anisotropy of the relic radiation as a test of theories of the early Universe", Proceedings of the 3rd seminar on "Quantum Gravity", (1984), Moscow, p.675, (1985). (With V.N. Lukash, P.D. Naselsky).
- "The problem of the origin of primordial perturbations and modern cosmology" in

"Galaxies, axisymmetric systems and relativity", ed./ M. MacCallum, Cambridge University Press, 23, (1985). (With V.N. Lukash).

- "The possible mechanism of the formation of the hard spectrum of active galactic nuclei" in *"Structure and Evolution of Active Galactic Nuclei"*, ed. G. Giurichin et al., D. Reidel Publ. Comp., p.149, (1986). (With B.E. Stern).
- "Obvious, Incredible ...", *"Foundations of Physics"*, **16**, 209, (1986). (With V.N. Lukash).
- "The electron-positron cauldron in AGN's", *Plasma Astrophysics*, Sukhumi (USSR), abstract, p.69, (1986). (With B.E. Stern).
- "Small-scale anisotropy of the microwave background in neutrino decay models of the Universe", *Astron. Zh.*, **63**, 1057, (1986). (With P.D. Naselsky, L.I. Reznicsky).
- "The electron-positron cauldron and the formation of the hard radiation of quasars and AGN's" in *"Quasars"*, eds. G. Swarup, V.K. Kapahi, IAU, p.383, (1986). (With N.S. Kardashev, B.E. Stern).
- "Evolution des Universums", (2nd edition), "Mir", Moscow, Leipzig, (1986).
- "Large distances, hidden mass and fluctuations of the relic radiation" in *"Observational cosmology"*, eds. A. Hewitt et al., p.73, (1987). (With V.N. Lukash).
- "Observational cosmology and cosmological models", *Uspekhi Fiz. Nauk (USSR)*, **51**, N1, (1987). (With N.S. Kardashev, V.N. Lukash).
- "Cosmology, theory and observations" in *"Evolution of galaxies"*, eds. J. Palous, L. Perek, Czechoslovak Academy of Sciences, Prague, Vol.4, p.401 (1987). (With V.N. Lukash).
- "The formation of the hard spectrum of the center of the Galaxy and active galactic nuclei" in *"HE-UHE behavior of accreting X-ray sources"*, eds. F. Givannelli, G. Mannocchi, p.165, (1987).
- "The baryonic islands cosmological model", *JETF*, **94**, 1, (1988). (With A.D. Dolgov, N.S. Kardashev, A.F. Illarionov).
- "Mathematical models for fitting the reflecting surface of the main mirror of the radiotelescope RT-70 to the form of an ideal paraboloid", *Publications of Space Research Institute*, N1369, (1986). (With N.S. Kardashev, D.A. Kompaneets, A.F. Illarionov, T.V. Perevodchikova, E.N. Sedih, B.I. Slish).
- "The baryonic islands model of large structures" in *"Large Scale Structure and Motions in the Universe"*, ed. M. Mezette et al., Kluwer Academic Publishers, p.159, (1989).
- "Spectral distortion and angular anisotropy of the relic radiation due to interaction with pregalactic dust", *Astron. Zh.*, **66**, 680, (1989). (With P.D. Naselsky).
- "An analysis of the operation of a time machine", *JETF*, **95**, 769, (1989).
- "Physical properties of the time-machine", *Uspekhi Fiz. Nauk (Russia)*, **157**, 549, (1989).
- "Spectral distortions of the relic radiation from decays of massive relic particles in the Universe", *Astron. Zh.*, **67**, 257, (1990). (With P.D. Naselsky).
- "The Cauchy problem in space-times with closed timelike curves", *Phys. Rev. D*, **42**, 1915, (1990). (With J. Friedman, M. Morris, F. Echeverria, G. Klinkhammer, K.S. Thorne, U. Yurtsever).
- "Propagation of electromagnetic radiation in a random field of gravitational waves and space radio interferometry", *Il Nuovo Cimento*, **1058**, 1141, 1990.
- "Physical effects in wormholes and time-machines", *Phys. Rev. D*, **42**, 1057, 1990. (With V.P. Frolov).
- "Explosion of a low mass neutron star", *Astron. Zh.*, **67**, 1181, (1990). (With S.I. Blinnikov, V.S. Imshennik, D.K. Nadezhin, T.V. Perevodchikova, A.G. Polnarev).

- "Black holes and the Universe", Cambridge University Press, Cambridge, 1990.
- "The microwave background radiation and the thermo-chemical history of a pregalactic Universe", Nordita report 90/41 A, (1990). (With P.D. Naselsky, B.V. Vainer).
- "On the gravitational field of an arbitrary axisymmetric mass endowed with magnetic dipole moment", *"Gravitation and Modern Cosmology"*, eds. A. Zichichi et al., Plenum Press, New York, p.121, (1991). (With V.S. Manko).
- "Generalization of the Kerr and Kerr-Newman metrics possessing an arbitrary set of multipole moments", *"Classical and Quantum Gravity"*, **9**, 2477-2487, (1992). (With V.S. Manko).
- "The appearance of highly relativistic, spherically symmetric stellar winds", *ApJ*, **369,** 175, (1991). (With M.A. Abramowicz and B. Paczynski).
- "A lucky chance for cosmology", Invited talk at 12-th European Regional Astronomy Meeting of IAU, Davos, Switzerland, October, 1990. (With V.N. Lukash). See also: *"Astrophysics on the Threshold of the 21-st Century"*, ed. N.S. Kardashev, Gordon and Breach Science Publishers, p.277, (1992)
- "Lectures on the early Universe" in *"Observational and Physical Cosmology"*, eds. F. Sanchez, M. Collados, and R. Rebolo, Cambridge Univ. Press, p.1, (1992). (With V.N. Lukash).
- "The Jinny of the time machine: nontrivial self-consistent solutions", *Classical and Quantum Gravity*, **9**, 1, 1992. (With A. Lossev).
- "Tidal capture of stars by a massive black hole", *MNRAS*, **255**, 276 (1992). (With C.J. Pethick and A.G. Polnarev).
- "Time machine and self-consistent evolutions in problems with self-interaction", *Phys. Rev. D*, **45**, 1989, (1992).
- "Nonlinear effects at tidal capture of stars by a massive black hole-I. Incompressible affine model", *MNRAS*, **258**, 715, (1992). (With A. Kosovichev).
- "Microwave background anisotropy on scales $10' - 2^o$ as an indicator of the cosmological model", *Ap.J.*, **417**, 14, (1993). (With P. Naselsky).
- "COBE data, microwave background anisotropy on scales $10' - 2^o$, and acoustic modes of perturbations", International School of Astrophysics, 2nd Course, *"Current Topics in Astrofundamental Physics"*, ed. D. Chalonge, World Scientific, p.195, (1993), (With P.D. Naselsky).
- "Kosmiske fingeraftyr på himlen og naturen af det mørke stof i Universet" (1993), Forsknings- og Teknologiministeriet Astronomisk udvalg, page 4, (1993). (With H.E. Jørgensen).
- "The Jinnee of the time machine" in *String quantum gravity and physics at the Plank energy scale*, International Workshop on Theoretical Physics, ed. N. Sanchez, World Scientific Publishing, p.480, (1993).
- "Wormhole as a device for studying a black hole's interior", *Phys. Rev. D*, **48**, 1607, (1993). (With V. Frolov).
- "Dynamical origin of the entropy of a black hole", *Phys. Rev. D*, **48**, 4545, (1993). (With V. Frolov).
- "Long distance correlations in the galaxy distribution and the nature of dark matter", *MNRAS*, **265**, 261-170, (1993). (With H.E. Jørgensen, E.V. Kotok, P.D. Naselsky).
- "Weak tidal encounter of a star with a massive black hole", *ApJ*, **418**, 163-180, (1993). (With A. Khohlov and C. Pethick).
- "Strong effects during close encounters of stars with a massive black hole", *ApJ*, **418**, 181-196, (1993). (With A. Khohlov and C. Pethick).
- "Inelastic billiard ball in a spacetime with a time machine", *Phys. Rev. D*, **47**, 1432, (1993). (With E. Mikheeva).

- "Sakharov Modulation of the Spectrum of Initial Perturbations and its Manifestation in the Anisotropy of Cosmic Microwave Background and Galaxy Correlation Function" in *"Present and Future of the Cosmic Microwave Background*, eds. J.L. Sanz, E. Martinez-Gonzales, l. Cayon. Springer-Verlag, p.146 (1994). (With H.E. Jørgensen, E.V. Kotok, P.D. Naselsky)
- "Relativistic tidal interaction of a white dwarf with a massive black hole", *ApJ*, **432**, 680, (1994). (With V.P. Frolov, A.M. Khohlov, C.J. Pethick).
- "Long Distance Correlations of Matter Distribution in the Universe" in *"Cosmology and Particle Physics"*, eds. V. de Sabbata and H. Tso-Hsiu, Kluwer Academic Publisher, p.159, (1994). (With H.E. Jørgensen, P.D. Naselsky, D.I. Novikov).
- "Inflation and Primordial Black Holes as Dark Matter", *Phys. Rev. D*, **50**, N12, 15 December, p.7173, (1994). (With P. Ivanov, P. Naselsky).
- "Evidence for Sakharov oscillations of initial perturbations in the anisotropy of the cosmic microwave background", *Astron. Astrophys.*, **294**, 639-647, (1995). (With H.E. Jørgensen, E. Kotok, P. Naselsky)
- "Dark Matter and Sakharov Oscillations" in *"Current Topics in Astrofundamental Physics: The Early Universe"*, eds. N. Sanchez and A. Zichichi, Kluwer Academic Publishers, 391-402, (1995).
- "MACHO photometry and astrometry", *Astron. Astrophys.*, **294**, 287-294, (1995). (With E. Hog, and A.G. Polnarev).
- "Black Hole Astrophysics" in *"Particle and nuclear Astrophysics and Cosmology in the next Millennium"*, eds. E.W. Kolb and R.D. Peccei, World Scientific, p.185, (1995).
- "Cosmology, Dark Matter and Sakharov Oscillations" in *"Frontier Objects in Astrophysics and Particle Physics"*, Conference Proceedings Vol.47, eds. by F. Giovanelli and G. Mannochi, SIF, Bologna, 79-88, (1995).
- "Concluding remarks" in *"Frontier Objects in Astrophysics and Particle Physics"*, Conference Proceedings Vol. 47, eds. F. Giovanelli and G. Mannochi, SIF, Bologna, 79-88, (1995).
- "General formula for radiative cooling and global structure of the accretion disk" in *"Frontier Objects in Astrophysics and Particle Physics"*, Conference Proceedings Vol.47, eds. F. Giovanelliand G. Mannochi; SIF, Bologna, 269-275, (1995). (With J. Artemova, G. Bisnovatyi-Kogan, G. Bjornsson).
- "Non-linear effects at tidal capture of stars by a massive black hole - II. Compressible affine model and tidal interaction after capture", *MNRAS*, **275**, 498-506, (1995). (With P. Diener, A.G. Kosovichev, E.V. Kotok, and C.J. Pethick).
- "Black-Hole Astrophysics" in *Particle and Nuclear Astrophysics and Cosmology on the Next Millennium (Snowmass 94)*, Proceedings of the 1994 DPF Summer Study on High Energy Physics, eds. E.W. Kolb and R.D. Peccei, World Scientific, p.377, (1995). (With P. Bender, E. Bloom, Lynn Cominsky, H. Ford, A. Harmon, P. Michel, K. Postnov, J. Scargle, J. Swank, J. Craig Wheeler, K. Woo).
- "Astrophysics of Black Holes" in *"Frontiers of Astrophysics"*, Proceedings of the Roseland Centenary Symposium, Oslo, 16-17 June, 1994, eds. P. Lilje and P. Malby, pp. 215-225, (1995).
- "Wormhole, Time Machines and the Principle of Minimal Action", Proceedings of the Markov's Physical Conference, Moscow, (1995). (With A. Carlini, V.P. Frolov, M.B. Mensky, H.H. Soleng).
- "Optically Thick - Optically Thin Transition in Accretion Discs" in Proceedings of Röntgenstrahlung from the Inoverse (Würzburg, Germany, 1995). (With J.V. Artemova, G.S. Bisnovaty-Kogan, G. Bjornsson)
- "Time Machines: the principle of self-consistency as a consequence of the principle

of minimal actions", *International Journal of Modern Physics D*, **4**, n5, 557, (1995). (With A. Carlini, V. Frolov, M. Mensky, H. Soleng).

• "Black Holes, Wormholes and Time Machines" in Proceedings of International Workshop Held in Rome, Italy, 18-21 May 1994, ed. F. Occhionero, Springer, p. 373, (1995).

• "Modified Newtonian Potentials for the Description of Relativistic Effects in Accretion Discs Around Black Holes", *ApJ*, **461** 565, (1996). (With J.V. Artemova, G. Bjornsson).

• "Structure of Accretion Disks with Optically Thick-Thin Transitions", *ApJ*, **456**, 119, (1996). (With J.V. Artemova, G.S. Bisnovaty-Kogan, G. Bjornsson).

• "Nature of Dark Matter and Some New Methods in Observational Cosmology" in Third Paris Cosmology Colloquium, eds. H. de Vega and N. Sanchez, World Scientific, p. 33, (1996). (With H.E. Jørgensen, E.V. Kotok, P.D. Naselsky and D.I. Novikov)

• "Decoherence Caused by Topology in a Time-Machine Spacetime", *International Journal of Modern Physics D*, **5**, 1, (1996) (With M. Mensky).

• "Cosmology Dark Matter and Sakharov Oscillations", Collected Papers in occasion of Prof. C. Castagnoli last teaching lecture, Editrice Composition Bologna, p. 121, (1996).

• "Big Bang Scenario and Nature of Dark Matter" in *"Examining the Big Bang and Diffuse Background Radiations"*, eds. M. Kafatos and Y. Kondo IAU, p. 289, (1996).

• "Three -Dimensional Billiards with Time Machine", *International Journal of Modern Physics D*, **5**, 179, (1996), (With M. Mensky).

• "Time Machines and the Principle of Self-Consistency as a Consequence of the Principle of Stationary Action (II): the Cauchy Problem for a Self-Interacting Relativistic Particle", *International Journal of Modern Physics D*, **5**, 445, (1996). (With A. Carlini).

• "Black Holes" in *Stellar Remnants*, 25th Saas-Fee Advanced Course, eds. G. Meynet and D. Schaerer; Springer-Verlag, Berlin, pp. 237-334, (1996).

Books

• "Relativistskaya Astrofisica" (Relativistic Astrophysics). Nauka, Moscow, (1967). (With Ya.B. Zeldovich). [Translation "Relativistic Astrophysics", Vol. I, The University of Chicago Press, eds. by K. Thorne and W. Arnett, (1971).]

• "Gravitational theory and stellar evolution", (in Russian), Nauka, Moscow (1971). (With Ya.B. Zeldovich).

• "Structura i evolutcia Vselennoy" (Structure and Evolution of the Universe), Nauka, Moscow, (1975). (With Ya.B. Zeldovich). [Translation "Relativistic Astrophysics, The Structure and Evolution of the Universe", v.2, ed. G. Steigman, University of Chicago Press, Chicago, London, (1983).]

• "Evolutcia Vselennoy" (Evolution of the Universe), Nauka, Moscow, (1979). [Translation "Evolution of the Universe", Cambridge University Press, Cambridge, (1983).]

• "Fisika chernikh dir" (Physics of black holes), Nauka, Moscow, (1986). (With V.P. Frolov). [Translation "Physics of Black Holes", Kluwer Academic publishers, (1989).]

• "E. Hubble, life and work, the man who discovered the explosion of the Universe", Nauka, Moscow, (1989). (With A.S. Sharov). [Translation "Edwin Hubble. The discoverer of the Big Bang Universe", Cambridge University Press, (1993).]